2nd International Conference on Advance & Scientific Innovation (ICASI 2019)

Journal of Physics: Conference Series Volume 1424

Medan, Indonesia
19 - 20 July 2019

Editors:

Robbi Rahim
Dodi Siregar
M. Mesran
Supriyanto
Rahmat Hidayat
Tengku Mohd Diansyah

ISBN: 978-1-7138-1328-6
ISSN: 1742-6588

Printed from e-media with permission by:

Curran Associates, Inc.
57 Morehouse Lane
Red Hook, NY 12571

Some format issues inherent in the e-media version may also appear in this print version.

Institute of Physics
Dirac House, Temple Back
Bristol BS1 6BE UK

Phone: 44 1 17 929 7481
Fax: 44 1 17 920 0979

techtracking@iop.org

Additional copies of this publication are available from:

Curran Associates, Inc.
57 Morehouse Lane
Red Hook, NY 12571 USA
Phone: 845-758-0400
Fax: 845-758-2633
Email: curran@proceedings.com
Web: www.proceedings.com

TABLE OF CONTENTS

EDUCATION

ELECTRICAL ENGINEERING

ENGINEERING

ENVIRONMENTAL ENGINEERING

HEALTH MANAGEMENT SYSTEM

INDUSTRIAL ENGINEERING

INFORMATION TECHNOLOGY

MATERIAL ENGINEERING

PHYSICS

WEATHER TECHNOLOGY

Preface

We are delighted to introduce the proceedings of the second edition of *International Conference on Advance and Scientific Innovation* (ICASI) 2019. The technical program has brought researchers and practitioners around the world to a good forum for discussing, leveraging and developing all scientific and technological aspects that are relevant to digital technology, computer science, engineering and education. This *ICASI 2019* acquired 210 full papers submission and only accept 128 papers that match with IOP JPCS scope. We hope that the future *ICASI* will be as successful and stimulating, as indicated with the contributions presented in this volume.

General Chair
Dr(c) Robbi Rahim, M.Kom

The Editors
Robbi Rahim
Dodi Siregar
M Mesran
Supriyanto
Rahmat Hidayat
Tengku Mohd Diansyah

Peer review statement

All papers published in this volume of *Journal of Physics: Conference Series* have been peer reviewed through processes administered by the proceedings Editors. Reviews were conducted by expert referees to the professional and scientific standards expected of a proceedings journal published by IOP Publishing.

RECENT ADVANCES IN EXTRACTION OF CLINACHANTHUS NUTANS: A REVIEW

Mustafa Khalid Abduljabar[1], Khairul Farihan Kasim[1,2*], Akmal Hadi Ma'Radzi[1,2], Noor Soffalina Sofian Seng[3]

[1]School of Bioprocess Engineering, Universiti Malaysia Perlis, 01000 Kangar, Perlis, Malaysia.
[2]Centre of Excellence for Biomass Utilization, School of Bioprocess Engineering, Universiti Malaysia Perlis, Kompleks Pusat Pengajian Jejawi 3, Arau, 02600 Perlis, Malaysia.
[3]School of Chemical Sciences and Food Technology, Faculty of Science and Technology, Universiti Kebangsaan Malaysia, 43600 Bangi, Selangor, Malaysia.

*Correspondence Author: mustafakhalid89@gmail.com

Abstract. Interest in the development of extraction of phytochemicals compounds from *Clinacanthus nutans* has increased in recent years due to its potential applications in food, chemical, and pharmaceutical industries. The quality and yield of the phytochemicals are greatly influenced by the extraction method employed as well as other factors. This paper provides an up-to-date overview of traditional (maceration, infusion, and soxhlet), with modern methods (pressurized hot water extraction, supercritical fluid extraction, ultrasound-assisted extraction, and microwave-assisted extraction) to extract phytochemicals compounds from *C. nutans*. Problems associated with conventional extraction methods have led to a demand for alternatives and indeed several new extraction methods have been developed and refined in the last two decades that can efficiently extract a great variety of valuable bioactive compounds.

1. Introduction

Recently, there is a growing consumer interest towards medicinal plants since it is perceived to be healthier and free from side effects unlike synthetic drugs. The majority of the world population still rely on plants and their extracts for primary health care and needs. Medicinal plants as herbal remedies maintain their use for therapeutic effect despite tremendous advances in modern medicine. The high efficacy, low cost, and low toxicity of medicinal plants have popularized them in the treatment of diseases even in modern times [1]. *Clinacanthus nutans* belongs to the Acanthaceae herbal family and grows exceedingly well in equatorial regions, especially in Southeast-Asia. It is traditionally used as a remedy for herpes, snake and insect bites, and inflammation [2]. Many studies have shown *C. nutans* extracts to have anti-inflammatory, anti-microbial, and anti-viral activity against herpes simplex virus (HSV) and varicella- zoster (VZV) [3]. *C. nutans* extracts also contain natural antioxidants that prevent oxidative damage caused by free radicals in the body [4]. Natural antioxidants are preferable to artificial antioxidants as they are non-toxic and are more easily decomposed by the body. The leaves of *C. nutans* are rich in natural antioxidants; phenolic and flavonoid compounds [5], that scavenge free radicals by way of acting as reducing agents, hydrogen donors, or oxygen singlet quenchers [6].

There exist numerous studies on the extraction of these bioactive compounds from plants. Conventional methods like maceration, infusion, and ordinary reflux have been the industry staple for many decades but they suffer from the disadvantages of being time- and solvent- consuming. More recently, modern methods such as supercritical fluid extraction (SFE), ultrasound-assisted extraction (UAE), microwave assisted extraction (MAE), and pressurized liquid extraction (PLE) have been developed and they have higher efficiencies, lower solvent consumption, and give better yields of phytochemicals compared to conventional methods [7]. The current literature body lacks a comprehensive study on the various methods to selectively extract phytochemicals from *C. nutans*. Therefore, the present mini review aims to provide an up-to-date technical discussion on current extraction techniques and their advantages and disadvantages, as well as an overview on the employment of cutting edge green extraction methods and the yield and quality of phytochemicals compound extracted from *C. nutans* using these methods.

2. Extraction methods of *C. nutans*

The choice of extraction method is important if one is to get a high quality and yield of bioactive compounds from *C. nutans*. Extraction methods can be classified into two groups; traditional or conventional extraction methods, and modern extraction methods. Conventional extraction methods include soxhlet, infusion, and maceration among others. A general overview has been provided by Khoo et al [8] on the extraction of phytochemicals compounds from *C. nutans* using conventional extraction methods. Conventional extraction methods often involve lengthy extraction times and excessive use of expensive organic solvents which are also bad for the environment. For example, to treat 20 g of plant material, the amount of solvent required is at least 250 mL while the extraction time may exceed 24 hrs [9]. Furthermore, toxics fumes are emitted during the extraction process as it is generally carried out at the effervescence point of the solvent used [10].

Previous studies have investigated the soxhlet extraction of phytochemicals from *C. nutans* using different solvents such as methanol, chloroform, ethanol, and hexane. Depending on the solvent used, *C. nutans* extracts are found to contain stigmasterol, C-glycosyl flavones, β- sitosterol, lupeol, vitexin, iso vitexin, betulin, shafto side, iso mollupentin, orientin, iso orientin, sulphur containing- glucosides, glycoglycerolipids, a mixture of-nine-cerebrosides, 7-O-β- glucopyran-oside, and mono acylmono galactosyl glycerol [11]. The yield for phytochemicals extracts from *C. nutans* using conventional methods is shown in Table 1.

In one study, soxhlet extraction was carried out on 3 g of dry *C. nutans* using 80 mL of absolute ethanol for 8 hrs. Then, the solvent was removed using a rotary evaporator at 45°C and the product was transferred into an amber glass bottle, weighed, and stored in a freezer at -7°C until use. The moisture content was 12.17% of the dry weight and the yield of *C. nutans* extract obtained was 21.28% [13]. Tiew et al [14] extracted flavonoids from *C. nutans* using the maceration method and got a yield of 0.05 ± 0.002 mg quercetin equivalent (QE)/g of dried plant material total flavonoids content (TFC). They had used methanol as the solvent and ran the process for 3 days. Furthermore obtained 0.22 ± 0.006 mg QE/g of dried plant material TFC also with the maceration technique but using a different solvent. The difference in TFC yield is mainly due to the choice of solvent used and therefore it is imperative that a suitable solvent is chosen when developing a protocol for the extraction of phytochemicals. Additionally, TFC will degrade to some degree during the extraction process and this must also be taken into account. Although the number of studies using this method to extract phytochemicals from *C. nutans* is significant, fact remains that the method is plagued by low yields, high organic solvent consumption, and low extraction efficiency [15].

Table 1. The yield of phytochemicals from *C. nutans* using conventional extraction methods.

Extraction Parameters							
Method	Plant Part	Solvent	Temperature (°C)	Time (h)	Amount of solvent (mL)	Phytochemical Compounds	Reference
Maceration	Leaves	Methanol	30	48	50	Total phenolic content, total flavonoid content, antioxidants. Analysed using GC-MS	[12]
Infusion	Whole plant	Methanol: water	40	6	150	DPPH and ferric reducing antioxidants. Analysed using (FRAP) assays	[2]
Soxhlet extraction	Leaves and stems	Ethanol	40	6	100	Phytosterols, phenols contents, flavonoids contents, and sitosterol	[12]

3. Novel extraction techniques

The major disadvantages of conventional extraction techniques are that they are time consuming, require a lot of expensive solvents, and have low extraction efficiency due to the thermal degradation of phytochemicals. To tackle these issues, a whole host of new extraction techniques have been developed in recent years [15]. Some of the most promising advanced extraction techniques are discussed below.

3.1. Supercritical fluid extraction (SFE)

Supercritical fluid extraction (SFE) is one of the new methods developed to extract phytochemicals compounds from plants [16]. Carbon dioxide (CO_2) is a typical solvent in the SFE process and it is non-toxic, non-flammable and safe enough to use even in food applications. The CO_2 gas is brought to the supercritical region mainly through the use of high pressures and mild heating where it then behaves like both a liquid and a gas. This fluid can then be used to extract thermo-labile compounds because its critical temperature is low, at only 31°C. In addition, the low boiling point of carbon dioxide at atmospheric pressure means the CO_2 can then be easily removed without leaving any unwanted residue [16]. Another advantage of using SFE is that the CO_2 fluid can be optimised for the extraction process by adjusting its temperature and pressure. Since supercritical CO_2 is a non-polar solvent, using it in tandem with amphipathic solvents such as ethanol can improve its ability to extract polar compounds [17].

Mustapa et al [15] used supercritical CO_2 at a pressure of 350 bars and temperature of 60°C to obtain 0.83 ± 0.10 mg sitosterol and 1.35 ± 0.12 mg phytosterols after 120 min. The SFE process was found to be especially unfavourable for the extraction of chlorophyll and phenols due to the non-polar nature of CO_2. The slow rate of extraction and the solutes being tightly bound to the plant matrix also contribute to the low extraction efficiency of phytochemicals from *C. nutans* using SPE.

3.2. Ultrasound-assisted extraction (UAE)

The UAE method uses ultrasonic waves to destroy the walls of the plant cells so that the phytochemical compounds within can be extracted [18]. Temperature, pressure, and extraction time are all factors that affect UAE efficiency. The UAE method is relatively easy, versatile, and low in cost compared to most other methods. Baharuddin et al [7] successfully extracted phenols and flavonoids from *C. nutans* leaves using 28 kHz frequency wave. He found the optimum temperature to be 55 °C and at higher temperatures, the phytochemicals will degrade. The maximum yield was obtained after 25 min at this temperature. Despite its utility, speed (5–30 min for one sample), high yield, and relative low cost, this method suffers from many problems, namely the lack of uniformity in the distribution of ultrasound energy and decline of power with time. Filtration is also required after the extraction step making the process labour intensive [19].

3.3. Microwave assisted extraction (MAE)

Microwaves are used to heat up the water content within the plant cell until it boils into steam. When it does, the water expands, rupturing the cell from within to release the phytochemicals. The process is quick, requires only a small amount of solvent, and is energy efficient. The MAE process is influenced by factors such as solvent type, compound type, and dielectric constants of the plant material [20]. The advantages of using the MAE process to extract phytochemicals from *C. nutans* include high extraction efficiency, low solvent consumption, good antioxidant-activity of the extracts, and short extraction times [21].

Ismail & Kasim [2], had used 80% methanol as the solvent and microwave heating at frequency of 2.45 GHz for 20 minutes. Theye found the TPC and TFC content to be 15 mg GAE/g DM and 15 mg Que/g DM respectively and they contribute to an antioxidant activity of 80%. Mustapa et al [15], considered the MAE to be a green process because it is safe, quick, easy to use, result in high extraction yields and can work with or without the use of solvent. Despite its good qualities, the high energy involved with the use of microwaves causes the oil itself to oxidise

3.4. Pressurized hot water extraction techniques

PHWE, also known as pressurized solvent extraction (PSE) or accelerated solvent extraction (ASE) uses high temperatures and pressures to improve the extraction of phytochemicals. The high pressure is used to maintain the solvent in the liquid phase even when heated above its boiling point. The process needs only a small amount of solvent, but still requires labour intensive washing and filtration after the extraction step [22]. Extraction efficiency can be improved by selecting the appropriate solvent or mixture, pressure (usually 3-20 MPa) [23], temperature (usually 100-200 °C), and extraction time. High temperature leads to an increase in the conductivity of the water as well as a decrease in surface tension and thus an increase in propagation of phytochemical molecules through the water. This method can be used to extract non-polar, moderately polar and polar compounds [24]. A brief summary of the extraction yields of phytochemicals from *C. nutans* using conventional extraction methods shown in Table 2.

Ismail & Kasim [2], studied the extraction of TPC, TFC, DPPH scavenging activity of different parts of *C. nutans* (stem, leaves, and mixture of both stem and leaves) by using PHWE technique with extraction condition at temperature of 121°C, pressure of 4 bars and extracted for 20 min. When compared to the control (maceration using water at 60°C), PHWE show a very significant effect. The use of water as solvent can be exploited with the used of pressurized hot water extractor. Leaves and mixture of *C. nutans* contain high TPC and TFC but lower DPPH activity. Howevr, the stem of *C. nutans* has high DPPH activity but lower TPC and TFC.

Table 2: A brief summary of the extraction yield of phytochemicals from *C.nutans* using novel extraction methods.

			Extraction Parameters				
Method	Part	Types of Solvent	Tmperature (°C)	Time (min)	Volume of solvent (mL)	Phytochemicals compounds	Reference
Pressurized hot water extraction	Leaves	Water	121	20	20	TPC, TFC, and DPPH antioxidant activity	[2]
Supercritical fluid extraction	Leaves-stems	Ethanol–water	60	120	50	Chlorophyll, TPC, TFC, and phytosterols	[15]
Ultrasound assisted extraction	Leaves-stems	Water	35	5-30	45	Phenolic, flavonoid contents, and vitamin C	[21]
Microwave assisted extraction	Leaves	Ethanol–water	186	5-20	14	Phytosterols, TPC, TFC, and sitosterol	[15]

4. Comparison of conventional and novel extraction methods

In general, conventional extraction methods have been used extensively and are rather effective in extracting compounds from plant. However, the main disadvantages of these methods are the need of large amounts of solvents and low extraction rates. On the other hand, modern methods can extract phytochemical compounds from *C. nutans* at a significantly higher rate. Mustapa et al [15], investigated and optimised modern methods for the extraction of phytochemicals compounds from *C. nutans*. In terms of extraction time, UAE and MAE were found to be the quickest, only requiring 5 and 10 min respectively but the extraction yield was lower than that obtained using PHWE and the extracts needed to be purified which is labour intensive. SFE resulted in the lowest level yield of polar phytochemicals because the supercritical carbon dioxide solvent used is non-polar and cannot extract polar compounds such as chlorophylls and polyphenol [25]. After taking into consideration the several modern extraction methods discussed above in terms of extraction time, temperature, and the ability to carry out sequential extractions using different solvents, it can be concluded that the PHWE method is the best method to use as it results in the highest yield of phytochemical compounds in the shortest time. The factors that influence the yield of phytochemicals extracted through the PHWE method are the temperature and pressure used, and the number of cycles and its duration [22].

5. Factors effecting extraction methods

Although most modern extraction techniques are quick, reliable, and highly efficient, there are several factors that influence phytochemical extraction efficiencies regardless of the method used. The solvent choice is the most important factor to consider in the extraction step and in general, the solvent used will depend on the desired phytochemical content to be extracted from the plant. There are several studies that report high recovery of phytochemicals using solvents provided that the other experiment parameters are optimized for the sample. Usually, pure solvents only result in low extraction efficiencies and 40 to 60% methanol, ethanol or acetonitrile with the remainder being water is necessary to achieve high recoveries. Also, due to environmental and economic concerns, it is preferable to use a lesser toxic solvent like ethanol or methanol instead of acetonitrile. The use of acids (especially HCl) in the extraction solvent should be avoided since there are evidence that indicate certain phytochemicals in *C. nutans* degrade under acidic conditions [7].

In non-conventional methods, both polar and non-polar solvents have been used. Extraction efficiency also depends on the solubility of the desired phytochemicals in the extraction solvent. In general, the like dissolves like principle applies and polar solvents will solubilize non-polar phytochemicals while non-polar solvents will solubilize non-polar phytochemicals. Lowly polar solvents have been reported to successfully extract flavonoids aglycones and an addition of a polar solvent allows the extraction of flavonoids glycosides and anthocyanins. Benzene, chloroform, ether, and ethyl acetate are typical solvents for the extraction of lowly polar compounds such as isoflavones, flavanones, dihydroflavonols, flavones, and flavunols whose chemical structures are very methylated [26]. Very polar flavonoids glycosides like hydroxylateflavones, flavonoles, biflavonyls, and chaliounes have been conventionally extracted using acetone, alcohols, and water. In fact, water is a typical solvent to use in the PHWE method to extract phenolic compounds [27].

In the MAE method, the extraction of quercetins and rutin is also affected by the concentration of solvent used. Mlyuka et al [28] observed that phytochemical yield increases with increase in ethanol concentration from 30% to 50% and then decreases with dilution of the ethanol. The same study also reported that the pH of the solvent affects extraction efficiency. Solvent volume is another important factor to consider and generally the PHWE method is advantageous in this aspect as the amount of solvent required is relatively low [25]. Because a high pressure is used, even large volumes of solvent can be heated up to the desired temperature quickly [25].

Extraction time is another important parameter to consider in the extraction of the phytochemicals from *C. nutans*. Conventional methods tend to require lengthy extraction times (2–24 hours) which will also degrade the phytochemicals during the extraction process. For this reason, the PHWE method is preferable because since higher temperatures can be achieved in the pressurized vessel, the extraction time can be kept short.

Temperature can greatly impact extraction efficiency as it affects the solubility of major phytochemical compounds, the average rate of diffusion of analytes through the solvent, and the proportion of the phytochemicals extracted. Some phytochemicals will degrade at high temperatures and therefore the selection of an appropriate temperature is important, especially if the extraction time is long. According to a study by [29], the yield of phenolic and flavonoids, and their DPPH scavenging activity when using PHWE technique was observed to increase with temperature from 100 until 180°C, however at above 180°C, the extract is expected to scorch. The increase of the extracted compound is believed due to the amplified mass transfer of the compound with the water temperature. At higher temperature, the compound degraded or completely decomposed due to thermal activity and/or Maillard reaction [30].

6. Conclusion

C. nutans extracts have great potential in medicinal research and therefore, efficient and commercially viable methods to extract them are well developed. This mini review presents a brief comparison between the conventional and non-conventional methods to extract phytochemicals from *C. nutans*. Major advantages of the non-conventional techniques include short extraction time, low solvent consumption, low generation of toxic pollutants and high extraction yields. The right selection of method results in good yield and quality of the extracts. There is always room for improvement in developing modern extraction method to cater to the demanding needs of commercial applications that use plant derived products. Improvement of existing extraction methods enhances extraction efficiency, yields, quality, and the eco- friendliness of the process.

7. Acknowledgement

The author would like to acknowledge the support from the Fundamental Research Grant Scheme (FRGS) under grant number FRGS/1/2016/TK02/UNIMAP/03/2 from the Ministry of Education Malaysia.

8. References

[1] Deb, J., and Dash, G. K., 2014 Der Pharm. Lett., **6**(3), pp. 61–66
[2] Ismail, N., and Kasim, K., 2018 J. Eng. Res. Educ., **10**(1), pp. 1-8 Special Issue on Bioprocess & Biosystems.
[3] Fazil, F., Azzimi, N., and Zubairi, S., 2019 *AIP Conference Proceedings*, Selangor, Malaysia, p. 50003
[4] Tee, L. H., Ramanan, B., Tey, C., Azrina, A., Bao, I., and Prasad, L., 2015 Med. Chem. (Los. Angeles)., **5**(1), pp. 23–27
[5] Hasmadi, M., Lusia Barek, M., Zaleha, A. Z., and Mohd Fadzelly, A. B., 2015 **22**(2), pp. 661–670
[6] Ghasemzadeh, A., Nasiri, A., Jaafar, H., Baghdadi, A., Ahmad, I., and Syarhabil, M., 2014 Molecules, **19**(11), pp. 17632–17648
[7] Baharuddin, N., Nordin, M., Morad, N., Aris, N., and Yunus, M., 2018 Malaysian J. Anal. Sci., **22**, pp. 659–666
[8] Khoo, L., Audrey, S., Lee, M., Tan, C., Shaari, K., Tham, C., and Abas, F., 2018 Evidence-Based Complement. Altern. Med., **2018**, p. 39
[9] Yadav, R. N. S., and Agarwala, M., 2011 J. Phytol., **3**(12)
[10] Ghude, K. R., Ayre, A., Mane, P., Nemade, M., Gosavi, S., and Pathare, A., 2013 Asian J. Biomed. Pharm. Sci., **3**(23), p. 1
[11] Yeo, B., Yap, Y., Koh, R., Ng, K., and Chye, S., 2018 Trop. J. Pharm. Res., **17**(2), pp. 375–382
[12] Majid, M., Laurencea, Y., Khana, M., and Sarjadia, M., 2018 Borneo Sci., **2**, p. 39
[13] Solibun, A., and Sivakumar, K., 2016 *IOP Conference Series: Earth and Environmental Science*, Kota Kinabalu, Malaysia
[14] Tiew, W., Png, X., Chin, J., Akowuah, G., Xiu, W., and Akowuah, G., 2014 J Appl Pharm, **6**, pp. 77–86
[15] Mustapa, A. N., Martin, A., Mato, R. B., and Cocero, M., 2015 Ind. Crops Prod., **74**, pp. 83–94
[16] Sanjaya, E., Tedjo, Y., Kurniawan, A., Ju, H., Ayucitra, A., and Ismadji, S., 2014 J. CO2 Util., **6**, pp. 26–33
[17] Tyśkiewicz, K., Konkol, M., and Rój, E., 2018 Molecules, **23**(10), p. 2625
[18] Rosello, S. E., Galanakis, C., Brncic, M., Orlien, V., and Trujillo, F., 2015 Trends Food Sci. Technol., **42**(2), pp. 134–149
[19] Teo, C. C., Tan, S. N., Yong, J. W. H., Hew, C. S., and Ong, E. S., 2010 J. Chromatogr. A, **1217**(16), pp. 2484–2494
[20] Angiolillo, L., Del Nobile, M. A., and Conte, A., 2015 Curr. Opin. Food Sci., **5**, pp. 93–98

[21] Yu, Q., Li, C., Duan, Z., Liu, B., Duan, W., and Shang, F., 2017 Czech J. Food Sci., **35**(1), pp. 89–94
[22] Nieto, A., Borrull, F., Pocurull, E., and Marcé, R. M., 2010 TrAC Trends Anal. Chem., **29**(7), pp. 752–764
[23] Mustafa, A., and Turner, C., 2011 Anal. Chim. Acta, **703**(1), pp. 8–18
[24] Kovačević, D. B., Barba, F. J., Granato, D., Galanakis, C. M., Herceg, Z., Dragović-Uzelac, V., and Putnik, P., 2018 Food Chem., **254**, pp. 150–157
[25] Ameer, K., Shahbaz, H., and Kwon, J., 2017 Compr. Rev. Food Sci. Food Saf., **16**(2), pp. 295–315
[26] Harborne, J. B., and Williams, C. A., 2000 Phytochemistry, **55**(6), pp. 481–504
[27] Suwal, S., and Marciniak, A., 2018 Nepal J. Biotechnol., **6**(1), pp. 74–91
[28] Mlyuka, E., Zhang, S., Zheng, Z., and Chen, J., 2016 African J. Biotechnol., **15**(22), pp. 1041–1049
[29] Nur, A. J., Khairul, F. K., Nuradibah, M. A., and Noor, S. S. S., 2018 IOP Conf. Ser. Mater. Sci. Eng., **429**, p. 012064
[30] Klungboonkrong, V., Lamsal, B. P., and Phoungchandang, S., 2019 J. Sci. Food Agric., **99**(2), pp. 933–940

Utilization of Tofu Dregs as Goat Fodder and Utilization of Nano Particles for Skin Care Products

Wahyu Lestari*, Ratu Fazlia Inda Rahmayani, Adhisyahfitri Evalina, and Mulia Saputra

Universitas Syiah Kuala, Banda Aceh, Indonesia

*wahyu_lestari2000@yahoo.com

Abstract. The purpose of this community service is to improve the welfare and income of partners through the use of tofu dregs into animal fodder. Then, it will correlate to the quality of milk which produced by the goat. Finally, this goat's milk will be processed through nanoparticles become a moisturizer for skin care. The location of community service is dedicated for small businesses in Lamnyong Village, Kuala Syiah sub-district, Banda Aceh. In this activity, a method was introduced tofu dregs could be used as quality food for goats which effects on goat milk were highly nutritious. In addition, nanoparticle processes will be introduced on the process of goat's milk which can be a moisturizer for skin care. This will be a new product that can be mass produced and marketed by partners and continuous promotion, thus increasing consumer interest in buying, which will increase partner income. The result of this service is the addition of new fodders resulting in better quality milk products, this is evidenced from laboratory results and other results is the creation of a skin care product that functions as a skin moisturizer. Finally, this product can be used to soften the skin and protect the skin from the sun.

1. Introduction

In order to be healthy and good, everyone needs milk as a source of energy. Milk can be easily found in various types and brands in the market. Foods that are circulating in the market usually come from the results of animal husbandry, sheep, goats, etc. and the results of processing agricultural products such as soybeans. Because of different sources, even though it is equally beneficial for the body, each milk produced from milk content is also different.

Today, the milk production process is managed in a traditional and modern way. In modern milk is produced with the help of technology such as the procurement of high-capacity and sophisticated machines. Although it uses many technologies, there are some businesses which still used traditional milk-squeezing to maintain the nature and freshness of the milk.

Utilization of Goat Milk is widely known for maintaining skin beauty and smoothness. Goat milk contains many vitamins, especially vitamins A and D. These vitamins will make the skin soft but strong. Goat milk in this service contains beta hydroxy acid which makes the dead skin cells peel off and replace them with new skin cells so that the new skin is fresher.

George Dermit, a doctor from Ohio University has conducted research and therapy on Goat Milk for children with skin diseases. From the results of the study, he concluded that goat milk is very good for sufferers of skin diseases such as eczema, zits, and itching. Moreover, the dermatologists from New Zealand also apply this to any country in the world.

Goat Milk which is rich in natural hydroxy beta acid is now often used as a mixture for scrubs. This mixture not only functions as a peeling that erodes dirt and dead skin cells but is able to brighten the skin to look smoother and not scaly. Additionally, the protein content in milk is also useful as a nutrient supply that functions to moisturize while coating the surface of the skin to make it more smooth and supple.

Besides the benefits of goat's milk, through nanoparticle processes, it can produce skin care moisturizers. This process must use the quality goat milk to produce good and healthy products. This moisturizer is useful for various skin problems.

1.1 Current Conditions and Partners Problem

a. Production Aspect

In the business processing, several obstacles faced by UD. Atjeh Livestock farm in producing milk from the aspect of production, there are:

1) Livestock change because looking for other goats takes a long time and requires a large amount of money. The type of goat that produces quality milk is the result of crossbreed originating from India. This type of goat, it can only be found in Medan and Central Java. It is because of the kingdom in the past, Indian traders came and traded in Indonesia with traders from Batavia, Medan, and Java. They crossbreed their livestock to produce quality milk and there was a crossing which produced a type of crossbreed goat.
2) Quality of human resources in the field of animal husbandry is still limited. Then, the number of workers at the production site is very small. The goat milk business is run by an owner with two workers. However, the milking goat process, there is more labor is needed which is caused business becomes lighter.
3) The amount of production is still limited which make them hard to establish a partnership or become a supplier with other businesses. This causes production to be limited to consumers who only need goat milk to drinks, it cannot reach other aspects such as cheese, yogurt, soap, and so on.
4) Lack of adequate production equipment such as goat milk equipment that uses existing machinery and equipment is limited, such as milk cooking tools that use machines as well, so that group members can more easily produce goat milk, and the value added can be enjoyed by group members.

b. Management Aspect

In business management, the goat milk business group experienced several obstacles, there are:

1) Lack of workers who are experts in their fields in conducting business activities such as feeding and drinking livestock, milking, cleaning cages, and others.
2) Weak in marketing. The lack of market information and skills in marketing products has resulted in the low sales which can be obtained by the goat's milk business, because they market their products without special promotions, only rely on mouth to mouth, a sales system that only relies on customer trust so that one customer will tell the experience of buying products and get health benefits from consuming these products to other prospective customers.

1.2 Problem of Partner

Priority Problem

We draw several priority issues to be resolved so that the goat milk business group becomes more developed in aspects of milk production and business management. Some of the priority issues are:

a. Production Aspect

1) Keep increasing the crossbreed goat quantity to produce quality goat milk.
2) Keep increasing the quality of human resources in the field of animal husbandry and squeezed milk.

3) Keep increasing the production capacity by adding milk machines, to create the faster and more numerous products, and also the profits obtained can increase with the milk produced.

b. Management Aspect

1) Increased number of workers who are skilled in the field of goat farming in managing goat milk business.
2) Increased marketing capacity and capability, in terms of analyzing share expansion and market information.
3) Increased capacity of actors in the business of goats in terms of organization/ management.

1.3 Solutions Offered

Regarding the main problems of the partners that have been stated, the solutions offered are:

a. Production Aspect

From the aspect of production, the solutions offered by the Science and Technology program for society include:

1) Increased capacity for the number of crossbreed dairy goats to produce quality goat milk by collaborating with suppliers of dairy goats. Then, when livestock changes it becomes easier.
2) Increased number of qualified human resources in the field of animal husbandry by employing graduates in the field.
3) Increased goat milk production capacity by increasing the number of dairy goats so that they can collaborate or become suppliers with other businesses.
4) Provision of goat milk milking machines to further accelerate production, which is accompanied by a mentoring program to use the tool.
5) Increased product marketing capacity, so partners can market their products with their own brand and packaging and also make consumers find out about these products.

b. Management Aspect

From the management aspect, the solutions offered from science and technology programs for members of this group include:

1) Training in Human Resource Management, which is the improvement of self-development management expertise, which is related to the production process of goat milk.
2) Increased marketing capacity and capabilities, in terms of expanding market share and market information.

2. Methodology

The community services program has been conducted include product creativity and capacity building for partner business organizations. Therefore, the application of science and technology is carried out through a comprehensive and multidisciplinary approach. The program must also involve the perpetrators of the goat milk group as beneficiaries of the program.

The training method is designed using a participatory approach with reference to the adult learning process, through an approach: Presentation of material equipped with PowerPoint Presentation and simulation, Discussion: limited, group, Practice, and Feedback from business actors in the form of understanding the material provided.

Increased production capacity carried out through two fields includes (1) training in marketing goat milk, (2) procurement of dairy machines and other processing equipment, both of which are accompanied by assistance. Mentoring is done by involving parties who are experts in their fields.

To improve capabilities in the marketing field, partners will be given material related to marketing techniques such as determining target consumers, analyzing the market or consumer needs, product marketing through the internet (web), making brand and product logos, as well as market development and breakthrough techniques market.

Development in the field of human resources is held through training and development of workers' expertise, increased collaboration among group members, and strategies for creating a group environment situation to brainstorm or design thinking in a short time, and provide motivation to the goat milk business group to encourage quality dairy products, beneficial to health and have a high market value.

3. Community Services Result

Processing of Animal Fodder into High Quality of Goat Milk

This community service focuses on how to increase farmers' income with various innovations, especially regarding goat feeding to improve the quality of goat milk so that goat milk produced is of high nutritional value and is favored by the community. There are many things have been discussed together with farmers. In the discussion, there were still many shortcomings that needed to be addressed. Among other things is equipment that is still very lacking in handling goat's milk. In addition, in this program, it tries to market dairy products to more extensive things, such as online systems. In the following picture, a discussion of the work plan of service and discussion in terms of overcoming the problems that have occurred so far. Discussions with Pak Bambang in terms of this service can be seen in Figure 1 below:

Figure 1. Road Map of Community Services

Additionally, there are other innovations that devotees try to do, namely by utilizing goat milk into skin moisturizers. Goat milk can be made into a skin moisturizing product through the nanoparticle system. If this product is finished, then this skin moisturizing product can be bought and sold to the public and it can increase farmers' income. However, to produce a good moisturizing product, go through good goat milk. Quality goat milk can be obtained from good goat fodder. Therefore, one of the food sources that must be considered is tofu dregs. It is the best and most nutritious food for goat.

The disadvantage of tofu dregs is that tofu dregs products cannot be stored for a long time, so every purchase of tofu dregs must be calculated when it can run out, so the tofu dregs are not wasted. Tofu dregs are mixed foods where the main food is grass. Mixed foods are very influential on goat's milk resulting from. Therefore, assistance from community service is very helpful for farmers in developing their business. The table below, assistance provided when visiting a goat farm. Giving out the tofu before serving is that they only buy tofu dregs and then give it to the goat. Then, another innovation which can be applied is to provide better quality food. The method used is by the addition of feedings from before. The table below will illustrate the prior fodder of the business, as it showed in the table:

Table 1. Fodders of Goat before the Program

No	Goat Fodder
1	Tofu Dreg
2	Sago Dreg
3	Rice Bran
4	Corn Bran
5	Corn Cinder
6	Mineral Salt

7	Grass
8	Corn Stem

In the implementation of the Grant program, additional food was provided to increase the production and quality of the milk. Giving these foods is mixed with other foods. The types of food additives can be referred in the following table 2:

Table 2. Additional Fodders of goat livestock

1	Sago flour Second Quality
2	Coconut Meal

The addition of these foods is based on the high protein from the feed. Sago flour is high in energy so it will increase milk production and viscosity of milk. While coconut meal is high in vegetable protein which it can increase the milk production. The availability and continuity are very good for milk production. The amount of food given per day per goat livestock is:

a. Feed the concentrate of 3% from the weight of livestock life = 0.9 kh/head/day
b. Forage fermented 10% of livestock weight = 3kg/tail/day

The results obtained after getting additional food from the program, there are as follows:

a. Increasing the amount of goat milk production per day. The previous increase in milk production from 1 liter/head/day to 1.3 liters /head/day.
b. Increased levels of milk nutrients that are better than before
c. The weight gain of dairy goats for 1 month averages 0.1 kg /head/day

The comparison of the analysis of milk quality before and after the addition of food is: After doing analysis in the laboratory, the quality of milk after adding the food ingredients is an increase in milk protein levels, milk fat, and milk lactose. Laboratory testing is carried out after 1 month of observation.

The lack of tools in the process of milking and milk production is also very pronounced in this effort, therefore we need the help of machines to launch this business process. The tools that exist before the programs are:

a. Clean water is used to clean the udders of the main livestock
b. A clean washcloth is used to clean and dry washed udders
c. Portable milking machine
d. The power cord as a source of electric current for the appliance
e. Glass liters are used for milked milk containers
f. Milking milk storage containers
g. Milk production logbook
h. Stationery for recording
i. Gauze for milk filters
j. Stainless pan for pasteurization

The equipment available is limited, therefore it needs help to add the tools for the business to smoothen again. The assistance of these tools includes:

a. Portable Milking machine for milking equipment
b. The CMT tool is used for mastitis disease testing in the main livestock
c. Milk fat test kits are used to test milk fat levels
d. The pH meter is used to see the pH of milk
e. Bio-lactation flour is used for additional milk production minerals in the main livestock
f. Drums are used to store and ferment tofu dregs mixed with concentrate

The provision of feeding assistance and tools can be seen in figure 2 below:

Figure 2. Provision of dedication assistance

All assistance provided has the goal to the small and medium entrepreneurs in order to get better milk production than before so they can increase their income, it can expand their business and the can add more workers. Other analyzes conducted to see the quality of milk are analyzes of the microbial composition. This analysis has been carried out and is still awaiting laboratory results from BPOM and it is expected that in the near future the results will be obtained. Figure 3 below, is the process of mixing food for dairy goats and the process of milking goats.

Figure 3. Mixing of Tofu Dregs for Fodder of Goat

However, after obtaining good quality goat milk, then this service will also process goat milk into skin moisturizing products for skin health through particle nano processes. The manufacture is carried out in a chemical laboratory to produce good products. In addition, these skin products will be examined by dermatologists and chemists to ensure that the process of converting milk into skin moisturizing products is correct. The skin moisturizing product is expected to be a product that is beneficial for skin farmers and can be marketed to the general public which it can improve the visibility of entrepreneurs.

The Processing of Goat Milk into a Skin Care Product

Hypoallergenic moisturizers (Hypo-Allergenic) are moisturizers that do not contain substances that can cause reactions allergies to the skin. This type of moisturizer is suitable for sensitive skin. Hypoallergenic means "minimizing the appearance of allergies". The term Hypo-Allergenic for skin care products means that the product uses as little as possible raw materials that can cause allergic reactions.

The use of Hypo-Allergenic products does not mean 100% free from allergy risks, but the reactions that occur due to some very sensitive raw materials on the skin can be limited to a minimum level. A product needs to be tested to prove that the product does not cause skin irritation or allergies. Products that have been tested by a dermatologist and proven to not cause skin irritation allergies are called dermatologist tested products.

To produce Hypo-Allergenic products can be done by removing allergens. In addition, the used of natural ingredients that function as an antioxidant and anti-bacterial, anti-inflammatory and others into skin moisturizers can also produce Hypo-Allergenic moisturizers.

CONCLUSION

From the community service program carried out to get a positive response for farmers and hopefully this collaboration can increase knowledge of farmers in continuing to manage goat's milk in accordance with the knowledge obtained from the service system. Then, the skin moisturizing products that are produced can also be produced in large quantities and can be marketed online and also manually. Dairy products produced in this service, become healthier which can increase the purchasing power of other communities. In addition, effective marketing, such as online, can introduce goat's milk products to be better in marketing, so that communities outside Aceh can buy and order online.

The skin care moisturizing products produced are also very helpful to partners in terms of new product introduction. This moisturizing product is much favored by women, especially those whose skin is often exposed to sunlight. So, what exactly is the superiority of goat's milk compared to other milk. Basically, both goat milk, cow's milk, and camel milk are rich in protein, vitamin A, vitamin D, and other minerals. Goat milk has a higher protein and vitamin A content than cow's milk, as well as smaller fat molecules. Goat's milk is also usually richer, making it suitable for pampering the skin like a soap and milk bath product.

References

1. Arikunto, Suharsimi. 2013. Research Procedure A Practice Approach, Fifteenth Published. Jakarta: Rineka Cipta.
2. Dedy, H (2005), Marketing Business, University of Sebelas Maret Surakarta.
3. Handoko (1999). Personnel & Human Resource Management. Second Edition. Yogjakarta
4. Lee, J.D.(2002), "Concise Inorganic Chemistry", McGraw-Hill Company. NewYork.
5. Iswari, R. dan F. Latifah. (2007). "Handbook of cosmetic science" Jakarta: Gramedia Pustaka Utama.
6. Majalah Health, (2011) "Natural Herbs for Skin Health" http://majalahkesehatan.com. Accessed on 1 October 2017.
7. Purwati, E. (2010). "Molecular Characterization of Lactic Acid Bacteria Isolated From Blondo (Waste of Virgin Coconut Oil) Biscuit Which Potential to Prevent Pathogen". Presentation of 2nd International Conference on Chemical, Biological and Environmental Engineering. Egypt, Kairo.
8. Pusat Bahasa. 2012. Indonesian Dictionary. Third Edition. Fourth Published. Jakarta: PT Gramedia Pustaka Utama.
9. Purwaningsih, Salamah, E., & Budiarti, T.A. (2014) "Good eating for livestock". Journal of Aquatica. 5(1) : 55-62.
10. Purbani, D.(20010). "The Process of Forming Salt Crystallization. Marine research center and non-biological resources". Jakarta.
11. Schmitt, W.H. (1996). "Skincare products. Cosmetic and Toiletries Industry". London. Blackie academic and professional.

Utilization Of Cow And Organic Waste As Organic Waste Empowerment Solution

Anggraeni Yunita[1], **Karmawan**[2], **Sumar**[3]

[1]Accounting Department, Economy Faculty, Bangka Belitung University, Integrated Campus of UBB Balunijuk, Bangka Belitung, Indonesia, 33172
[2]Accounting Department, Economy Faculty, Bangka Belitung University, Integrated Campus of UBB Balunijuk, Bangka Belitung, Indonesia, 33172
[3]Management Department, Economy Faculty, Bangka Belitung University, Integrated Campus of UBB Balunijuk, Bangka Belitung, Indonesia, 33172
Email: anggi21.ay@gmail.com

Abstract. *Bukit Kijang Village is one of the villages that is classified as young in the Namang Bangka Tengah District. one of the flagship programs of the Bukit K.ijang village, which is farm-based agriculture on ex-mining land. To achieve the success of the program, it is necessary to make appropriate solutions for waste management to support the progress in agriculture. Village farms have 9 adult cows and 1 turtle calf so that cow dung produced is quite a lot. To overcome the problem of cow waste and household waste, one step is by making biogas from cow dung which can also produce solid and' liquid fertilizer and utilization household organic waste in making biofertilizers.* Mentoring activities were attended by *Ladies of Family Welfare Empowerment*, villagers and village youth groups. *Both products produced are less than perfect due to equipment leakage and lack of shelf life so that the biogas and biological fertilizers that are formed are not perfect. So can't to test the products.*

Keywords: Cow Dung; Organic Waste; Biogass; Biological Fertilizers

1. Introduction

Bukit Kijang Village is one of the villages in Namang Subdistrict, Central Bangka Regency which has an area of 813,188 ha with a population of 1,145 people and is still classified as a young village or a new village. The village is divided into two hamlets, namely Hamlet of Bukit Lesung and Hamlet of Batu Kijang. The majority of communities in the two hamlets work as farmers and miners. The large amount of mining activity in the Bukit Kijang Village area causes unproductive land so that one of the village's flagship programs is developing agriculture that correlates with livestock in post-mining areas. The correlation is intended as a step to repair the soil due to mining activities.

The program is supported by the existence of village livestock facilities, namely cattle farming, but the cow manure is still not optimally utilized. Cow manure is only managed by one person in charge and is only used in making compost to improve the texture of the former land of the mine before planting. In some cases cow manure has many benefits apart from being a basic ingredient in composting, cow manure can also be used as an alternative energy-producing material such as electricity and gas. The gas produced from cow dung or other organic material is called biogas. Biogas is a gas produced from the fermentation process of organic materials. This gas comes from various types of organic waste such as biomass waste, human waste, animal waste can be used as energy through anaerobic digestion process [8].

The next target is the utilization of household organic waste in the manufacture of liquid compost or biological fertilizers aimed at housewives as a step to develop household scale farming or home garden gardens. Making this biological fertilizer can play a role in reducing household waste pollution while changing the mindset of the community in the utilization and empowerment of organic materials. The activity of utilizing organic waste has been running in several advanced villages in Indonesia, especially in the Java region and has a positive impact of creating chemical-free vegetable farming which can also reduce production costs in the agricultural sector [5].

The lack of awareness of Bukit Kijang Village community towards the empowerment of organic waste both livestock waste and organic waste encourages the idea of inviting the Bukit Kijang Village community to develop village potential in agriculture and animal husbandry so that synergies between rural agriculture and livestock may result in an economic impact village. Especially the development in agriculture and the utilization of cow manure so that it can produce economically valuable products.

Biogas is an alternative energy source. Biogas is a gas formed because of anaerobic fermentation process from waste or waste materials and other organic materials. To produce biogas, a biogas reactor (digester) is needed which is an airtight installation so that the process of decomposition of organic matter can run optimally [10]. Biogas is a mixture of gases produced by methanogenic bacteria that occur in materials that can decompose naturally under anaerobic conditions. Biogas is odorless and colorless and when burned, it will produce a bright blue flame like LPG gas [1].

Biogas is a collection of several gases produced by organic materials through anaerobic fermentation process. Biogas has the main content of methane and carbon dioxide but also contains other gases. Natural gas is a gas consisting of several gas elements that have different chemical compositions. The biogas composition produced by anaerobic processes includes methane (CH_4), carbon dioxide (CO_2), several inert gases, and sulfur compounds. Methane is a gas derived from natural gas that can be burned [3]. Biogas technology with the concept of zero waste (not produced by waste) is expected to help slow the pace of global warming. In addition to being able to become alternative energy, biogas can also reduce environmental problems, such as air pollution, soil pollution and global warming [9].

The utilization of manure can be an easy alternative for the Bukit Kijang Village community in overcoming the need for electrical energy and also the need for LPG gas. Three cows can produce biogas which can be used in cooking, heating water and also as a source of electricity for petromarks. In addition, the use of biogas is also safer than using synthesis gas. This is because biogas has lighter and volatile molecules that reduce the risk of fire and explosion [8].

Biogas production is also relatively easy, the main ingredients used are cow dung and water with a ratio of 1: 1 and fermented in anaerobic digester. Digester is an oxygen impurities compost device so that methanogenic bacteria can produce methane gas from cow dung. Besides being able to produce biogas, the use of biodigester can also produce byproducts in the form of liquid compost and solid compost, so that it can support in agriculture [8].

Waste is waste material that is not wanted after the end of a process or activity. Waste is a source of environmental pollution because it causes unpleasant odors, can pollute water, soil and is seen as aesthetically reducing the beauty of the environment. Household waste such as leftover vegetables and rice water laundry can be used as basic ingredients for making compost and biological fertilizers. The remains of vegetables in household waste are nutrient-rich organic waste needed by plants. Therefore, the remaining vegetables can be used as basic ingredients for making biological fertilizers. Vegetable waste is also a material that is easily decomposed by microbes so that the required process does not take long. Vegetable waste in biological fertilizers acts as a source of microbes which is one of the important components in organic fertilizer solutions (Retno, 2009). The use of waste is done by making it a new thing that is very useful. One effort to utilize the waste is to make it compost. The composting process in order to run faster and more efficiently is done by adding microorganisms to the organic matter or activator and supporting materials by utilizing rice washing water [7].

Waste of rice water laundry itself plays a role in producing carbohydrates in making biological fertilizer solutions. Milk-white rice washing water, it indicates that the protein and vitamin B1 that is abundant in rice are also eroded. And then vitamin B1 is used to reduce stress on plants after transferring media and spurring plant root growth [6]. This is in line with the results of previous studies which stated that used rice washing water can stimulate the growth of Adenium roots. The benefits of rice washing water have also been studied by Leonardo (2009), the first washing water of rinse rice has an effect on increasing the number of leaves and height of tomato and eggplant plants. One of the contents of rice laundry water, namely phosphorus which is a macro nutrient that is needed by plants [4].

Brown rice washing water contains calcium, iron and vitamin B1 nutrients which are greater than white rice washing water, while white rice washing water has higher nutrient content of nitrogen, phosphorus, magnesium and sulfur than brown rice washing water. The difference in nutrient content is noticeable in sulfur (S) nutrients. The content of S in brown rice washing water is 0.005% while in white rice washing water is 0.027%. Sulfur in plant metabolism has a role in protein synthesis and part of the amino acids cysteine, biotin and thiamin. Sulfur helps stabilize protein structure, helps oil synthesis and chlorophyll formation, and reduces disease attacks in plant bodies [11].

The use of organic waste or organic household waste can encourage housewives in Bukit Kijang Village to make a product that has economic value and also supports agricultural activities at household scale. The other positive impact is to provide selling value to organic waste which has been made as liquid organic fertilizer, adding value to organic waste, and can reduce the dependence of farmers on the use of chemical fertilizers [2]. The making of this biological fertilizer is also relatively simple and easy, the organic waste used will be fermented in an anaerobic composter with the help of the Effective Microorganism EM-4 activator. Simple and easy making is very possible to be carried out by the community including housewives [11]. The empowerment of organic waste in fertilizer production also contributed to the development of household agricultural programs in Bukit Kijang Village and reduced production costs in rural agriculture.

2. Methodology

The activity was held in Bukit Kijang Village which lasted for 4 weeks and was divided into 4 stages of activities, namely:

1. The first stage is making a digester and composter tool
2. The second stage of socialization
3. The third stage is assistance with direct practice methods
4. The fourth stage of evaluation of results

3. Results and Discussion

3.1. Tools and Materials

The tools used are simple biodigester consisting of gallons, 1/2 inch paralon pipes, ¼ inch water hoses, gas taps, clamps, inner tubes, plastic glue, black paint, and order. Another tool is the composter which consists of a water drum, 1 inch paralon pipe, 1/2 inch paralon pipe, filter / viber, connecting pipe Y pipe cover, water tap, and plastic glue.

The materials used are cow dung waste, water, household organic waste (leftover vegetables, banana peels, bamboo humps, eggshells, fish scraps, etc.), rice washing water and EM-4 activator (Effective Microorganism).

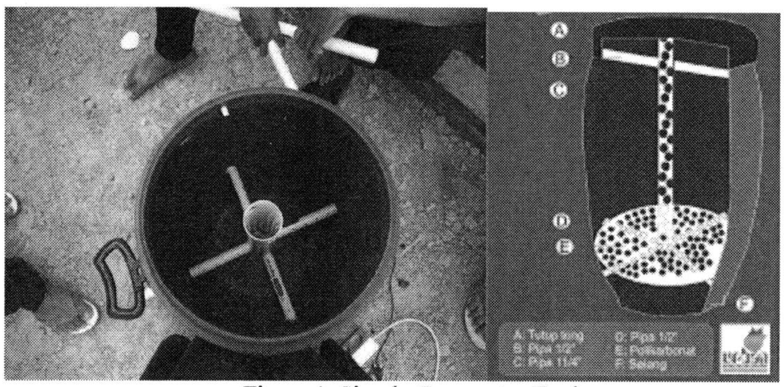

Figure 1. Simple Composter Tools

3.2. Dissemination and Direct Making of Biofertilizer by Ladies of Family Welfare Empowerment in Bukit Kijang Village

The dissemination was held in front of the village office, with a total of 20 participants who were members of the Ladies of Family Welfare Empowerment in Bukit Kijang Village. The dissemination began with the presentation by the presentator about the steps and importance of the separation of organic and inorganic waste. This dissemination invites Ladies of Family Welfare Empowerment to foster a sense of care in processing waste according to the type of waste by increasing important functions in sorting waste before being utilized. Even though the organic and non-organic waste systems have been implemented, however, in reality residents, especially the women of Bukit Kijang Village, still find it difficult to separate or dispose of waste according to the type of waste. They tend to still mix organic and inorganic waste so that the waste is difficult to recycle. The activity continued with the delivery of material that was more focused on organic waste, namely the use of organic waste in making fertilizers. In addition to socialization, participants were also invited directly or directly involved in making biological fertilizers so that there was interaction between participants and speakers.

Journal of Physics: Conference Series **1424** (2019) 012003 doi:10.1088/1742-6596/1424/1/012003

Figure 2. Making Biofertilizers That Directly Engage Participants

The making of biofertilizer starts with making an activator by mixing EM-4 (Microorganism Effectivity) as much as 2 tablespoons into 1 liter of rice washing water then adding 250 gr of white sugar as a source of energy or microbial glucose. To be more effective activators should be stored first for 3-4 days until the color of the water turns clear and smells of tape, but can also be used directly. Household vegetables or organic waste are then chopped to speed up the rate of fermentation or composting by bacteria and put into the composter. The waste that is already inside the drum is then doused with rice washing water which has been evenly mixed with EM-4. The drum is closed tightly and stored in a cool place for up to 2 weeks.

The direct interaction or involvement of participants in making biofertilizers shows that participants gave a positive response to this activity, and also with the involvement of participants in making fertilizers directly can make it easier for participants to understand the steps to make biological fertilizers so that they can be practiced directly at home. Liquid biofertilizer produced at home can be applied to household crops or home gardening so as to reduce dependence on chemical fertilizers, harvested vegetables are free of chemicals and reduce production costs.

3.3. Making Simple Biodigester Tools

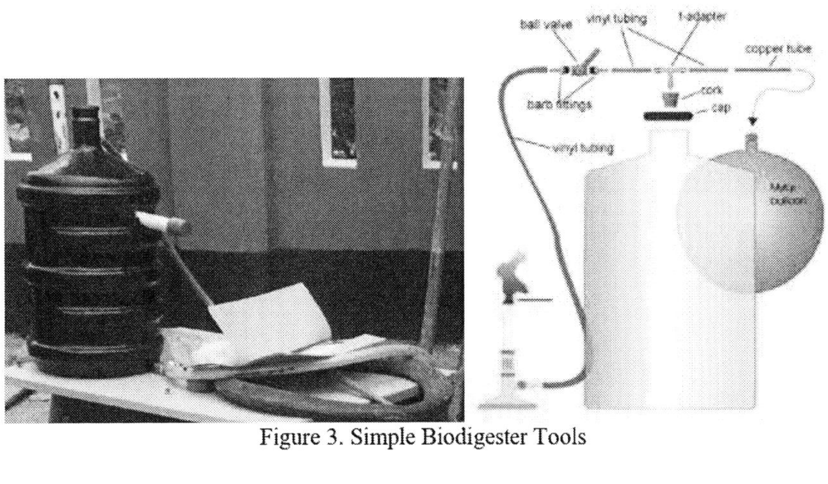

Figure 3. Simple Biodigester Tools

Making biodigester in producing biogas is quite easy, first of all gallon is given a hole in the side of the gallon neck and 2 holes in the side of the gallon body are 1 cm from the hole in the gallon neck. Each hole faces vertically and horizontally which will be filled with 1/2 inch paralon pipes which inside the gallon form the letter T. The hole in the gallon neck is then inserted an ¼ inch water hose connected to the Y connecting pipe and reinforced with a clamp. The left side of the hose is connected to the inner tube and the other part is connected to the gas tap. The digester that has been finished is then painted black to block the light from entering the gallon.

3.4. Socialization and Making Simple Biogas by Residents and Youth Organization in Bukit Kijang Village

The socialization of biogas was carried out in the village yard with 30 people consisting of ordinary residents and village cadets. This socialization was carried out with the aim of first presenting the targets from preparation to biogas production. In the socialization, it was also explained how to make a digester and the type of digester. In addition, the advantages of biogas and benefits for the community are also explained.

Figure 4. Biogas socialization in Bukit Kijang Village

The activity continued with the manufacture or direct practice of making simple biogas. Cow manure that is used is cow dung that comes from village farms which is as much as 3 kg. Cow dung used is cow dung that is still wet or still mixed with urine. Cow dung is then mixed with water with a ratio of 1: 1, so that 3 Kg cow dung is mixed with 3 liters of water. After that, the two ingredients are stirred until smooth before finally being put into the biodigester. Biodigester is then tightly closed and stored in a cool, low-light room. Fermentation time of about 1 - 2 weeks new gas can be produced. Besides being able to produce biogas fermented cow manure can also produce solid and liquid compost as a by-product. The benefits obtained are not only one but can also be beneficial in the field of agriculture, especially in accordance with one of the village programs, namely livestock-based agriculture. If 3 cows have been able to produce enough biogas for cooking and power generation, village farms that have 9 adult cows and 1 calf will not rule out the village cow dung can be one source of energy for residents and the byproducts are solid compost and liquid can be distributed to surrounding farmers. Based on the results of checking after the waiting period both of them experience imperfections due to the leakage of the composter and digester devices so that the fermentation and composting process takes place imperfectly so that further testing cannot be carried out.

4. Conclusions

Based on the activities that have been carried out both in making biological fertilizers and biogas conclusions can be drawn as follows:

1. The resulting product is not perfect, this is due to tool leakage and limited storage time. Whereas in the biogas on the 6th day a small amount of gas is formed but afterwards it decreases or the gas is not formed again.
2. Ladies of Family Welfare Empowerment are enthusiastic in making biological fertilizers.
3. Biogas that has been produced has not yet been tested on a flame. Biogas formed has a by-product in the form of solid and liquid fertilizer.

Therefore, then:

1. There needs to be guidance from the government related to the utilization of organic waste both in making biogas and biological fertilizers.
2. It is necessary to have a correlation that can facilitate the making of complex digester around the farm as a village cow manure fermentation tool so that it can be fully utilized.
3. Training and coaching still need to be done in the village community both for Ladies of Family Welfare Empowerment and youth groups.

5. Acknowledgments

The author expresses his deepest gratitude to the University of Bangka Belitung for funding this activity through the Real Work Lecture program, which one of the outputs is this paper. Furthermore, the authors would like to thank the parties who helped so that this paper was presented at the National Symposium on Community Service 2018 held by HSBC Indonesia in collaboration with Sampoerna University supported by Indonesian Journal Volunteers.

6. References

[1] A Ayub, A. Haryanto, S. Prabawa. 2015. Produksi Biogas dari Rumput Gajah (Pennisetum Purpureum) Melalui Proses Fermentasi Kering.Artikel Ilmiah Teknik Pertanian Lampung: 33-38.

[2] Asngad, A., Astuti, P., dan Rahmawati, I.N. 2013. Pemanfaatan Limbah Air Cucian Beras IR-36 dan IR-64 (Air Leri) Untuk Pembuatan Sirup Melalui Fermentasi Dengan Penambahan Bunga Rosella Sebagai Pewarna Alami. *J. Agrociltural*. FKIP UNS 10(1).

[3] Deublein, D., and A. Steinhauser. 2008. Biogas from Waste and Renewable Resource. Wiley-VCH Verlag GmbH & Co. KgaA. Weinheim. 443 hlm.

[4] Istiqomah, N. 2012. Pengaruh Pemberian Air Cucian Beras Coklat terhadap Pertumbuhan dan Hasil Tanaman Seledri (*Apium graveolens* L.) pada Tanah Rawa Lebak. *Agroscientiae*, 3 (17): 152-155.

[5] Nurdiyanti D, Utami A.S, Bastian N, Johan J. 2017. Pemanfaatan Limbah Organik Pasar Sebagai Bahan Pupuk Kompos Untuk Penghijauan Di Lingkungan Masyarakat Kota Cirebon. *The 5th Urecol proceeding*. UAD Yogyakarta.

[6] Purnami, WG. Yuswanti dan Astiningsih. 2014. Pengaruh Jenis dan Frekuensi Penyemperotan Leri terhadap Pertumbuhan Bibit Anggrek (*Phalaenopsis* sp) Pasca Aklimatisasi. *Agroekoteknologi Tropika*, 3(1) : 22-31.

[7] Retno S. 2009. Kajian pemanfaatan pupuk organik cair mikroorganisme lokal (MOL) dalam priming, umur bibit dan peningkatan daya hasil tanaman padi (*Oryza sativa* L.) (Uji coba penerapan *system of rice intensification* (SRI)). *Tesis*. Universitas Sebelas Maret Surakarta.

[8] Sulistyanto Y, Sustiyah S, Zubaidah S dan Satata B. 2016. Pemanfaatan kotoran sapai sebagai sumber biogas rumah tangga di kabupaten pulang pisau provinsi Kalimantan tengah. J. udayan mengabdi 15(2):153-158.

[9] Wahyuni, S. 2011. Menghasilkan Biogas dari Aneka Limbah. Edisi Pertama. PT Agro Media Pustaka : Jakarta. Hlm. 96.

[10] Wibowo, T.S., A, Dharma, dan Refilda. 2013. FermentasiAnaerob dariCampuran Kotoran Ayam dan Kotoran Sapi dalam Proses Pembuatan Biogas. Jurnal Kimia Unand. 2 (1): 113-118.

[11] Zakaria. 2013. Pemanfaatan Kulit Telur dan Air Cucian Beras dengan Penambahan CMA pada Media Tanaman untuk Pertumbuhan Tanaman Tomat (*Solanum licopersicum*). *Skripsi*. Fakultas Keguruan dan Ilmu Pendidikan Universitas Muhammadiyah Surakarta.

The Analysis of Attribution Reduction of *K-Nearest Neighbor* (KNN) Algorithm by Using *Chi-Square*

Muhammad Danil[1], Syahril Efendi[2], Rahmat Widia Sembiring[3]

[1,2,3]Department of Computer Science, Faculty of Computer Science and Information Technology, Universitas Sumatera Utara, Medan, 20155, Indonesia

*daniel.fikom@gmail.com

Abstract. Data reduction is one of the applicable techniques used to obtain the reduction representation from the data whose volume is much smaller, but still retains the original integrity of the data. Attribute reduction is a process to identify and eliminate an attribute with irrelevant or excessive values. In this study, attribute reduction was carried out using the Chi-Square Algorithm implemented in *K-Nearest Neighbor* (KNN) for classifying the objects based on the closest data to the objects.The test was carried out on the Pima Indians dataset with the total of 768 data. The Chi-Square method was used to reduce the dimensions of large datasets and to improve the accuracy of the prediction of the closest *K-Nearest Neighbor* results.*K-Nearest Neighbor* using *Chi-Square* as the choice of features proved to be accurate and effective in reducing data. The results of this study show that *K-Nearest Neighbor* using *Chi-Square* as the choice of features was accurate and effective in reducing data without reducing the integrity of the original data and reducing the quality of the information produced.

Keywords: *K-Nearest Neighbor*, KNN, *Chi-square*, *classification*, data reduction, Pima Indians dataset

1. Introduction

Analysis of complex and large amounts of data can take a long time to proceed and often leads to impractical or inappropriate analysis. Many researchers have conducted research on data reduction methods. Such technique is useful in reducing complex data without reducing the integrity of the original data and reducing the quality of the information produced. [4]

This selection process is implemented using the *K-Nearest Neighbor* algorithm. *K-Nearest Neighbor* is a method for for classifying the objects based on the closest data to the objects. *K-Nearest Neighbor* (KNN) can also be defined as a classifier used to classify data based on the comparison of the nearest K value. K parameter on *K-Nearest Neighbor* (KNN) has a large influence on the final outcome of the prediction produced. [5]

The method used for attribute selection in this study was the *Chi-Square* method, one of the feature selection methods which include the filter method. *Chi-Square* is a type of non-parametric comparative test performed on two variables, where the scale data of the two variables is nominal. If one of the two variables is variable with a nominal scale, Chi-Square test must be carried out at the lowest degree. [5]

2. Methodology

To obtain accurate, relevant, and valid data, the data was collectedusing the documentary research method. This method is a type of data collection examining various kinds of documents which are useful for analyzing the materials. The stages carried out in the implementation of this research are as follows:

a. Study of literature

In order to explore the previous studies discussing the process of data reduction in the *K-Nearest Neighbor* (KNN) algorithm by using *Chi-Square*, the study of literature was carried out. The data of this study were obtained from various books, journals, articles and several other references.

b. Research Analysis

The analysis carried out in this study was to evaluate the level of accuracy of the *K-Nearest Neighbor* algorithm by combining the *Chi-Square* algorithm in order to reduce the large data so that the accurate results can be gained.

Workflow diagram and Flowchart of Algorithms Work System undertaken in this study are illustrated in figure. 1 and figure. 2:

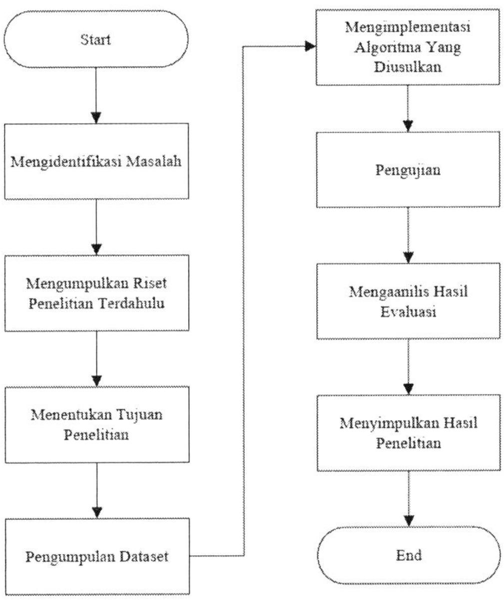

Figure 1. Research Work Flow Chart

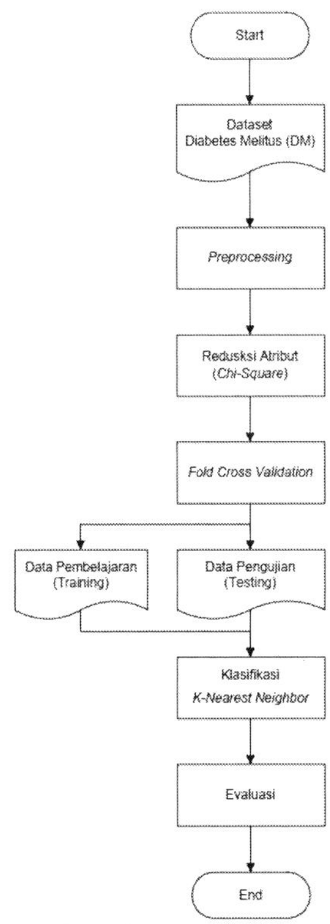

Figure 2. Flowchart Work System Flow Algorithms

Figure 1. Workflow diagram

Start – Problem identification – Reviewing the literature- determining the purpose of the study- dataset collection- implementing the proposed algorithms- testing- analyzing the evaluation results- summarizing the result of the study- end.

Based on figure 1, it can be explained as follows:

a. Pre-processing Stages

Pre-processing is a preparation process that is carried out on the dataset so that it is ready to be processed. All documents including training data, test data, and new data will be subject to pre-processing stage first.

In the preprocessing stage, the data is uniformised and reduced. This stage is done by selecting, cleaning, and then transforming the data into the desired form so that the model can be prepared. The purpose of this stage is to prepare valid data before being processed in the next stage. Thus, cleansing, transformation, and attribute reduction are carried out in this stage. The data obtained is processed to gain relevant and appropriate attributes. However, in this study, pre-processing data stage was not carried out.

b. Attribute Reduction (*Chi-Square*)

In this study, *Chi-Square* was used to reduce the data. It checks the null hypothesis which states that there is no relationship between attributes. Based on this null hypothesis, the model for distributing data is categorized with the assumption that there is no relationship between attributes. This test is undertaken based on the comparison between the actual data distribution and the expected data distribution. Contingency tables for calculating *Chi-Square* are also desirable. The frequency observed for c_{ij} cell is n_{ij}. The expected frequency for C_{ij} cell is e_{ij}.

$$e_{ij} = \frac{n_i n_j}{n} \quad \text{.......................................(1)}$$

The *Chi-Square* test formula can be calculated as follows:

$$(x^2) = \sum_{i=1}^{r} \sum_{j=1}^{c} \frac{(n_{ij} - e_{ij})^2}{e_{ij}} \quad \text{.......................................(2)}$$

Information:
n: Expected frequency
x2: *Chi-square* value
e: Frequency obtained

c. Fold Cross Validation

In this stage, the data sharing stage is carried out to divide the training data and the testing data. The amount of training data is 90% whilethe test data is 10%. Such distribution is used based on the suggestion of k-fold cross validation method so that the performance of the classification model can be measured. 10-fold cross validation is the basic form of cross validation. In this study, the 10 fold cross validation was used because the data is quite large which is 768 datasets. It means that the testing data used in this study was 70 datasets for each fold.

In cross validation, the training and testing data must be crossed in a row so that each data has the opportunity to be validated. According to Refaeilzadeh et al. (2008), there are two main objectives using cross-validation:

1. To measure the performance on the provided training modelusing one algorithm. In another word, to check the general performance of the algorithm.
2. To compare the performance of two or more different algorithms and gain the best algorithm for the available data. Moreover,it also compares the performance of two or more variations of the parameter model.

d. *Classification* (*K-Nearest Neighbor*)

K-Nearest Neighbor (K-NN) is a classifier that is used to classify data based on the comparison of the nearest K value and the K parameter on KNN which has a large influence on the final outcome of the prediction. K-NN measures the similarity of data using distance measurements. In this study, Euclidean distance is used to calculate the similarity of two or more data.

$$d(x, y) = \sqrt{\sum_{a=1}^{m} (x_a - y_a)^2} \quad \text{.......................................(3)}$$

Variable d is the distance of both data. K-NN is known as a method that requires time and large computational costs because the testing data and all available training data need to be matched first. Thus, there is a need for a data reduction process in order to reduce attributes that are not needed in the data so that computing time can be accelerated.

e. Evaluation

The final step is to do an evaluation. This stage aims to determine the accuracy of the results from the datasets classification using the classification method of the test data. Evaluation is performed to find the optimal solutions resulting from various and complex classification methods that are carried out repeatedly. [13] In this study, evaluation was conducted to determine the level of accuracy of the results in the preprocessing data, attribute reduction, and classification of test data stages.

Novakovic et al. (2017) mention that there are several standard measurements used to measure classifications with two class labels: accuracy, true positive rate, false positive rate, true negative rate, false negative rate, and precision. [13] In this study, the accuracy standard was used because it is considered adequate. Accuracy can be formulated using the following equation:

$$Accuracy = \frac{jumlah\ nilai\ benar}{jumlah\ data\ keseluruhan} x\ 100\% \quad \text{.......}(4)$$

f. Testing

The dataset that has been collected is then classified using the *K-Nearest Neighbor* (K-NN) algorithm and reduced by using *Chi-Square*. At this stage, the system that has been implemented is tested. After testing the system, the stage of measuring system performance was performed in order to draw conclusions on the results of the research conducted.

g. Analysis of the Evaluation Results and Conclusion of the Research Results

The next step is analyzing the results of the two algorithms in order to reach the conclusion of the study. From the results, it is expected that *K-Nearest Neighbor* and *Chi-Square* can predict a person's tendency to develop diabetes mellitus (DM). Furthermore, through the data reduction, it is also expected that the results and the accuracy of the study can be optimized.

3. Result and Discussion

The selection process in this study was implemented using the *K-Nearest Neighbor* algorithm. It is a method for *classification* on objects based on training data that is the closest to the object in terms of distance.

By carrying out the first test towards all attributes, there were 445 correct data out of 538 total data. The accuracy equation of the K-NN classification uses the following equation:

$$Accuracy = \frac{jumlah\ nilai\ benar}{jumlah\ data\ keseluruhan} x\ 100\% \quad \text{.......}(5)$$

By using the above accuracy equation for k = 10, the result for the accuracy of the *classification* using KNN was 82.71%. The classification using the KNN method was carried out with 7-times-trial using different Ks for each experiment. The Ks used were 5, 10, 15, 20, 25, 30, and 35.

The results of the accuracy obtained are shown in table 1 below:

Table 1. Results of experimental accuracy of different K

The value of k used	Correct data	Accuracy Results
k-5	430	**79,93%**
k-10	445	**82,71%**
k-15	408	**75,84%**
k-20	428	**79,55%**
k-25	405	**75,28%**
k-30	413	**76,77%**
k-35	404	**75,09%**

The results of the accuracy in the form of a diagram can be seen in Figure 3 below:

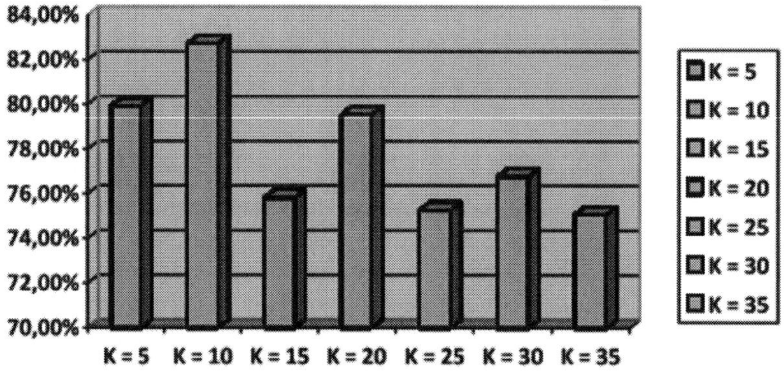

Figure 2. Comparison of different k accuracy

Used K value/ correct data/ accuracy results

The table above shows the results of accuracy of different k. From the table, it can be concluded that k = 10 has the highest accuracy of 82.71% with 445 true data out of 538 total data.

Then, the second test is carried out on all attributes resulting in 89 correct data out of 90 total data.

By using the above accuracy equation for k = 6, the results obtained from the accuracy of classification calculations using KNN is 98.89%. Classification using the KNN method is performed with 5 trials using different k for each experiment. The Ks used were 3, 5, 6, 10, and 15. The results of the accuracy obtained are shown in table 2 below:

Table 2. Results of experimental accuracy of different K

The value of k used	Correct data	Accuracy Results
k-3	87	**96.67**
k-5	88	**97.78%**
k-6	89	**98.89%**
k-10	88	**97.78%**
k-15	86	**95.56%**

Used K value/ correct data/ accuracy results

The table above shows the results of accuracy of different K. From the table, it can be concluded thatthe highest accuracy for k = 6 was 98.89%. The results of the accuracy in the form of a diagram can be seen in Figure 4 below:

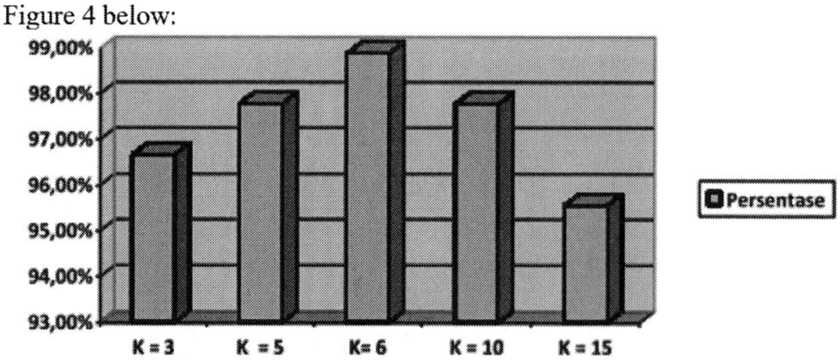

Figure 3. Comparison of different accuracy of k

From figure 3, it can be seen that k-6 has the highest accuracy. Therefore, it was used for the next process to retrieve the value of accuracy.

After conducting the classification and normalizationprocess using the *K-Nearest Neighbor* algorithm, attribute reduction was performed by using the *Chi-Square* method, one of the attribute reduction methods including the filter method. It is also one type of non-parametric comparative test performed on two variables where the scale data of the two variables is nominal.

K-Nearest Neighbor can solve data class imbalance problems and the selection feature of *Chi-Square* is one way to reduce large dataset dimensions. Chi-Square has an important role in choosing the appropriate subset of the original feature setsince not all attributes are relevant to the problem. Hence, the use of *K-Nearest Neighbor* in this study can help reducing complex data without reducing the integrity of the original data and reducing the quality of the information produced. From the result of the study, it was proven the use of *K-Nearest Neighbor* and *Chi-Square* stabilized and even increased the accuracy.

4. Conclusion

Based on the testing and evaluation of the *K-Nearest Neighbor* (KNN) method in attribute reduction using the *Chi-Square* algorithm, it was concluded that the *K-Nearest Neighbor* algorithm showeda stable and increased accuracy from datasets with large data dimensions. The use of Chi-Square method can also reduce the dimensions of large datasets and help improving the results of *K-Nearest Neighbor* prediction accuracy. In this case, K-Nearest Neighbor utilizes attribute reduction function from *Chi-Square* to select data attributes with the characteristics of the data itself, and increases the accuracy of *K-Nearest Neighbor* predictions.

Chi-Square based on *K-Nearest Neighbor* only considers a small subset of all possible models so that the risk of skipping or losing the best model will increase, along with the addition of independent variables. This research can be further developed by using other Data Mining *classification* methods, the use of attribute reduction methods or other optimization methods that can overcome large dimensional data problems without reducing the integrity of the original data and reducing the quality of the information produced.

References

[1] Alpaydin, Ethem. 2010. Introduction to Machine Learning Second Edition.

[2] Ardy Wibowo Haryanto, Edy Kholid Mawardi, & Muljono, 2018. Influence of Word Normalization and Chi-squared Feature Selection on Support Vector Machine (SVM) Text Classification. IEEE International Seminar on Application for Technology of Information and Communication (iSemantic).

[3] D. Hand, H. Mannila & P. Smyth, 2001. Principles of Data Mining. MIT Press. Florin Gorunescu, Data Mining Concept, Models and Techniques, 12th ed., Prof. Lakhmi C. Jain Prof. Janusz Kacprzyk, Ed. Craiova, Romania: Springer, 2011.

[4] Han, J., & Micheline K., 2012. Data Mining: Concepts and Techniques Second Edition, Elsevier.

[5] Indah Listiowarni & Eka Rahayu Setyaningsih, 2018. Feature Selection Chi-Square dan K-NN pada Pengkategorian Soal Ujian Berdasarkan Cognitive Domain Taksonomi Bloom. Sekolah Tinggi Teknik Surabaya.

[6] Jatmiko Indriyanto, Purwanto, & Catur Supriyanto, 2014. Algoritma K-Nearest Neighbor Berbasis Chi-Squared Untuk Prediksi Nasabah Asuransi. Pascasarjana Teknik Informatika Universitas Dian Nuswantoro.

[7]Liu & Motoda, 199. Feature Selection for Knowledge Discovery and Data Mining. Kluwer Academic, 1998.

[8] Maimoon, Oded & Lior Rokach. 2010. Data Mining and Knowledge Discovery Handbook.

[9] M. Arinal Ihsan, 2018. Reduksi Atribut Pada Algoritma K-Nearest Neighbor (KNN) Dengan Menggunakan Algoritma Genetika. Tesis. Universitas Sumatera Utara

[10] Miles, B. Mathew & Michael Huberman. 1992. Analisis Data Kualitatif Buku Sumber Tentang Metode-metode Baru. Jakarta: UIP

[11] Muhammad Ejazuddin Syed, 2014. Attribute Weighting In K-Nearest Neighbor Classification.

University of Tampere School of Information Sciences Computer Science
[12] Novakovi´c J. 2011. "Toward Optimal Feature Selection Using Ranking Methods and Classification Algorithms" dalam: Yugoslav Journal of Operations Research21(1).[Internet].[Diakses tanggal 19/03/2019]. http://yujor.fon. bg.ac.rs/index.php/yujor/article/ download/364/255.
[13] Novakovi´c J et al. 2017. "Evaluation of Classification Models in Machine Learning" dalam: Theory and Applications of Mathematics & Computer Science 1 (1), pp.39-46. [Internet]. [Diunduh tanggal 19/03/2019]. Dapat diunduh dari: http://uav.ro/stiinte_exacte/journal/index.php/TAMCS/article/viewFile/158/126.
[14] Nachirat Rachburee & Wattana Punlumjeak, 2015. A Comparison of Feature Selection Approach Between Greedy, IG-ratio, Chi-square, and mRMR in Educational Mining. IEEE International Conference on Information Technology and Electrical Engineering (ICITEE), Chiang Mai, Thailand.
[15] Refaeilzadeh P, Tang L, dan Liu H. 2009. Crossvalidation. Springer, pp. 532–538. [Internet]. [Diakses tanggal 19/03/2019]: http://leitang.net/papers/ency-crossvalidation.pdf.
[16] Runxiu Wu, 2010. Improved K-Modes Clustering Method Based on Chi-square Statistics. IEEE International Conference on Granular Computing.
[17] Said Bahassine, Abdellah Madani & Mohamed Kissi, 2016. An Improved Chi-Sqaure Feature Selection For Arabic Text Classification Using Decision Tree. IEEE 11th International Conference on Intelligent Systems: Theories and Applications (SITA).
[18] Witten & Frank, 2005. Data Mining – Practical Machine Learning Tools and Techniques with Java Implementations, second edition, Morgan Kaufmann Publishers, USA, 2005.

The Amino Acid of Sandfish Sea Cucumber (Holothuria scabra): Dry Method with Various Feeding Enzyme

[1]Kristhina P. Rahael, [2]Santi Penina Tua Rahantoknam, [3]Syahibul Kahfi Hamid

[1]Post-Harvest Processing Fisheries Study Program, Tual State Fisheries Polytechnic, Langgur-Sathean Ringroad Km 6, Southeast Maluku District 97611, Indonesia

[2]Marine Culture Engineering Study Program, Tual State Fisheries Polytechnic, Langgur-Sathean Ringroad Km 6, Southeast Maluku District 97611, Indonesia

[3]Agribusiness Fisheries Study Program, Tual State Fisheries Polytechnic, Langgur-Sathean Ringroad Km 6, Southeast Maluku District 97611, Indonesia

Email: syahibul_kahfi@polikant.ac.id

Abstract. Sand sea cucumber processing stages require the release of calcium on the surface of the skin, it is necessary for the addition of the enzyme papain from papaya latex to facilitate the process. The purpose of this study is to determine the type and composition of amino acids in dried sea cucumber due to the effect of the enzyme papain from papaya leaves with different methods. The results showed that dried sea cucumbers from the two methods approaches produced the same amino acid profile, even with different composition values. Where for sea cucumber type A obtained 9 essential amino acids and type B, obtained 8 non-essential amino acids. Thus, in order to maintain the amino acid content in dried sea cucumbers, it is better that when boiling sea cucumbers, the water temperature and the slices of papaya leaves to be controlled at a temperature of 500 C ± 50 C in order to minimize damage to amino acids.

1. Introduction

Sea cucumbers are economically valuable commodity that the potential is quite high in Indonesia. According to data from fisheries in 2001, production reached 3.517 tons of sea cucumbers. In the world there are 1,250 species of sea cucumbers that have been described, while in Indonesia there are 25 species that have commercial value derived from water corals in Indonesia [1]. Sandfish sea cucumbers (Holothuria scabra) is an economically valuable species of sea cucumber and is one of the leading commodities in Southeast Maluku. Sea cucumbers is included in the Holothuridae class. Elongated rounded body shape like an elongated cylindrical or slightly flattened with a mouth and anus located on the opposite corners. The outer skin of the sea cucumber consists of a layer of sand attached to strong and rough with needle-shaped frame or small pieces of lime spread in body tissues [2].

In general, processing raw sea cucumber is not fully utilized due to production of dried sea cucumber is only temporary and fluctuate only of there is demand from customer for sea cucumber. One characteristic of this sea cucumber that is on the surface of the body covered with lime so it tends to require special treatment to remove the lime substance. The release of calcium usually has to use a certain material to soften it and using the enzyme papain is a way to do so. The enzyme papain in papaya latex is widely available in the region. This enzyme is classified as protolithic enzyme that can hydrolyze the protein. Enzymes are biocatalyst that can speed up a reaction without co-react with the material.

Although in terms of aesthetics, these species are unattractive but fairly complete in nutritional value. Proximate analysis of the results obtained by the protein composition of sea cucumber meat 43%, 2% fat, 17% water content, mineral ash content of 21% and 7% [3]. Nutrient content can be altered by the treatment process. Protein is one nutrient component which are crucial for the human body. These types of nutrients is essential for the growth and replacement of human body cells that are damaged also supporting the body's metabolism. Sea cucumber protein content in wet conditions is 44-55% [4]. According [5], protein dry conditions was 82%, 1.7% fat, 8.9% moisture content, ash content of 8.6% and 4.8% carbohydrate.

Amino acids are the monomeric building blocks of protein, plays an important role in the metabolism of living cells. Acids of amino acids can be classified based on the ability of the synthesis of human and animal bodies [6]. Essential amino acids are amino acids that cannot be made in the body and must be obtained from food which also called exogenous amino acids. Amino acids are often called and known as a builder substance which is the end product of protein metabolism [7]. Non-essential amino acid is an amino acid that can be made in the body is also called endogenous amino acids [8]. These amino acids function of repairing damaged tissue after injury, protect the liver from toxic substances, lowering blood pressure, regulate cholesterol metabolism, promote the secretion of growth hormone, and reducing ammonia levels in the blood [9]. The existence of the essential amino acids is often associated with the quality of the protein. This study aims to determine the type and content of amino acids in dried sandfish sea cucumber due to the effect of the enzyme papain from papaya leaves with different methods.

2. Material and Method

2.1. Material

Materials used in this research was sandfish sea cucumber obtained from the cultivation farm owned by our partners in Langgur Village, Southeast Maluku, Indonesia.

2.2. Equipment

Cool box, knives, pots boiling, conventional curing tool, agitator. The tools used to analyse the amino acid analytical balance, oven, syringe filter paper and High-Performance Liquid Chromatography (HPLC).

2.3. Work Procedure

Cucumbers harvested from pen-culture put cool box and transported to mills. Sea cucumbers are weighed to determine the initial weight. Furthermore, the sea cucumber entrails removed (eviscerate) by piercing from the anus to the head using parts of the bamboo skewers and sharpen like a jab. Sea cucumbers pressed to exit the entire contents of the stomach. Furthermore, the boiling, the boiling is used two methods. Method A is based on study by [10], where started with boiling the sea cucumbers mixed with sliced papaya on the water temperature of 60 0 C ± 50 ° C for 30 minutes, while the second method (B) by using traditional methods used by the Southeast Maluku, where sea cucumber processing boiled with water for ± 30 minutes until it was chewy and then proceed with stirring along with slices of papaya leaves. Afterwards, added with brushing with a soft brush and wash cucumbers, then braised sea cucumbers being wash twice. So, the first method, proceed by boiling the slices of papaya leaf and sandfish sea cucumber, whereas the second method uses only boiling water. Subsequently cooled, and waited for 24 hours of curing process by using a conventional dryer with coconut shell fuel. Initially using cold fogging subsequently using heat curing. Furthermore, sandfish sea cucumbers dried using the sunlight for 3 days. Sandfish sea cucumbers are being further analysed the content of amino acids in the Environmental Biotechnology Laboratory PT. Biotechnology biodiversity of Indonesia, Bogor. The results were tested by descriptive (Fig 1).

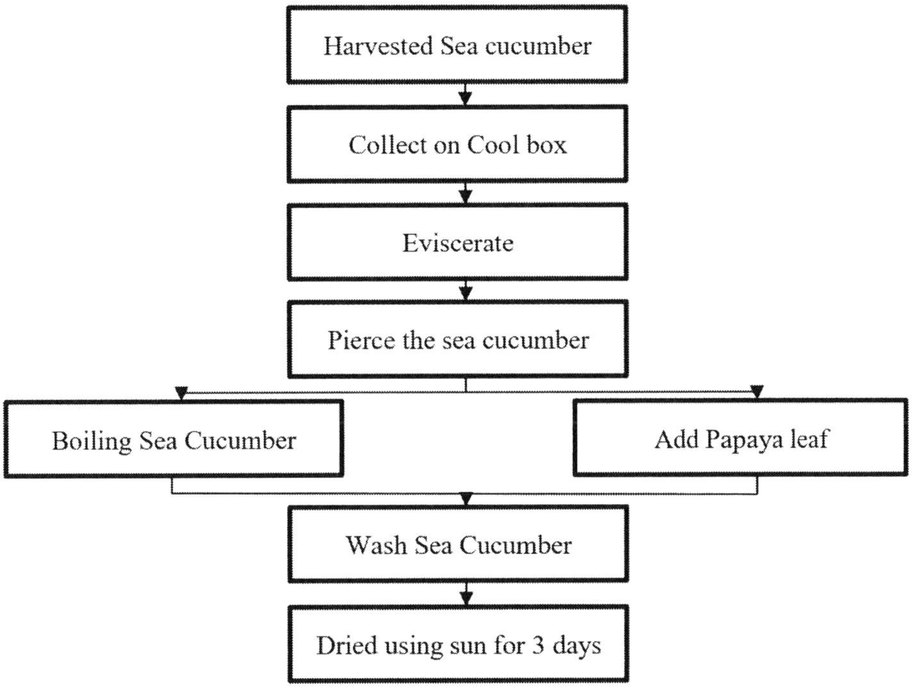

Figure 1. Work Procedure

3. Result and Discussion

The results of the analysis of amino acids in dried sea cucumber is based on two methods that can be seen in Table 1 below:

Table 1. Amino Acid Profile Dried Sea Cucumber

No	Amino Acid Type	Unit	Sample Code	
			A	B
1	Aspartic acid	%	4.22	3.72
2	Acid glutamate	%	8.00	7,11
3	Serin	%	2.52	2.05
4	Glycine	%	7.90	8.04
5	Histidine	%	1.41	0.96
6	Arginine	%	1.25	1.19
7	Threonine	%	1.72	1.75
8	Alanine	%	1.51	1.41
9	Proline	%	0.63	0.62
10	Tyrosine	%	1.37	1.39
11	Valin	%	3.10	3:37
12	Methionine	%	1.14	1.96
13	Cysteine	%	1.07	0.96
14	Isoleucine	%	2.16	1.72
15	Leucine	%	6.02	5.75
16	Phenylalanine	%	1.38	1.23
17	Lysine	%	3.81	4.34
		Total	49.21	47.57

Based on the analysis of amino acids in above shows that there are about 17 types of amino acids contained in the dried sea cucumber produced both a sea cucumber A and cucumbers B, consisting of nine essential amino acids are: histidine, arginine, threonine, valine, methionine, isoleucine, leucine, phenylalanine, and lysine and 8 non-essential amino acids are: aspartic acid, glutamic acid, serine, glycine, cysteine, alanine, proline, and tyrosine. The results of this study are not much different from the results of the study of [11] and [12] who argued that the sand sea cucumbers (Holothuria scabra) contains almost all the essential amino acids except tryptophan. The difference lies in the composition of each amino acid. Total amino acid produced by the treatment A is 49.21% higher than treatment B, namely 47.57 in accordance with the results of this research study [13] states that the average amount of amino acids contained in the sea cucumber ranges from 33.32 - 54.13 g / 100 g bk with the highest amino acid is glycine, glutamate, aspartic, alanine and arginine. The high amino acid composition of trepang B were related to the method used. [10] method using controlled temperature is 500C ± 5 0C at boiling 1 so as not to damage the amino acid composition contained in it besides the temperature is the optimal temperature for the enzyme papain work.

Boiling along with slices of papaya leaves at this temperature causes the release of calcium from the surface of the sea cucumber skin but damage to the nutritional components can be minimized. In contrast to the method [10], This method of temperature control is not done because the goal is only to be rubbery meat with a sea cucumber that is inconsistent, causing breakdown of amino acids can occur. [13], stating that all the amino acids in food, especially lysine, threonine and methionine is sensitive to dry heat and radiation. Furthermore [14], states that heating meat at a temperature of 700C would reduce the amount of lysine contained in it to be 90%, while heating at a temperature of 1600C would reduce levels of lysine up to 50%. Differences in amino acid composition of these two methods can be seen in Figure 1.

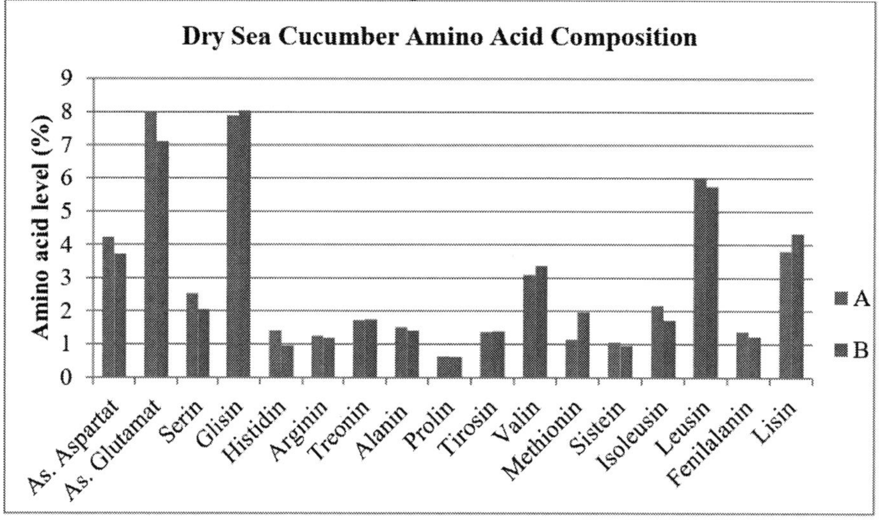

Figure 2. Amino Acid Composition

From the chart above showed that the highest content of essential amino acids namely leucine of cucumbers A amounted to 6.02%. Leucine works with the amino acid isoleucine and valine in repair muscles, regulate blood sugar, and provide energy reserves. Leucine also serves to increase the production of growth hormone and helps burn visceral fat located in the deepest layers of the body. Essential amino acid histidine lowest for the sea cucumber B of 0.96%. Histidine is an essential amino acid for children and babies. The lowest of the content of histidine so to say as the limiting amino acid. Non-essential amino acid content is highest with glycine of cucumbers B of 8.04%, followed by glutamate acid in the sea cucumber A while the lowest is proline of sandfish sea cucumber B at 0, 62% nevertheless not unlike the sea cucumber A. Glycine is an amino acid that can inhibit processes in the brain that causes stiffness of motion such as in multiple sclerosis [15]. Glycine and glutamic acid is an essential component for the cell to synthesize glutathione, which can stimulate the activation and proliferation. Arginine can enhance immunity by promoting cell activation and proliferation of T cells [16]. The average content of glutamate acid in both treatments was higher than that generated by the research of sea cucumbers in the islands of Belitung [17] while the lower glycine. Although glutamate acid instead of an amino acid essential but this amino acid is important in the formation of the flavour of the fishery, which can provide the savoury flavour of seafood products. Hydrogen group on glutamic acid can substance with sodium to form monosodium glutamate which has a high intensity savoury flavour so widely used as a flavour enhancer. Proline has a free amino group and form the aromatic structure. These amino acids can be obtained from the hydraulic casein [14]. All amino acids in food, especially lysine, threonine and methionine is sensitive to dry heat and radiation. [18] Differences in the womb These amino acids can also be caused by age, fishing season, and stages in the life cycle of the organism [19]; [20].

4. Conclusion

Based on research by amino acids contained in the dried sea cucumber from both methods have the same kind are 9 essential amino acids and 8 essential amino acids, but different composition. The highest total amino acids contained in the dried sea cucumber with the method of use of the enzyme by [10] is 49.21%. The content of the essential amino acid leucine highest at 6.02%. In the dried sea cucumber with the method of use of the enzyme by [10] while the lowest was 0.96 on a histidine dried sea cucumber with traditional methods of enzymes, non-essential amino acids, glycine as high as 8.02% on the dried sea cucumber with methods of use traditional enzyme [10], while the lowest was 0.62% proline in dried sea cucumber with traditional delivery methods of enzyme. Preferably during the first boiling water temperature and slices of papaya leaves to remain controllable at a temperature of 500 C ± 50 C in order to minimize the breakdown of amino acids.

Acknowledgement

We would like to thank the Directorate of Research and Community Ministry of Research Technology and Higher Education Republic o Indonesia to fund the study.

References

[1] Darsono, P 2007 Teripang (Holothuridea) kekayaan alam dalam keragaman biota laut. *Journal Oseana* 32: 1-10.
[2] Sudrajat, Y 2002 Teknik penghilangan lapisan kapur pada teripang pasir menggunakan enzim papain. Bulletin teknik pertanian, 7(2).
[3] James, D. B 1989 Beche-de-mer: Its resources, fishery and industry. *Marine Fisheries Information Service, Technical and Extension Series, 92*, 2-30.
[4] Dewi, K. H 2008 Kajian ekstraksi steroid teripang pasir (Holothuria scabra J) sebagai sumber testosteron alami.[Disertasi]. *Bogor (ID): Institut Pertanian Bogor.*
[5] Martoyo, J., Aji, N.,and Winanto, T 2006 Budidaya Teripang, Edisi Revisi. *Penerbit Penebar Swadaya-Jakarta, 75.*
[6] Toha, A. H. A 2001 Biokimia: Metabolisme Biomolekul. *Alfabeta, Manokwari.*
[7] Ningsih, P 2009 Karakteristik protein dan asam amino kijing lokal (Pilsbryoconcha exilis) dari situ gede bogor akibat proses pengukusan. *Skripsi. Institut Pertanian Bogor.(Tidak dipublikasikan).*
[8] Winarno, F. G 1997 *Kimia Pangan dan Gizi. Cetakan Kesembilan*. Jakarta: Gramedia.
[9] Kamiya, T 2002 Biological functions and health benefits of amino acids. *Foods and Food Inggredient Journl of Japan*, 33-44.
[10] Sasongko, B 2015 Kajian pengolahan dan mutu teripang asap. Skripsi. Program Studi Ilmu Kelautan, Fakultas Pertanian, Universitas Bengkulu. Bengkulu.
[11] Chen, S. G., Li, G. Y., Yin, L. A., Huang, W. C., Dong, P., Xu, J., ... and Xue, C. H 2010 Identification of eight species of sea cucumber chondroitin sulfates by high temperature~ 1H NMR. *J. Instrum. Anal, 29*(8).
[12] Nurjanah, Trilaksani, W., dan Kustiariyah 2004 *Bahan Ajar: Teknologi Preparasi Hasil Perikanan.*Departemen Teknologi Hasil Perikanan. FPIK-IPB. Bogor.

[13] Wen, J., Hu, C., and Fan, S 2010 Chemical composition and nutritional quality of sea cucumbers. *Journal of the Science of Food and Agriculture*, *90*(14), 2469-2474.

[14] Palupi, N. S., Zakaria, F. R., and Prangdimurti, E 2007 Pengaruh pengolahan terhadap nilai gizi pangan. *Modul e-Learning ENBP, Departemen Ilmu & Teknologi Pangan-Fateta-IPB.*

[15] Harli, M 2008 Asam amino esensial. http://www.supamas.com. [4 September 2019].

[16] Bordbar, S., Anwar, F., Nzamid, S 2011 *High-value components and bioactives from sea cucumbers for functional foods*—A review. *Mar. Drugs*. 9: 1761-1805.

[17] Ridhowati, S and Asnani 2015 Profil asam amino dan asam lemak pada teripang pasir (*Holothuria scabra*) olahan Belitung. Kepulauan Belitung.

[18] Hawab, H. M 2007 *Dasar-Dasar Biokimia*. Jakarta: Diadit Media.

[19] Okuzumi, M., and Fujii, T 2000 Nutritional and functional properties of squid and cuttlefish. National Cooperative Association of Squid processors, Tokyo

[20] Litaay, M. 2005. Peranan nutrisi dalam siklus reproduksi abalone. *Journal Experimental Oseana* 75(3): 1-7

Journal of Physics: Conference Series **1424** (2019) 012006 doi:10.1088/1742-6596/1424/1/012006

Design and Realization of A Low Cost Clinometer based on ADXL345 Sensor, Ultrasonic Sensor, and Android based Smartphone

A Najmurrokhman[1], Kusnandar[1], U Komarudin[1], A M Annas[1], R Rahim[2]

[1]Department of Electrical Engineering, Universitas Jenderal Achmad Yani, Cimahi, Indonesia
[2]School of Computer and Communication Engineering, Universiti Malaysia Perlis, Kubang Gajah, Malaysia

asep.najmurrokhman@lecture.unjani.ac.id

Abstract. Manually height measurement using a measuring instrument with a limited scale range can hamper the performance of building construction activities. To deal with this problem, a device is needed to be able to measure the height of the building more quickly, has a larger measuring limit, and is easy to use on the measurement of the inner building that is on a flat plane. This paper describes a prototype of clinometer using ADXL345 accelerometer sensors connected to Android-based smartphones. Other components that build this system are HC-SR04 ultrasonic sensor, HC-05 bluetooth module, and Arduino Uno microcontroller. The proposed system is able to connect with Android-based smartphones through the connection of the HC-05 bluetooth module, retrieving distance data through ultrasonic sensor readings and elevation angle data via the ADXL345 sensor, sending or receiving data from an Android smartphone to Arduino and vice versa, calculating the approximate height, and display it on the Android based smartphone. Experimental results show that applications installed in Android-based smartphone can display the results of approximate height. The proposed instrument can quickly calculate the approximate height of a building with an error percentage of 2.5 %. The maximum distance from the proposed instrument to the building is 1.75 m.

1. Introduction

Measuring the length or height of an object by using the conventional equipment is limited by its measuring scale. There are two possible approaches to cope such condition. One could employ the equipment with the larger scale and higher precision. However, it is simplest way but costly. The other way applies indirect measurement to render the actual measurement. To make it clear, imagine that a student is given a task to find the height of a building. It is widely known that one can use the theodolite to gain the height of a building. In principle, the theodolite is a device which consists of telescope placed on a rounded base (disk) that can be twisted around the vertical axis, allowing the horizontal angle to be read. The telescope is also mounted on the second disk and can be rotated around the horizontal axis, allowing vertical angles to be read. So far, the theodolite is an expensive equipment to measure the length or height of some objects [1]. To overcome the unavailability of an instrument, an indirect measurement could be done by using a protractor as illustrated in Fig. 1a. The measurement results are provided by such a way using trigonometric rule. If the dangling string is steady, then the angle between the midpoint of the protractor (90°) and the point where the string crosses the rim can be counted by subtracting one from the other as shown in Fig. 1b. In that case, the height of a building becomes the product between a horizontal distance from an observer to building and the tangent of the angle. Such equipment is called a protractor clinometer [2].

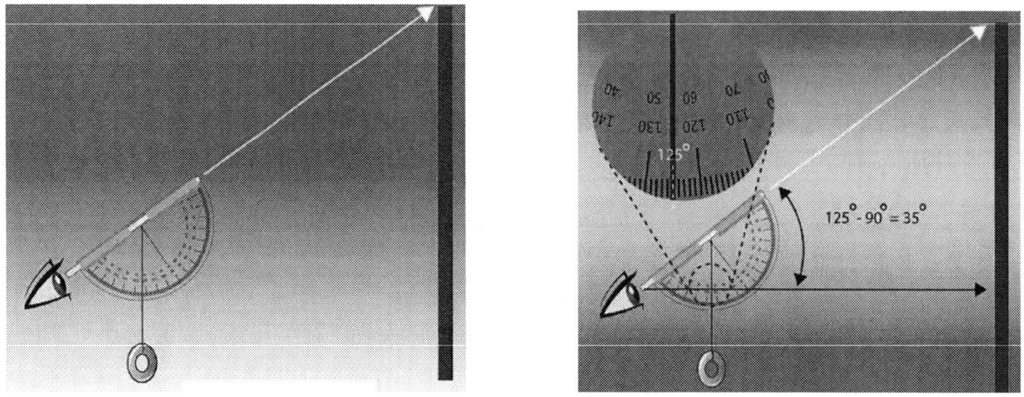

(a) (b)

Fig. 1. Indirect measurement of the height of building by a protractor [2]

The clinometer has been recently used in sport and health [3]–[5], structural health monitoring and geotechnical measurements [6], [7], automatic excavator [8], and so on. In medical field, clinometer was used to measure range of motion of a patient's body part such as shoulder [4], ankle [5], wrist of patients whose paretic hand [9], and so forth. In this recent our work, the clinometer is designed to measure an angle and calculate the approximate height of the building. As previously described, the information that needed to get the height of an object is the horizontal distance and an appropriate angle. With advances in microelectronics technology, especially fabrication of electronic sensor devices, measuring instruments can be made cheaper, smaller, and more practical. Along with this advancement, we could employ the electronic sensors to gain a horizontal distance and an appropriate angle in indirect measurement of the height of a building. It is widely known that the ultrasonic sensor was used to measure a range detection and obstacle recognition in robots [10], [11], to detect road surface for supporting an elderly and visually impaired people [12], to support vector machine–based fall detection [13], to calculate the liquid volume inside oil tank [14], and so forth. In that case, we employ the ultrasonic sensor to gain a horizontal distance between the instrument and the building. Meanwhile, we obtain an appropriate angle by using the ADXL345 accelerometer sensor. Such sensor has been used in a self-balancing robot [15], a car data recorder for recording the condition of four-wheeled vehicles [16], fall detection for elderly people [17], digital pedometer [18], for name a few.

Nowadays, smartphone user penetration is estimated to touch the two billions in 2015 [19] and there are approximately 2.7 billion smartphone users around the world in early 2019 [20]. In US, smartphone has become the second-most-common product with 87% penetration [21]. Smartphones with an Android-based operating system have been used extensively for applications that enhance the functionality of these smartphones [4], [7], [19], [22]. In line with this advancement, this paper describes the use of Android-based smartphones to realize an altitude measuring instrument with an ADXL345 sensor as the main component that can detect elevation angles and the ultrasonic sensor for gaining the horizontal distance. Furthermore, the data obtained from the sensor is then processed by the microcontroller and displayed in a smartphone via Bluetooth HC-05 module. Using data obtained from the two sensors, the height of objects or buildings is calculated by the principle of simple trigonometry. Furthermore, the measurement results are displayed visually in the smartphone.

2. System design

Block diagram of the proposed instrument is shown in Fig 2. Based on Figure 2, two sensors namely ADXL345 and HC-SR04 are employed in this instrument. An Arduino Uno R3 type microcontroller is used as the main data processor which collects and proceeds data from those sensors to be sent to the smartphone using the Bluetooth module HC-05. Meanwhile, the servo motor is installed to move the laser pointer as a marker of height to be measured.

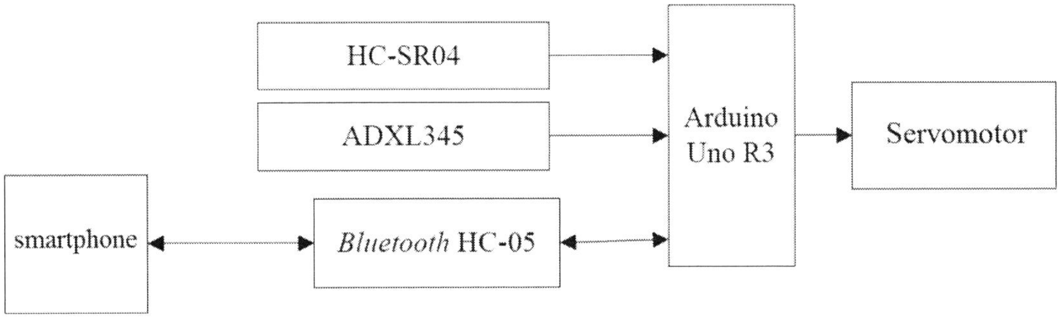

Fig. 2. Block diagram of the proposed instrument

ADXL345 sensor is an accelerometer sensor that is used to obtain an elevation angle between buildings measured by horizontal lines, while the HC-SR04 sensor is an ultrasonic sensor that is used to determine the horizontal distance between objects or buildings as measured by instruments. ADXL345 is a 13-bit resolution 3-axis motion sensor / acceleration module that can detect a pull at a range of up to 16 g (g = gravitational acceleration). The application includes detecting slope by monitoring changes in static attraction and dynamic acceleration arising from movements or collisions. With the high resolution it produces, this sensor is able to detect inclination smoothly less than one degree. It can detect movement activity status by comparing the acceleration on any axis with a sensitivity threshold that can be set through the programming code. In addition, this module provides tap sensing that can detect single or multiple beats in various directions. It can also detect free-fall sensing. These functions can be mapped separately on the two output pins contained in the device. The ADXL345 sensor has a 32-bit static internal memory management system with FIFO (first-in first-out) queue type that can be used to store measurement variables / temporary data so as to reduce microcontroller loads and generally reduce system energy consumption. This module has a sophisticated power management circuit where the modules are placed in a very small power consumption mode until detected movements that pass a certain threshold that reactivate normal mode. After the reading process is complete, the module is automatically returned to standby mode to save energy. The physical form of ADXL345 is shown in Figure 3.

Fig. 3. ADXL345 sensor

Meanwhile, the HC-SR04 sensor is a ready-to-use ultrasonic sensor that can be used to measure the distance of objects from 2 cm to 4 m with an accuracy of 3 mm. This sensor has 4 pins, namely Vcc, GND, Trig, and Echo. Trig pins are used to trigger signals sent from sensors, while Echo pins are used to capture reflected signals from objects to be measured. The physical form of the HC-SR04 sensor is depicted in Figure 4. In ultrasonic sensors, ultrasonic waves are generated through a material called piezoelectric with a certain frequency. This piezoelectric will produce ultrasonic waves (generally with a frequency of 40 kHz) when an oscillator is applied to the sensor. In general, this device will direct ultrasonic waves towards an area or a target. After the wave touches the target surface, the target will reflect back the wave. Reflection waves from the target will be captured by the sensor. Furthermore, the distance could be calculated based on the difference between the wave delivery time and the reflected wave time received.

Fig. 4. HC-SR04 ultrasonic sensor

As the main processor in this instrumentation system, the Arduino Uno R3 type microcontroller is used to process data and connect among components in the system. It is an ATmega328P chip-based microcontroller development board which has 14 digital input-output pins with six pins can be used as PWM output, six analog input pins, using 16 MHz crystal oscillator, USB connection, ICSP header, and reset button. The physical form of the microcontroller is shown in Figure 5.

Fig. 5. Arduino Uno R3 Microcontroller

All components used in the proposed instrument are placed in a black plastic box with dimensions of 5 cm high, 12 cm long, and 8 cm wide. The compact shape of the proposed instrument is given in Figure 6(a). At the front of the box there are two holes for placing the transmitter and receiver of ultrasonic sensor. In addition, a laser pointer is installed to give a marker of the top of the object or building to be measured in height. The pointer can move freely through the motion mechanism produced by the servomotor. The servomotor motion is adjusted by sending the instructions from the smartphone. Meanwhile, the application program created and displayed on the smartphone is given in Figure 6(b). In this view, section 1 is used to access the list of addresses and available Bluetooth names, while section 2 is enabled to activate the application with the chosen Bluetooth address and name. Section 3 is used to drive a servomotor in the proposed instrument. The servomotor is moved up and down by pressing parts 4 and 5, while the laser pointer is turned on by pressing section 6. Section 7 is the result of reading the ADXL345 sensor, while section 9 marks the results of the HC-SR04 sensor reading. Section 8 is filled with the height of the instrument from the ground. After all the required data filled in, a button whose number 10 in display is used to execute the calculation result and the approximate height of the building is shown in section 11. Meanwhile, realization of the proposed instrument is shown in Figure 7.

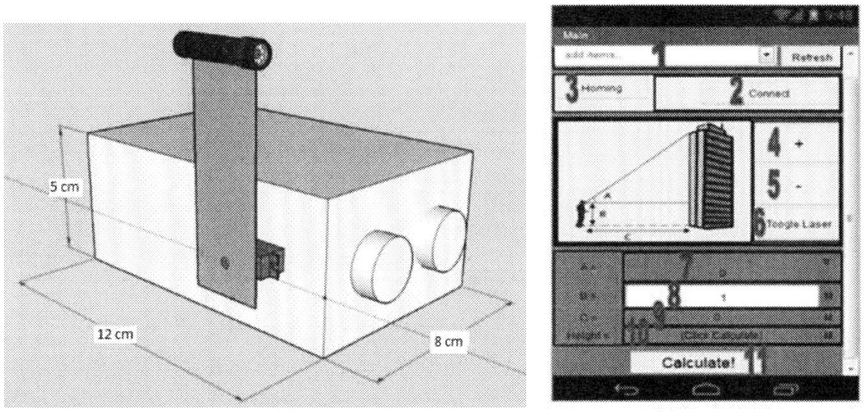

(a) (b)

Fig. 6. The compact shape of the proposed instrument (a) and display on the smartphone (b)

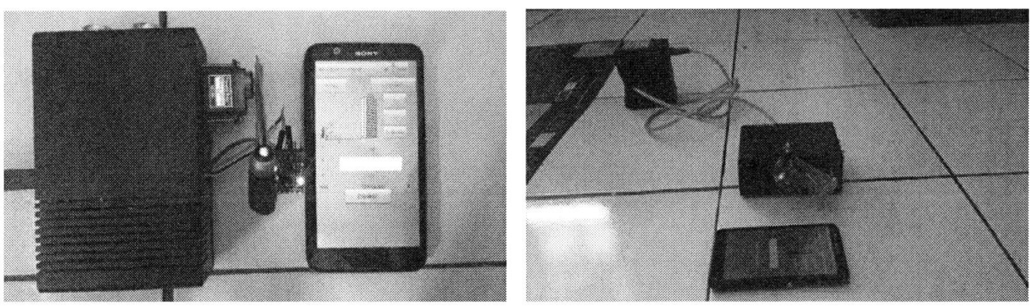

Fig. 7. Realization of the proposed instrument from front view (left) and side view (right)

3. Results and Discussion

In experimental setup, we firstly tested the accuracy of sensor readings by comparing the values that appear in the Arduino IDE serial monitor with those randomly sent to the Android interface via HC-05 Bluetooth module. The ADXL345 sensor will move along with the servomotor movement. Whenever the servomotor is actuated by pressing the button on the Android interface, the ADXL345 sensor will move up, down, or go to its initial position according to the instruction signal sent from an Android based smartphone. Experimental results show that the reading of the ADXL345 sensor is the same as that shown on the Android interface so that it can be used as measurement data. Meanwhile, ultrasonic sensor testing is done by comparing the sensor readings to the horizontal distance between the proposed instrument and the wall of the building to be measured in its actual distance. The results of ultrasonic sensor testing are given in Table 1. As observed from Table 1, the discrepancy between measurement data gained by sensor and its actual distance is below 0.001 % for horizontal distance of more than 0.15 m. Based on those results, measurement data obtained by employing ultrasonic sensors can be used as reference data for determining the horizontal distance between the proposed instruments and buildings to be measured.

After all the components are integrated, experimental testing of the proposed instrument is done by comparing the results of sensor measurements and height calculations by the application program installed in a smartphone with actual height measured manually. In this experiment, the proposed instrument is placed 1 m above the ground. Experimental results are tabulated in Table 2. The fist column on Table 2 indicates the readings of angle data in degree when the ADXL345 sensor is moved by the servomotor such that the laser pointer points to the tip of the measured building. Meanwhile, the second column of Table 2 shows the readings of horizontal distance in metres between the proposed instrument and wall of the measured building. The maximum distance from the proposed instrument to the building is 1.75 m. Based on the results of that experiment, the designed clinometer shows quickly calculate the approximate height of a building with an error percentage of 2.5 %.

Table 1. Comparison of HC-SR04 ultrasonic sensor data and actual distance

sensor data (m)	actual distance (m)
0	0.01
0	0.02
0.15	0.15
0.30	0.30
0.50	0.50
1.20	1.20
1.34	1.35
1.61	1.60
2.32	2.30
2.67	2.65

Table 2. Experimental results of the proposed clinometer

ADXL345 (deg)	HC-SR04 (m)	Approximate height displayed in *smartphone* (m)	Actual height (m)
30	0.25	1.14	1.14
30	0.50	1.28	1.29
30	0.75	1.43	1.43
30	1.00	1.58	1.58
30	1.25	1.72	1.72
30	1.50	1.87	1.87
30	1.75	2.50	2.01
45	0.25	1.25	1.25
45	0.50	1.50	1.50
45	0.75	1.75	1.75
45	1.00	2.00	2.00
45	1.25	2.25	2.25
45	1.50	2.50	2.50
45	1.75	2.75	2.75
60	0.25	1.43	1.43
60	0.50	1.87	1.87
60	0.75	2.29	2.30
60	1.00	2.73	2.73
60	1.25	3.17	3.17
60	1.50	3.59	3.60
60	1.75	4.03	4.03

4. Concluding remarks

This paper has discussed the design and realization of the clinometer using the ADXL345 accelerometer sensor as a slope detection sensor and HC-SR04 ultrasonic sensor as a horizontal distance measuring sensor between instruments with objects or buildings to be measured in height. Data from the sensor is then processed by the microcontroller and then sent via HC-05 Bluetooth module to an Android-based smartphone. Data from sensors and measurement results are displayed in the smartphone device. Experimental results show that the proposed instrument provides measurement accuracy with an error percentage of around 2.5 %.

Acknowledgments

We thank to the Institute of Research and Public Services of Universitas Jenderal Achmad Yani who already partially supported this work in Competitive Research Scheme for academic year 2018/2019.

References

[1] S. Gill and A. Aryan, "To Experimental Study for Comparison Theodolite and Total Station," *Int. J. Eng. Res. Sci.*, vol. 2, no. 3, pp. 153–160, 2016.

[2] B. Ruff, "Using a Protractor Clinometer," *https://www.wikihow.com/Make-a-Clinometer*, 2019.

[3] N. Romero-Franco, J. A. Montaño-Munuera, and P. Jiménez-Reyes, "Validity and Reliability of a Digital Inclinometer to Assess Knee Joint-Position Sense in a Closed Kinetic Chain," *J. Sport Rehabil.*, vol. December, pp. 1–5, 2017.

[4] P. Boissy *et al.*, "Trueness and Minimal Detectable Change of Smartphone Inclinometer Measurements of Shoulder Range of Motion," *Telemed. e-HEALTH*, vol. 23, no. 6, pp. 1–4, 2017.

[5] C. Junghyeon, C. Jongman, N. Taewoo, P. Junho, and P. K. Inseon, "Development of an automatic ankle stretching device for lower limb monoplegia patients," in *Proceedings of the 29th Annual International Conference of the IEEE Engineering in Medicine and Biology*, 2007, pp. 4779–4781.

[6] H. Gong, Z. Qian, X. Yang, C. Zhao, and X. Dong, "Optical Fiber Inclinometer Based on a Fiber Taper Cascading a Peanut-Shape Structure," *IEEE Sens. J.*, vol. 15, no. 7, pp. 3917–3920, 2015.

[7] S. Lee, J. Suh, and H. Park, "Smart Compass-Clinometer: A smartphone application for easy and rapid geological site investigation," *Comput. Geosci.*, vol. 61, pp. 32–42, 2013.

[8] J. Kim, S. Yoo, M. Kim, Y. Kim, and G. Lee, "Safety Control of Automatic Excavator for Swing Collision Avoidance," in *15th International Conference on Ubiquitous Robots (UR)*, 2018, pp. 758–762.

[9] J. Y. Kim, H. S. Sung, Y. K. Kang, I. Gaponov, and B. K. Jung, "Development and evaluation of a method to measure wrist range of motion on paretic hand rehabilitation device," in *IEEE International Conference on Rehabilitation Robotics*, 2017, pp. 1337–1342.

[10] J. Park, Y. Je, H. Lee, and W. Moon, "Design of an ultrasonic sensor for measuring distance and detecting obstacles," *Ultrasonics*, vol. 50, no. 3, pp. 340–346, 2010.

[11] A. Najmurrokhman, K. Kusnandar, G. I. Sofyan, E. C. Djamal, A. Munir, and B. Wibowo, "Design and Implementation of Fuzzy Logic Controller for A Class of Hexapod Mobile Robot," in *Electrical Power, Electronics, Communications, Controls and Informatics Seminar (EECCIS)*, 2018, pp. 269–273.

[12] S. Nakashima, S. Aramaki, Y. Kitazono, S. Mu, K. Tanaka, and S. Serikawa, "Application of Ultrasonic Sensors in Road Surface Condition Distinction Methods," *Sensors*, vol. 16, no. 1678, pp. 1–8, 2016.

[13] Z. Chen and Y. Wang, "Infrared-ultrasonic sensor fusion for support vector machine-based fall detection," *J. Intell. Mater. Syst. Struct.*, vol. 29, no. 9, pp. 2027–2039, 2018.

[14] M. Husni, D. O. Siahaan, H. T. Ciptaningtyas, H. Studiawan, and Y. P. Aliarham, "Liquid

volume monitoring based on ultrasonic sensor and Arduino microcontroller," in *IOP Conf. Series: Materials Science and Engineering 128*, 2016, p. 012026.

[15] H. Ferdinando, H. Khoswanto, and D. Purwanto, "Performance Evaluation of MMA7260QT and ADXL345 on Self Balancing Robot," *Telkomnika*, vol. 11, no. 1, pp. 1–10, 2013.

[16] S. A. Nugroho, E. Ariyanto, and A. Rakhmatsyah, "Utilization of Onboard Diagnostic II (OBD-II) on four wheel vehicles for car data recorder prototype," in *Proceedings of 6th International Conference on Information and Communication Technology (ICoICT 2018)*, 2018, pp. 7–11.

[17] X. Yuan, J. N. Li, and Z. Fang, "Hardware Design of Fall Detection System Based on ADXL345 Sensor," in *Proceedings of 8th International Conference on Intelligent Computation Technology and Automation (ICICTA 2015)*, 2015, pp. 446–449.

[18] S. Y. Kim and G. Kwon, "Interrupt-Based Step-Counting to Extend Battery Life in an Activity Monitor," *J. Sensors*, vol. 2016, no. ID 5824523, pp. 1–6, 2016.

[19] C. A. Kardous and P. B. Shaw, "Evaluation of smartphone sound measurement applications," *J. Acoust. Soc. Am.*, vol. 135, no. 4, pp. 186–192, 2016.

[20] G. Deyan, "60+ Smartphone Statistics in 2019," *https://techjury.net/stats-about/smartphone-usage/*, 2019.

[21] B. Markwalter, "The Consumer Technology Market-Past and Present [CTA Insights]," *IEEE Consum. Electron. Mag.*, vol. 8, no. February, pp. 108-C3, 2019.

[22] A. Najmurrokhman, K. Kusnandar, A. Budiman Krama, E. C. Djamal, and R. Rahim, "Development of a secured room access system based on face recognition using Raspberry Pi and Android based smartphone," *MATEC Web Conf.*, vol. 197, p. 11008, 2018.

Effect Of Groove Infill To The Compressive Strength Of Multiple Groove Concrete Block Masonry Unit Under Uniformly Distributed Load

Fauziah Aziz[1], Mohd. Fadzil Arshad[2], Hazrina Mansor[3]

[1]Faculty of Civil Engineering, University Technology Mara, Shah Alam, Selangor, Malaysia
[2]Associate Professor, Faculty of Civil Engineering, University Technology Mara, Shah Alam, Selangor, Malaysia
[3]Senior Lecturer, Faculty of Civil Engineering, University Technology Mara, Shah Alam, Selangor, Malaysia

* fauziah2876@gmail.com

Abstract. Interlocking concrete block is a new concrete masonry unit that is used in the construction industry. This masonry unit can improve the quality of the construction process; reduce the cost of labor, no need high skill workmanship and speeding the construction time. Normally, the interlocking concrete block masonry unit in the market place was designed in one way interlocking concept only either x or y-axis, shorter in length and low compressive strength value. However, the biaxial interlocking concrete block is a new concept being introduced in this research. Biaxial interlocking concrete block is a block with the interlocking system at both directions. Besides that, this masonry unit is offered the specialty compared to the normal interlocking concrete available in the market place due to its length and the geometry of the groove and tongue. This concrete masonry unit can be used as a non-load bearing wall or load-bearing wall depends on the application of the masonry. But, there is a lack of technical data was produced before. This paper presents a new finding on the compressive strength of biaxial interlocking concrete block masonry unit. Two series of biaxial interlocking concrete block unit namely BICB1 and BICB2 have tested the compressive strength. BICB1 is the masonry unit of a hollow biaxial interlocking concrete block meanwhile; BICB2 is the masonry unit of groove infill interlocking concrete block. The masonry unit sample tested under static compressive load. The results examine that BICB2 is higher in compressive strength compared to the BICB1. This result shows that BICB2 is better than BICB1, especially in load-bearing wall application.

1. Introduction

Nowadays, masonry structures are widely used in most building construction. It is a common component with low maintenance and good durability aspect. Normally, masonry structures can react as a load-bearing or non-load bearing structures. Basically, in masonry structures, it can be divided into beam structures, column structures or wall structures. The traditional masonry unit that has been used in the masonry construction are clay @ earth brick, cement sand brick, and concrete block. Typically this masonry unit has the weakness in their application such as environmental issues for clay brick[1], quality control during construction for cement sand brick and the self-weight load for concrete block. In view of this weakness, several new masonry units and modern building masonry technologies are now available in the building marketplace to speed up the masonry construction project.

Biaxial interlocking concrete block (BICB) unit is one of the new concrete masonry units that is used in masonry construction especially in a building project. This material was developed with the interlocking system in both direction, x and y-axis and purposely used as a walling system. The interlocking was reacted through the groove and tongue element of the interlocking concrete block unit. Furthermore, this element is good to resist the loading acting on the structure from both sides and reduce the buckling and shear effect of the structure from the compressive load and lateral load. Besides that, this new material also giving the industries benefit from its geometry. The larger of their length reduces the number of material used in masonry wall construction. BICB can improve the process of masonry construction by speed up the construction time, reduce the number of skilled workers and its good in mechanical properties such as high compressive strength. Regarding [2], hollow element in concrete block offers many benefits such as ventilation aspects, reduce the dead load compared to the solid concrete block and its require less amount of mortar. However, since this is a new product, there is a lack of technical data has been produced before.

On the other than, this material comes out with the new technique in masonry construction called mortarless or dry stacked system. Dry-stack system or mortarless is a new technique of masonry wall construction without traditional mortar application in bed and heads joint [3][4][5]. This system reacts through interlock part by groove and tongue to led the masonry unit connected tightly to be an entire system [6]. The material being used in walling system by stacked directly on top of each other likes a lego concept and used 2mm cement glue as a binder to increase the capacity of the masonry wall [4].

As a new material, there is a need to identify its capability to be used as a masonry unit to replace the traditional one. Biaxial interlocking concrete block (BICB) is a hollow interlocking concrete block that can be used as a non-load bearing masonry wall. However, the application of this material can be extended to the load-bearing masonry wall by putting groove infill in the multiple groove area. This method called as grouting procedures. Regarding [7] in their research found out that the insertion of grouting material in the hollow concrete walls increased about 80% of their resistance capacity. The most important mechanical properties of the concrete masonry unit are their compressive strength. It is because the capacity of the masonry wall highly depends on the masonry unit performance. By that's, the compressive strength of the interlocking concrete block plays a significant role to enhance the capacity of the masonry wall. It is supported by [8] that the capacity of masonry wall under axial compression is influenced by different parameters such as the compressive strengths of masonry units and mortar, the thickness of mortar binder, shape of the units, mortar bedding or binder methods (full or face shell), type of testing (prism or wallette) and masonry specimen aspect ratio.

This study aims to investigate the capacity of biaxial interlocking concrete block units under compression load due to its hollowness and groove infill material effect.

2. Methodology

2.1 Material Properties

Materials used in this study are a biaxial interlocking concrete block (BICB), sand, cement, and water. The details of the material used are per discussed as follow.

2.2 Biaxial Interlocking concrete block (BICB)

Interlocking concrete block namely Biaxial interlocking concrete block (BIPB) is a block produced and supply by Innovative Precast Builder Sdn.Bhd. The dimension of the IPB unit is shown in Figure 1.

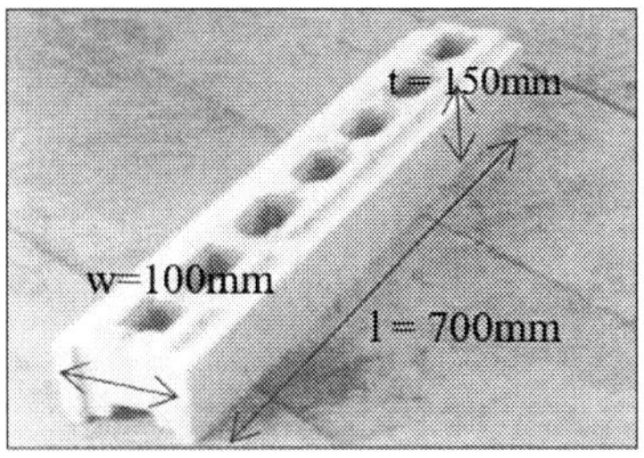

(a)

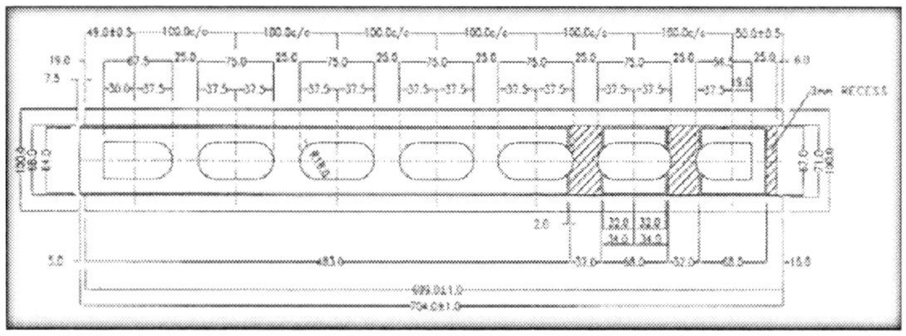

(b)

Figure. 1 (a) and (b) Dimension of the biaxial interlocking concrete block unit

This BICB unit is 700 mm x 100 mm x 150mm in size consist of seven (7) numbers of grooves and 2 sides of the tongue. For this research, 40 unit of IPB was involved in its physical and mechanical properties. The numbers of block required for the testing as follows in Table 1. Out of 40 units, 20 units group as a hollow unit and 20 units as a grouted.

Table 1. Numbers of BICB unit sample

BICB sample	No of sample	Type of testing
Hollow unit	10	Compressive strength test/modulus of elasticity
	10	Density test/Dimension check
Groove infill unit	10	Compressive strength test/modulus of elasticity
	10	Density test/Dimension check

2.3 Groove infill material

In this study, the groove infill material used as a reinforced method to enhance the capability of the block. Besides that, to compliance the requirement by Malaysian Standard as a number of percentages the void area is less than 35%, the grooves should be infill to become a solid. For the grouting process, it is involved the mixture between ordinary Portland cement, sand and water. The mix proportion of this material shows in Table 2.

Table 2. Design mix of groove infill material

Design mix	Cement	Aggregates	Water/cement ratio (w/c)
1:3 (grouting material)	Ordinary Portland Cement	Fine sand	0.5

Grouting material design mix was as shown above. Based on [9], the design mix 1:3 with 0.5 water-cement ratios produced enough compressive strength to fill in the grooves. Meanwhile, the chemical compositions of OPC Type 1 as shown below in Table 3.

Table 3. Chemical Compositions for ordinary Portland cement [10]

Chemical Compositions	%
CaO	63.00
SiO_2	20.00
Al_2O_3	5.70
MgO	0.99
Fe_2O_3	2.90
SO_3	3.50
K_2O	3.55
LOI	2.8

2.4 Sample preparation

The experimental program involved the preparation of BICB unit sample for determination of its mechanical properties and the preparation of cube sample for groove infill material. The samples divided into two groups as a hollow and grouted as shown in Table 4. The experimental program involved the preparation of cube sample for determination of the mechanical properties of the groove infill material. Twelve (12) numbers of a cube sample size of 150mmx150mmx150mm and 50mmx50mmx50mm prepared in the concrete laboratory. The sample cured by immersed in the water for 7 and 28 days of age. Meanwhile, twelve (12) numbers of the hollow and groove infill biaxial interlocking concrete block tested to certify the Malaysian Standard MS 1933-1-2007, Compressive strength of the unit. Blocks were cut and grind for surface preparation and capped on top and bottom to reproduce the same contact area to simulate the actual compression state of the block in the masonry.

Table 4. Numbers of BICB unit for testing

BIPB sample	No of sample	Type of testing	Standard compliance
Hollow and Groove infill unit	20	Compressive strength test/modulus of elasticity	MS 1933-1-2017: Methods of test for masonry units – Part 1: Determination of compressive strength
	20	Density test	MS 1933-13-2007: Methods of test for masonry units – Part 13: Determination of net and gross dry density of masonry units (except for natural stone)

(a) (b)

Figure 2.(a) hollow BICB unit, (b) groove infill BICB unit

2.5 Testing procedures

The testing procedures involved grouting material and BICB unit compliance with Malaysian Standard.

2.5.1 Compressive strength and density of grouting material

The compressive strength of mortar is determined by using 50mm cubes as per ASTM C109 / C109M – Standard Test Method for Compressive Strength of Hydraulic Cement Mortars. The test was conducted for six (6) cubes and report the average value as the test result for both 7day and 28-day compressive strength.

2.5.2 Density test of IPB unit

All the sample will follow the MS 1933-13-2007: Methods of test for masonry units – Part 13: Determination of net and gross dry density of masonry units (except for natural stone). Total 20 unit of IPB as a hollow or grouted used to determine the net and gross density of masonry units. For whole unit specimens, dry the specimens to constant mass, in a ventilated oven at a temperature of 70° ± C 5° C. Constant mass is reached, if during the drying process in two subsequent weighing with a 24h interval, the loss in mass between the two determination is not more than 0.2% of the total mass.

2.5.3 Compressive strength test of IPB unit

All the sample will follow MS 1934-1:2007(confirmed 2014) Method of test for masonry-part 1: Determination of compressive strength. A minimum number of sample required is six (6) per 1000 consignment. Firstly, immerse the specimens in the water at a temperature of 20 °C ± 5 °C for a minimum period of 15 h and subsequently allow them to drain for 15 min to 20 min. According to clause 7.4.2 where the net loaded area of masonry units with a frog which is intended to be filled with mortar in practice (see also 7.2.3) is not less than 35 % of the gross area then the compressive strength shall be calculated on the basis of the net loaded surface of the frogged bed face. Where the net loaded area of masonry units with a frog is less than 35 % of the gross area then the compressive strength shall be calculated on the basis of the gross area of the masonry unit.

3 Result and Discussion

3.1 Density value of the sample

Density is a physical property of the concrete block that plays a significant role in masonry construction. Based on the density values, the material can be classified as a normal concrete block or lightweight concrete block. The denser of the material, it will reduce the dead load of the structures. Figure 3 below, shown the density of the BICP1 and BICP2 sample. From the figure, it is shown that the density value of BICB1 unit is 2059kg/m^3 meanwhile BICB2 unit is 2200kg/m^3. The higher value density of the sample it might be due to the effect of groove infill material that fills in the area of the groove of the BICB unit. It is also regarding the mix ratio used. Based on [11], the weights of the solid blocks are higher than those of hollow blocks lead to higher in density. Besides that, [12] found out that the decrease in weight can be explained by the Law of Mixtures. The weight of sample influence the density of the sample as finding out by [13]. As a conclusion, the density of the sample will be influenced by the load resistance capacity of the structures.

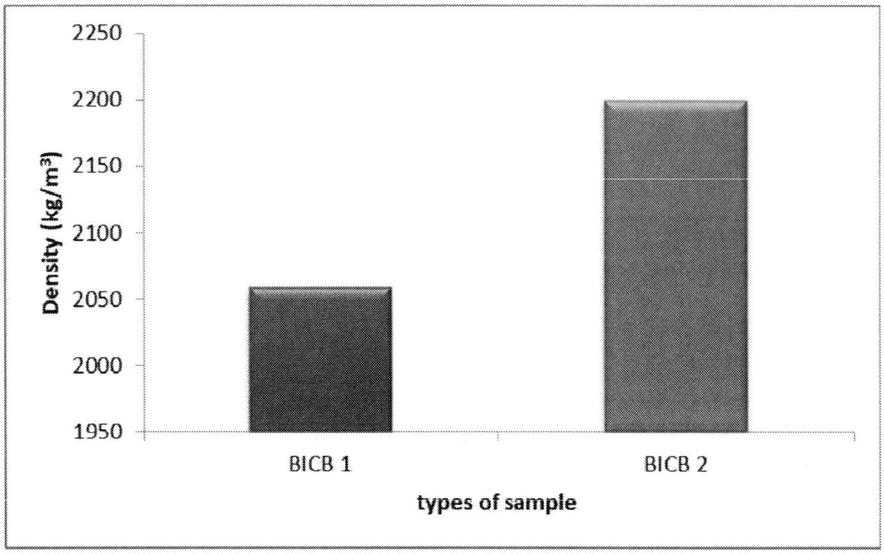

Figure 3. Comparison of the density value for BICB blocks

3.2 Compressive strength of the biaxial interlocking concrete block unit.

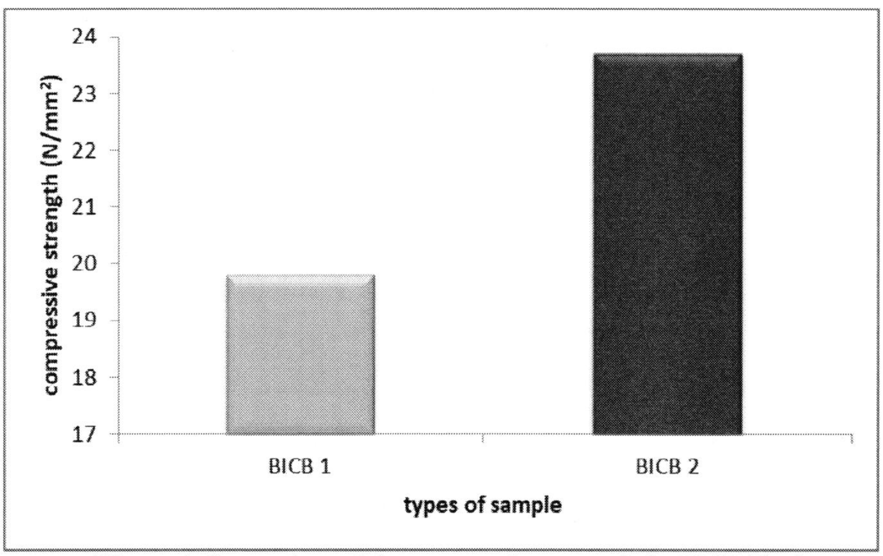

Figure 4. Comparison of the compressive strength for BICB blocks

Figure 4, presents the compressive strength of hollow BICB and groove infill BICB. From the figure, it is shown that the compressive strength of hollow BICB is 19.8 N/mm^2 meanwhile grouted BICB is 23.7 N/mm^2. It is found that the compressive strength of grouted BICB is higher than hollow BICB unit. The compressive strength of the grouted IPB is higher than the hollow unit might be due to the effect of groove infill material compressive strength when reacting with the hollow block. A study conducted by [14] mentioned that the voids number of block unit significantly affect the compressive strength value. Regarding [15], their study showed that the grout bond strength is improved with the decrease of the water/cement ratio.

3.3 Correlation of compressive strength with the density.

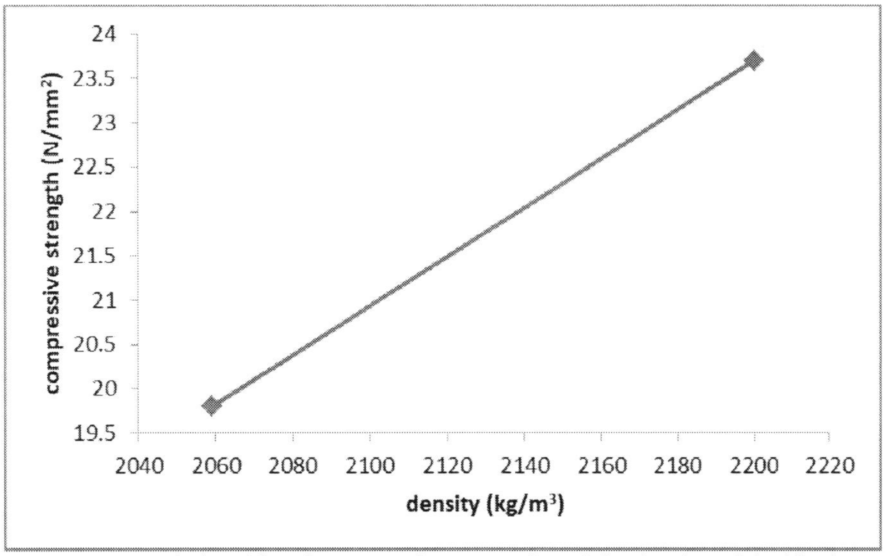

Figure 5. Relationship of compressive strength and the density of the blocks

Figure 5, presents the relationships of compressive strength of BICB 1 and BICB 2 with their respective density. From the figure, it is shown that the linear relationships between these two types of sample. The higher the density value leads to the higher in compressive strength and vice versa. It is found that the higher compressive strength with higher density leads to an increase in a dead load of the BICB unit. By that, these results show that to develop a new lightweight concrete block it must take into account the reliable density but produced the good compressive strength. The raw material that been used must lighten but have a good capacity.

4. Conclusion

This paper presented a comprehensive experimental program with the objective of assessing the compressive strength of the biaxial interlocking concrete blocks. Nominal resistance of blocks was 23.7 N/mm^2 with the groove infill material produced with shrinkage-compensating admixture. The main conclusions of the research presented here are:

a) The density of the sample depends on the weight of the sample
b) All the BICB 2 units showed a higher value in compressive strength compared to the BICB 1 unit;
c) There is a linear relationship between the density and the compressive strength of the sample
d) Experimental results show the significant result of compressive strength of BICB units due to the effect of the groove infill material.

Acknowledgment
The authors gratefully acknowledge the Innovative Precast Builder Sdn. Bhd. As a main supplier of the interlocking concrete block in this research, University Technology Mara, Malaysia and wish to thank the Ministry of Education, which provided the scholarship and also whoever is involved in this project.

References

[1] S. Liu, J. Zhang, L. Liu, H. Guo, and B. Wang, "Compressive performance of new types of load-bearing horizontal-hole interlocking concrete hollow blocks," *Adv. Struct. Eng.*, vol. 22, no. 6, pp. 1368–1383, Apr. 2019.

[2] T. S. V. M. Thorat, Mahesh Papal, Vivek Kacha, and S. Gaikwad, "Hollow Concrete Blocks-A New Trend," *Int. J. Mod. Eng. Res.*, vol. 5, no. 5, pp. 19–26, 2015.

[3] X. X. He and Y. Lin, "Research on Compressive Strength of Grouted Dry-Stack Concrete Block Masonry," *Appl. Mech. Mater.*, vol. 638–640, pp. 202–207, Sep. 2014.

[4] S. Atamturktur *et al.*, "Compressive Strength of Dry-Stacked Concrete Masonry Unit Assemblies," *J. Mater. Civ. Eng.*, vol. 29, no. 2, p. 06016020, Feb. 2017.

[5] N. A. N. A. Safiee *et al.*, "Behaviour of interlocking mortarless hollow block walls under in-plane loading," *Aust. J. Struct. Eng.*, vol. 7982, no. 2, pp. 1–9, Jan. 2018.

[6] H. C. Uzoegbo and J. V Ngowi, "Structural Behaviour of Dry-stack Interlocking Block Walling Systems Subject to In-plane Loading," *Concr. Bet.*, vol. 103, pp. 9–13, 2003.

[7] J. S. Camacho, B. G. Logullo, G. A. Parsekian, and P. R. N. Soudais, "The influence of grouting and reinforcement ratio in the concrete block masonry compressive behavior Influência do graute e da taxa de armadura no comportamento à compressão da alvenaria de blocos," vol. 8, no. 3, pp. 341–352, 2015.

[8] J. A. Thamboo and M. Dhanasekar, "Correlation between the performance of solid masonry prisms and wallettes under compression," *J. Build. Eng.*, vol. 22, pp. 429–438, Mar. 2019.

[9] S. B. Singh, P. Munjal, and N. Thammishetti, "Role of water/cement ratio on strength development of cement mortar," *J. Build. Eng.*, vol. 4, pp. 94–100, Dec. 2015.

[10] A. H. Abdul Ghani, M. Fadzil Arshad, N. A. Salehuddin, and A. R. Mohamad Ridzuan, "Hydration Kinetics and Strength Development of Waste Paper Sludge Ash (WPSA) with Different Particle Size as Binary Blended Cement," *Appl. Mech. Mater.*, vol. 754–755, pp. 382–388, 2015.

[11] M. M. K., "Load Carrying Capacity of Hollow Concrete Block Masonry Column," *IOSR J. Eng.*, vol. 2, no. 10, pp. 5–8, 2012.

[12] J. . Lasco, M. . Madlangbayan, and M. . Sundo, "Compressive Strength and Bulk Density of Concrete Hollow Blocks (CHB) with Polypropylene (PP) Pellets as Partial Replacement for Sand," *Civ. Eng. J.*, vol. 3, no. 10, pp. 821–830, 2017.

[13] A. A. Raheem, A. Kayode Momoh, and A. A. Soyingbe, "Comparative Analysis Of Sandcrete Hollow Blocks And Laterite Interlocking Blocks As Walling Elements," *Int. J. Sustain. Constr. Eng. Technol.*, vol. 3, pp. 2180–3242, 2012.

[14] N. Sathiparan, M. K. N. Anusari, and N. N. Samindika, "Effect of Void Area on Hollow Cement Masonry Mechanical Performance," *Arab. J. Sci. Eng.*, vol. 39, no. 11, pp. 7569–7576, 2014.

[15] O. S. Izquierdo, M. R. . Correa, and I. . Soto, "Study of the block / grout interface in concrete and clay block masonry structures Estudo da interface bloco / graute em alvenaria de blocos," *IBRACON Struct. Mater. J.*, vol. 10, no. 4, pp. 924–936, 2017.

Journal of Physics: Conference Series **1424** (2019) 012008 doi:10.1088/1742-6596/1424/1/012008

Development of Web-based Coffee Management Information System to support the Management of Regional Superior Products

Folkes E. Laumal[1*], Jhon A. Wabang[2], Raden S. B. Suharto[3], Paulus E. Plaimo[4], Lita A. Ndoloe[5]

[1,2,3,5]Politeknik Negeri Kupang, East Nusa Tenggara, Indonesia
[4]Universitas Tribuana, Kalabahi, East Nusa Tenggara, Indonesia

*folkeslaumal76@gmail.com

Abstract. One of the areas in Alor Regency which the majority of people like to drink coffee is South of Alor. Coffee is considered a reliever of fatigue and drowsiness and provides additional energy before activity. The management of coffee in South of Alor is carried out by the Koperasi Beringin using raw materials from farmer groups from surrounding villages and sold in the form of packaging under the name Lonsilar Coffee. So far, Lonsilar's coffee production process has been using roasted machines, but the administration of the management still uses cash books and sales are limited to cooperative showrooms and souvenirs. Even though in terms of Telecommunications and Information Technology support, telecommunications infrastructure has reached the South of Alor District, moreover the majority of the people have used mobile phones/smartphones as communication media. This condition allows the development of a web-based coffee management information system so that administrative, promotional and sales processes are carried out through the website. The system is built using PHP language structure with database management using MySQL. The interconnection between users, products, and Koperasi Beringin is organized using detailed and integrated diagram algorithms. The system page can be accessed by the admin and user. Users can process transactions when they have passed the registration/registration stage, while the admin has full rights in managing, changing and verifying all processes. The system test results can run correctly according to recommendations in the Data Flow Diagram.

1. Introductions

Coffee is one of the drinks favored by people in Alor Regency. The hobby of drinking coffee causes people to be able to make coffee as a favorite drink or continuously drink coffee because it is considered as a reliever of drowsiness and giving stimulus (additional energy) before carrying out strenuous activities during the day. One area that the majority of the people like to drink coffee is South of Alor District. This area is ± 30 km from Kalabahi, the capital city of Alor Regency. Some villages in the South of Alor area are mountainous areas with altitudes between 1500 - 2000 m above sea level so that besides vanilla, the community also grows coffee plants and is processed through farmer groups. The annual coffee harvest is processed into coffee and sold to the Beringin KUD in the sub-district capital (Apui) (Microhidro & Power, 2014; Wabang, Laumal, Suharto, & Plaimo, 2018). KUD Beringin then processes and sells in the form of 100 grams and 250 grams. Packaging which was originally only in the form of ordinary plastic packaging, but a touch of mechanization through the

Community Service Program of the Republic of Indonesia Ministry of Research and Technology has changed the coffee processing model in KUD Beringin using a roasting machine with a capacity of > 5 kg, grinding using grinding machines, packaging using electric packaging machines, so that currently the ground coffee products named Lonsilar Coffee have been processed with better, cleaner, faster mechanization and packaged in more attractive attractive packages (Laumal, Wabang, Hattu, & Plaimo, 2017).

The administration, distribution and sale of packaged coffee from Alor Regency, namely Kopi Lonsilar and Kopi Cita Rasa Typically, only rely on manual administration systems using cash books, sales in cooperative showrooms and coffee as souvenirs. Whereas in terms of taste, aroma and protein content, Lonsilar Coffee has a distinctive smell and delicious taste (soft), 14.6 % Protein content and 35.18 % Carbohydrate, the content of antioxidant compounds to ward off free radicals in the body (Drug and Food Testing Center, 2018). With sales in showrooms, there has been an increase in demand for Lonsilar coffee by 2.5 % from the total production of 2 tons from 2016 when it was still conventionally processing into 3.5 tons of production a year.

As one of the superior products from Alor Regency (especially from South of Alor District), the management of Lonsilar Coffee continues to be improved both from the quality of its coffee powder, packaging and promotion. However, locations far and mountainous topography have limited coffee sales. Lonsilar Coffee sales have not yet come out of Alor Regency. On the other hand, South Alor district has quite good telecommunications infrastructure because several Telecommunication towers are available and serve the communication needs of the community through the use of mobile / smartphone communication devices. The use of manual cash books has provided several problems when making recapitulation of production and sales for Annual General Meetings (RAT). Cooperative staff must take more time to summarize all records of production and sales, not to mention getting sheets that have been lost due to water, the paper is damaged or has been torn. In the end the recapitulation produced was also invalid because of these problems. In terms of the development of communication and informatics (ICT) techniques, with the support of telecommunications infrastructure and the ease of designing online information systems, a web-based coffee management information system was developed to change the management, sales, and conventional reporting mechanisms using books to be web-based management that can accessed online.

2. Methods

WEB-based management system development is carried out with several stages, namely designing Context Diagrams, Designing Data Flow Diagrams, Designing Entity Relationship Diagrams (ERD) and designing data dictionaries that will be utilized in making information systems. The first stage is designing a Context Diagram as part of a data flow diagram that functions to map the environmental model, which is presented with a single circle that represents the entire system. Context Diagram highlights a number of important characteristics of an information system, namely groups of users (users), organizations or other systems that make communication or as terminators. Entry data, namely data received by the system from the environment and must be processed in a certain way. Data comes out, namely data generated by the system and given to the outside world (Suryati, 2015). The Context Diagram mechanism in the Web-based Coffee management information system is given in Figure 1

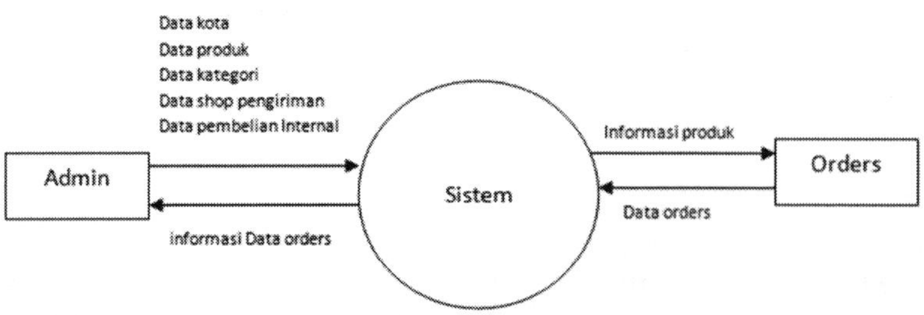

Figure 1. Context Diagram of a Web-based Coffee Processing System

In Figure 1, the Coffee management information system consists of 3 main parts, namely Admin as a system manager who can enter purchase data, sales and data recap, can see and change all data entered into the system. The second part is orders (parties), namely parties who will use the system to view the appearance of products, order, make payments and monitor the delivery process carried out by the system. The third part is the information system, in the form of an interface that displays information on products sold by cooperatives, prices and all stages of ordering, payment and monitoring, management of purchase and sale data and management process records that can be stored, displayed and printed. The second stage is designing Data Flow Diagrams (DFD) which is a logic data model or process that is made to describe where the data comes from and where the data that comes out of the system, where data is stored, what processes produce the data and the interaction between stored data and the process imposed on the data (Afyenni, 2014). DFD is generated in System Analysis and System Design, resulting in an Alternative Evaluation Hierarchy [7]. DFD for Coffee Management Information System is given in Figure 2. This DFD displays an overview of each process in more detail starting from the event of inserting coffee products as input, managing raw materials into packaged coffee (process) to packaging coffee distribution to buyers or customers / users (output).

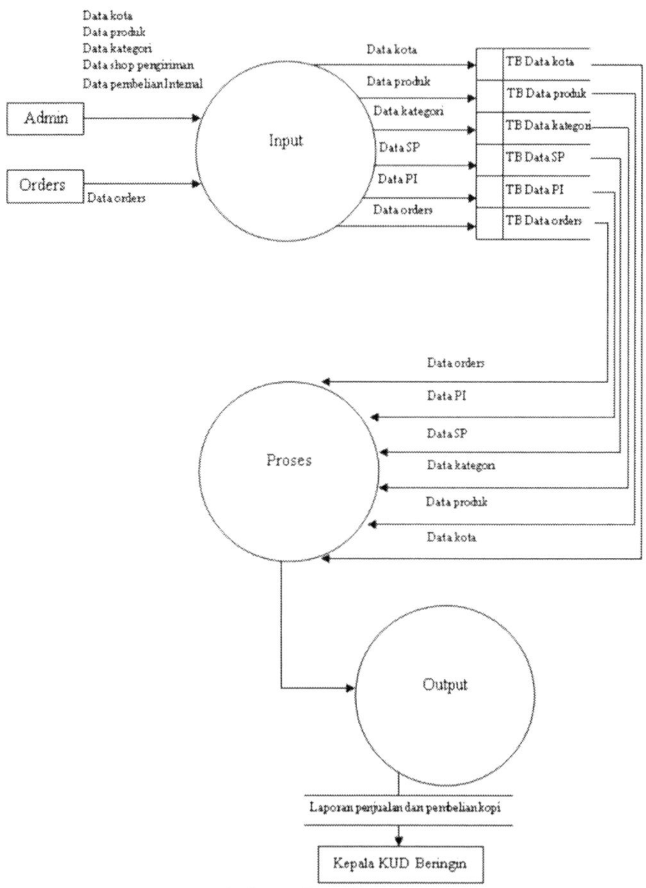

Figure 2. DFD Web-based Coffee Management System

DFD in Figure 2 consists of 3 parts, namely process input and output. The input data section consists of admin as inputting data, with inputted data in the form of coffee products in the form of packaged coffee and coffee beans purchased from farmers, data on categories of coffee beans, packaging of 100 g, 200g or 250 g, sales data , shipping data and purchase data carried out internally both coffee beans from peani and keamsan coffee purchases carried out in the showroom. Orders or buyers as inputting data on orders to be purchased. The Process section consists of product data manipulation, category data, shipping shop data, internal purchase data and data orders in the coffee management information system. The Output section is the result of manipulating data from the coffee management information system that produces reports according to the processing process and transactions that occur in the system. The third stage is designing the Entity Relationship Diagram (ERD) which is one technique to form an activity development through an action diagram (Brinkkemper, 1996; Teorey, T., Lightstone, S., Nadeau, T., & Jagadish, 2011). ERD is also a conceptual model that describes the relationship between data storage (data files) using a number of notations and symbols to describe the structure and relationships between data, or ERD shows the relation or relationship between the tables to be made [10]. The ERD design for the Coffee Management Information System consists of 6 entities, namely orders/products, products, internal purchases, shop purchases, cities and categories. Entities are individuals who represent something tangible (existence) and can be distinguished from something else. Each entity has a relation to aribut, namely the description of the characteristics of the entity, described in the form of a circle or ellipse and becomes the key to the entity or key that is underlined. These entities are then developed into data dictionaries as tables in the database that are used as data storage locations inputted to the coffee management information system. The fourth stage in the development of a lonsilar coffee management information system, namely the design of the Data Dictionary. Data Dictionary is a list of data elements (parts of a database) that are used in the application of an information system (Anwar, Nugroho, & Lestariningsih, 2013). Some of the Coffee Management system Data Dictionary schemes are given in Table 1 and Table 2.

Table 1. Data Dictionary orders

No	field	Data Type	Length	Description
1	*id_orders*	Int	(5)	Primary Key
2	nama_kustomer	Varchar	(100)	
3	address	Text		
4	telpon	Varchar	(20)	
5	email	Varchar	(50)	
6	status_order	Varchar	(50)	
7	date_order	Date		
8	hour_order	Time		
9	id_kotaordering	Int	(3)	

Thesection as in Table 1 has 9 fields, namely id_orders (as primary key), nama_customer, address, telephone , email, statur_order, tgl_order, jam_order, kota. Masiing-each also has a data type and length of data that is tailored to the data processing requirements in the information system.

Table 2. Product Data Dictionary

No	field	Data Type	Length	Description
1	*product ID*	Int	(5)	Primary Key
2	id_category	Int	(5)	Foreign Key
3	product_name	Varchar	(100)	
4	produk_seo	Varchar	(100)	
5	description	Text		
6	price	Int	(20)	
7	stock of	Int	(5)	
8	weight	Decimal	(5.2)	
9	date	Date		
10	picture	Varchar	(100)	
11	purchased	Int	(5)	
12	discount	Int	(5)	

Whereas the Parts of Product in Table 2 have more fields due to being objects that can be processed by information Systems. Twelfth product field includes id_product (as primary key), id_kategory (as a foreign key), nama_produk, produk_seo, description, price, stock, weight, date of entry, image, purchase, discount.

3. Result and Discussion.

3.1. Making a Database for a Web-based Coffee *Management Information System.*
The database created is a database management that will accommodate all the data in the information system. The database is blinded using the Xampp application. The database that has been created contains tables with parameters specified in the system creation method. Display of one of the tables in the database is given in Figure 3.

Figure 3. Display of tables in database management of coffee management information system

After filling in the table name and number of fields in the column provided, the creation is continued by filling in the teacher table fields according to the name fields, data types of fields and sizes of data from fields and stored. In the same way, the other tables are developed according to the design made. Tables in the database will correlate with each menu option that will be displayed on the system interface. Login menu with user table, product menu with tables of categories, descriptions, prices, stocks and other related tables. The ordering process also correlates with tables purchased, discounts, prices, banks and recapitulation.

3.2. Making and Implementing a Web-based Coffee Management Information System.
Lonsilar Coffee management information system was developed using the PHP programming language with the main display as in Figure 4.

Figure 4. The main view of the Coffee Management Information System

3.2.1. There are 3 main parts in Figure 4, namely the header section that contains a display menu that can be used to log in, registration, view shopping documents (basket) and search links. In the second part, the program body displays a list of products that can be ordered by choosing which type of product you want. Each product has included product photos, prices and information about the product. The third part, the footer, displays additional information about the lonsilar coffee management information system.
Customers who will place an order can order directly through the front page of the information system or can also register first through the available login menu. The Login page displays 3 parts, namely if you register as a new customer, then the buyer can register through the new customer section. If the buyer has already made a transaction / order, then he can continue through the loyal customer section. Accounts that have been owned can be used to log into the lonsilar coffee management information system. In the third part, displays additional information about the user including changes to passwords, order history, periodic payments, purchases and rewords. The programming footage for the login menu is given in Figure 5.

```
<h2>Pelanggan Setia</h2>
<p><strong>Saya adalah pelanggan setia</strong></p>
<form action="http://www.kopialor.com/new/index.php?route=account/login" method="post"
enctype="multipart/form-data">
<div class="form-group">
<label class="control-label" for="input-email">Alamat E-Mail</label>
<input type="text" name="email" value="" placeholder="Alamat E-Mail" id="input-email"
class="form-control" />
</div>
<div class="form-group">
<label class="control-label" for="input-password">Kata Sandi</label>
<input type="password" name="password" value="" placeholder="Kata Sandi" id="input-
password" class="form-control" />
<a href="http://www.kopialor.com/new/index.php?route=account/forgotten">Lupa Kata
Sandi</a></div>
<input type="submit" value="Log Masuk" class="btn btn-primary" />
</form>
```

Figure 5. Snapshot of the program language for the login menu

Figure 5 shows that there is one menu named "I am a loyal customer" which if clicked will run the login page and request to fill in the email address label and password. But if the user accidentally forgets his account, then the user can run the "Forgot Password" menu for the password request procedure. When a user makes a transaction through product ordering, the user will select photos and product information found on the front page. After selecting the product, the system will display information such as Figure 6.

Figure 6. Early Ordering Stage

Figure 6 displays the start page when the user orders the product. The user will be faced with the stage of filling the product into the basket and choosing the number of products to be purchased. If it has finished and ensures the product, the system will allow the user to reconcile the entire order, enter the identity (including the destination of the shipment) and validate the certainty of funds transfer in accordance with the amount calculated by the system. The system also provides an opportunity to have a coupon if provided by the seller. If not, the user can process payments to the bank account that is informed by the system, validate the transaction by uploading proof of delivery and the product will be sent to the user. The appearance of the recapitulation in the user's basket is shown in Figure 8.

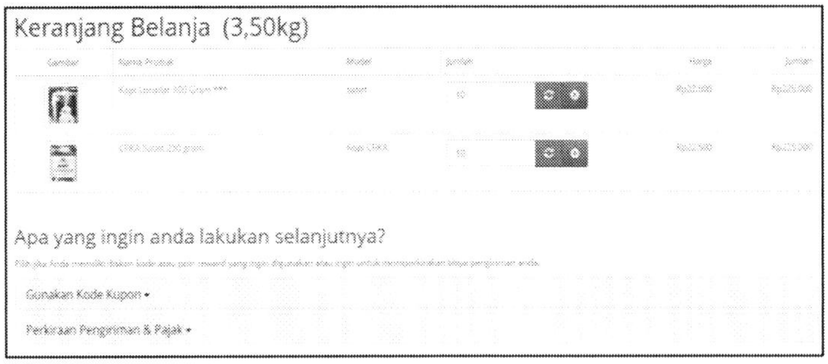

Figure 7. Recapitulation of Shopping (Basket)

Figure 7 displays one of the stages of the transaction when the customer orders the product. On the order page, a list of products that have been selected appears, but the customer can still change both the product category and type. If the customer makes sure the data is correct, then the system page will recommend to proceed to the payment process. The payment system regulated in this system only uses bank transformers, so customers need to make payment transactions first, then the transfer proof document is submitted to the system. The payment receipt status will be received by the Admin and confirmed to the customer via the customer's e-mail address as proof that the product being spent temporarily is processed to be sent to the customer's address.

Admin has its own management page in a web-based coffee management information system. The Admin page is shown in Figure 9.

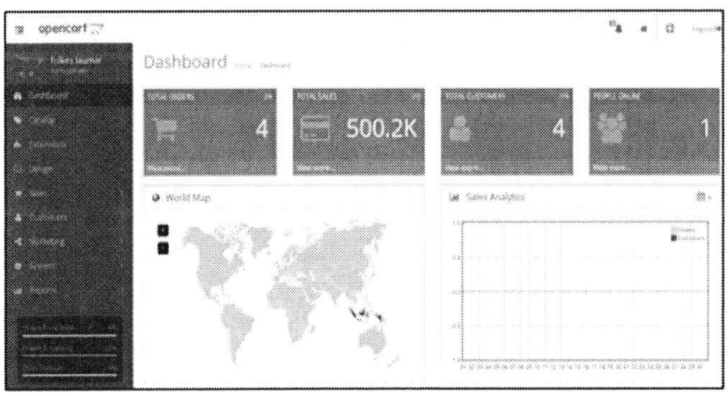

Figure 8. Periodic Coffee Management Recapitulation

Admin page designed consists of 2 parts, the left part displays menus that can be accessed for page management, while the right part displays the contents of each menu on the left . The Admin page has full access to a Web-based Coffee management information system. The Admin page as shown in Figure 9 displays the development of product introduction in graphical form that can be displayed weekly, monthly or yearly in preparation for the periodic evaluation. The Admin page also displays a catalog menu for entering products, reviewing products and customers, deleting customers and setting the layout of the webpage itself. The results of testing the implementation of a web-based coffee management information system show the data as in Table 3.

Table 3. Results of Testing the implementation of the use of a web-based coffee management information system.

No	Data Input	Recommended	Results achieved	
			True	False
1	Admin Login.	Admin page displayed	√	
2	Adding coffee products for sale.	All product categories can be input into the system.	√	
3	Add system and product descriptions.	The description results can be accessed by customers.	√	
4	Register as a customer and login.	The registration process is received and can log in to display the customer's page.	√	
5	Verify customer transaction data in the form of payment.	The transaction was successfully received and confirmed to the customer.	√	
6	Creating a transaction report is displayed on the Admin page.	Transaction reports appear on the admin page in the form of numbers and graphs.	√	

Table 3 shows that the web-based coffee management information system developed can work well when the admin and customer carry out transaction tasks. Admin managed to log in and manage admin pages, while customers also managed to register and make transactions from the account page.

4. Conclusion

The web-based coffee management information system is one of the efforts to develop superior products in the NTT area, especially processed coffee from Alor Regency. One of the processed coffee products from Alor is Lonsilar Coffee produced by the KUD Beringin in Apui and the typical Alor Coffee Taste produced by the Plantation Cooperative. Both are produced in the form of packaging. Coffee management in Alor Regency even though the processing has used machines, but the administration and sales management are still manual, so to avoid damage to management data and increase sales and promotions, a web-based management information system is developed. This system consists of 2 user bases, namely admin and customer. Admiin has the right to full access to change and regulate the information system, while the user is only limited to seeing, ordering and making product shopping transactions. The test results show that the 6 parts of this information system testing can work well when accessed by the admin or by the customer.

References

[1] Afyenni, R. (2014). Designing Data Flow Diagrams for School Information Systems (Case Study of High School UNP Laboratory Development). Teknoif Journal, 2 (1), 2338–2724. https://doi.org/10.1093/jme/tjv089

[2] Anwar, SN, Nugroho, I., & Lestariningsih, E. (2013). MOBILE APPLICATION DESIGN AND IMPLEMENTATION OF SEMARANG, 5 (2), 135–145.

[3] Drug and Food Testing Center. (2018). Report on External Sample Testing Results. Kupang. Retrieved from https://osf.io/preprints/inarxiv/w7u5v/

[4] Brinkkemper, S. (1996). Method engineering: Engineering of information systems development methods and tools. Information and Software Technology, 38 (4 SPEC. ISS.), 275-280. https://doi.org/10.1016/0950-5849(95)01059-9

[5] Laumal, FE, Wabang, JA, Hattu, EP, & Plaimo, PE (2017). INCREASING THE PRODUCTION OF LONCILAR COFFEE IN THE WINTERED KUD. In FE Laumal, JA Wabang, EP Hattu, & PE Plaimo (Eds.), Vocational and Technology National Seminar (Semnasvoktek) 2017 (pp. 366–373). Denpasar. https://doi.org/10.17605/OSF.IO/YFXJ8

[6] Lutfi, HW, & Riasti, BK (2011). Maintenance Information System and Laboratory Inventory in

Middle School 1 Rembang Based on the Web. Journal Speed - Engineering and Education Research Center, 10 (1), 83–91. https://doi.org/10.1145/1518701.1518726

[7] Microhidro, C., & Power, E. (2014). Rural productive economic development through high-tech cooperatives, 24 (2), 116–130.

[8] Suryati, BEP (2015). Development of the Poor People's Data Collection Information System for the Poor (Raskin) Rice Program in Mantren Village, Kebonagung District, Pacitan Regency. Communication, 6 (1). Retrieved from http://portal.ejurnal.net/index.php/speed/article/view/406

[9] Teorey, T., Lightstone, S., Nadeau, T., & Jagadish, HV (2011). The Entity-Relationship Model. In Database Modeling and Design. https://doi.org/https://doi.org/10.1016/B978-0-12-382020-4.00002-1

[10] Wabang, JA, Laumal, FE, Suharto, RSB, & Plaimo, PE (2018). Synergy of Farmers, UKM, Higher Education and Government Groups in processing Regional Superior Products and Online-based Information System Implementation. Radar NTT. https://doi.org/10.31227/osf.io/6ehfk

[11] Wang, L., & Raz, T. (1991). Analytic hierarchy process based on data flow diagram. Computers & Industrial Engineering, 20 (3), 355–365. https://doi.org/https://doi.org/10.1016/0360-8352(91)90007-S.

RSA Cryptographic Algorithm using Cubic Congruential Generator

Nurul Khairina[1], Muhammad Khoiruddin Harahap[2]

[1] Informatics Engineering Department, [Universitas Medan Area]
[2] Informatics Engineering Department, [Politeknik Ganesha Medan]

E-mail : nurulkhairina27@gmail.com[1], choir.harahap@yahoo.com[2]

Abstract. In the field of cryptography, the key and procedures for converting messages into secret messages have a very important role. The RSA key is built with a random number generator. The RSA key is made as strong as possible to increase ciphertext robustness. A key that is difficult to crack and requires a lot of time certainly has higher robustness. Two initial variables that are very important in building RSA keys are prime number p and prime number q. In this research, the p and q prime numbers are generated by Cubic Congruential Generator algorithm. The results of this study can be seen by testing the ciphertext robustness. Ciphertext test results that are resistant to cryptanalytic attacks certainly show the strength of the key that is built, including the selection of primes p and primes q. The ciphertext test shows that 90% of the greater primes p and primes q will achieve a higher percentage of ciphertext robustness, and 10 % of the combination from small prime number p with large prime number q is also able to achieve the highest percentage of ciphertext robustness. This proves that the selection of prime number p and prime number q in generating RSA keys really can affect the ciphertext robustness.

1. Introduction

Cryptography is the science used to protect the integrity of data [1]. Cryptography is very closely related to the confidentiality [2], integrity and availability of confidential data [3]. Cryptographic algorithms have had rapid development from year to year. Since the RSA algorithm was first introduced by three researchers Ron Rivest, Adi Shamir, and Len Adleman [4] in 1977 - now, many researchers have been interested in continuing to develop, modify, and combine these RSA algorithms with other algorithms to obtain strong keys and robustness to attacks by a cryptanalyst [5].

On the RSA algorithm, choosing prime number p and prime number q [6] can affect the key and also the ciphertext. Ciphertext that is not resistant to cryptanalyst attacks will have a bad impact on data confidentiality [7]. This research uses Cubic Congruential Generator (CCG) to generate random numbers, where two primes will be taken which will be the p-value and q value to generate the key. The researcher was interested in combining the RSA algorithm with the Cubic Congruential Generator (CCG) algorithm to increase the ciphertext robustness to cryptanalyst attacks.

There are several studies related to this research, among others: Previous research has been carried out by [8]. This study generates prime numbers on a large scale using Mersenne Prime Number and tests them with Rabin Miller Primality Test. In the results of this study, it can be seen that the security of the RSA algorithm is higher when using prime numbers on a large scale.

Research on RSA with the Linear Congruential Generator (LCG) algorithm has been carried out by [9]. The results of this study indicate the existence of double-layer security as indicated by the presence of new tables arising from the results of randomizing ASCII tables and the process of encryption and decryption such as RSA in general.

Research conducted by [10] combines the AES (Advanced Encryption Standart) algorithm with RSA. The results showed that the results of encryption have high security and are difficult to solve.

Research conducted by [11] modifies RSA to be more unique. Modification of the RSA is done by using two different public keys and a private key obtained from a large factor N. This research also does double encryption and decryption, so that ciphertext will get better security.

Research conducted by [12] compares performance between RSA-based cryptosystem and ECG (Elliptic Curve Cryptography) – based cryptosystem. From the results of the study, ECG-based cryptosystem is better than RSA - based cryptosystem.

Research conducted by [13], The researchers conducted a combination of MD5 Hash, RSA and DSA algorithm. The results of the study show that researchers can achieve efficiency, authentication, and data integrity that are quite good.

Research on cryptography always prioritizes the robustness of keys and ciphertext. The results of this study will show several things, like what is the role of prime number p and prime number q for ciphertext robustness, how much the percentage of ciphertext robustness and how long is the estimated cracking times.

2. Methodology

In this study, the researchers want to combine the RSA algorithm and Cubic Congruential Generator algorithm. Here are the steps in the process of generating the key, encryption [14], and decryption [15] [16]:

2.1 Generating The Key :

a. Generate two different random numbers p and q with the Cubic Congruential Generator algorithm [17] :

$$X_n = (a*X_{n-1}{}^3 + b*X_{n-1}{}^2 + c*X_{n-1} + d) \bmod m \qquad (1)$$

The following are the results of random numbers obtained with a = 1; b = 3; c = 4; d = 3; X_{n-1} = 3; and m = 14.

Table 1. Random Number with CCG

No	Xn	Xn mod n
1	69	13
2	2759	1
3	11	11
4	1741	5

b. Test two numbers p and q with Fermat Primality Test Algorithm whether the numbers are primes or composite numbers. We take value p = 13 and q = 11.
   - $p = 13 \Rightarrow 2^{(13-1)} \equiv 1 \pmod{13} = 2^{12} \bmod 13 = 1$
     Test result : p = prime
   - $q = 11 \Rightarrow 2^{(11-1)} \equiv 1 \pmod{11} = 2^{10} \bmod 11 = 1$
     Test result: q = prime

c. Calculate n = p . q, make sure p is not equal to q
   - n = p * q = 13 * 11 = 143

d. Calculate $\Phi(n)$ = (p-1) * (q-1) = (13-1) * (11-1) = 12 * 10 = 120

e. For the public key, select the value of e which is relatively primed to $\Phi(n)$ where $1 < e < \Phi(n)$ and gcd (e, Φ (n)) = 1, we take e = 7.
   - e evidence is relatively prime and $\Phi(n)$:
     7 mod 120 = 7
     120 mod 7 = 1 => gcd = 1

f. For the private key, d obtained by equation $d \equiv 1 \pmod{\Phi(n)}$, where $(0 \leq d \leq n)$. After conducting the experiment, get it d = 223.
   - Proof : d * e mod $\Phi(n)$ = 1 => 223 * 7 mod 120 = 1

g. We obtain the public key of the couple from e and n.
   - Public key = (e, n) => (7, 143)

h. We obtain the private key of the couple from d and n.
   - Private key = (d, n) => (223, 143)

2.2 Encryption Process :

a. Take the public key *e* and *n,* where e = 7 and n = 143.
b. Convert plaintext letter to ASCII Code
c. Perform the encryption process with $c_1 = m_i^{\,e} \bmod n$, so obtained the ciphertext
   - First Ciphertext :
     $C_1 = m_1^{\,e} \bmod n$
     $C_1 = 110^{\,7} \bmod 143 = 33$ => (Ciphertext_1 = !)
d. Repeat and continue the steps above until C_5.
   The results of the encryption process can be seen in Table 2 below:

Table 2. Encryption Process

No	Plaintext	ASCII	C_i	Ciphertext
1	N	110	33	!
2	U	117	39	‘
3	R	114	49	1
4	U	117	39	‘
5	L	108	4	EOT

2.3 Decryption Process :

a. Take the private key *d* and *n,* where d = 223 and n = 143.
b. Convert each ciphertext letter to ASCII Code
c. Perform the decryption process with $m_1 = c_1^{\,d} \bmod n$ so obtained the first plaintext
   - First Plaintext:
     $M_1 = c_1^{\,d} \bmod n$
     $M_1 = 33^{\,223} \bmod 143 = 110$ => (Plaintext_1 = N)
d. Repeat and continue the steps above until M_5.
   The results of the decryption process can be seen in Table 3 below:

Table 3. Decryption Process

No	Ciphertext	ASCII	M_i	Plaintext
1	!	33	110	N
2	‘	39	117	U
3	1	49	114	R
4	‘	39	117	U
5	EOT	4	108	L

3 Result and Discussion

In this chapter a test will be conducted to find out the relationship between the values of prime number p and the prime number q to the ciphertext robustness and the estimating cracking times, which can be seen in detail in the table below:

Table 4. Relationship between the Values of Prime Number p and Prime Number q for The Ciphertext Robustness and Estimating Cracking Times

No	Prime Number p	Prime Number q	Ciphertext Robustness	Estimating Cracking Times
1	13	11	18 % (Dangerously Low)	19 hours
2	29	37	44 % (Medium)	4 months
3	71	73	59 % (Medium)	39 thousand years
4	61	67	63 % (Fairly Good)	386 thousand years
5	113	103	63 % (Fairly Good)	386 thousand years
6	17	71	72 % (Fairly Good)	120 million years
7	191	373	78 % (Fairly Good)	4 billion years
8	271	317	82 % (Excellent)	39 billion years
9	503	541	89 % (Excellent)	4 trillion years
10	19	83	100% (Excellent)	27 quadrillion years

Visualization of table 4 above can also be seen in the graph below:

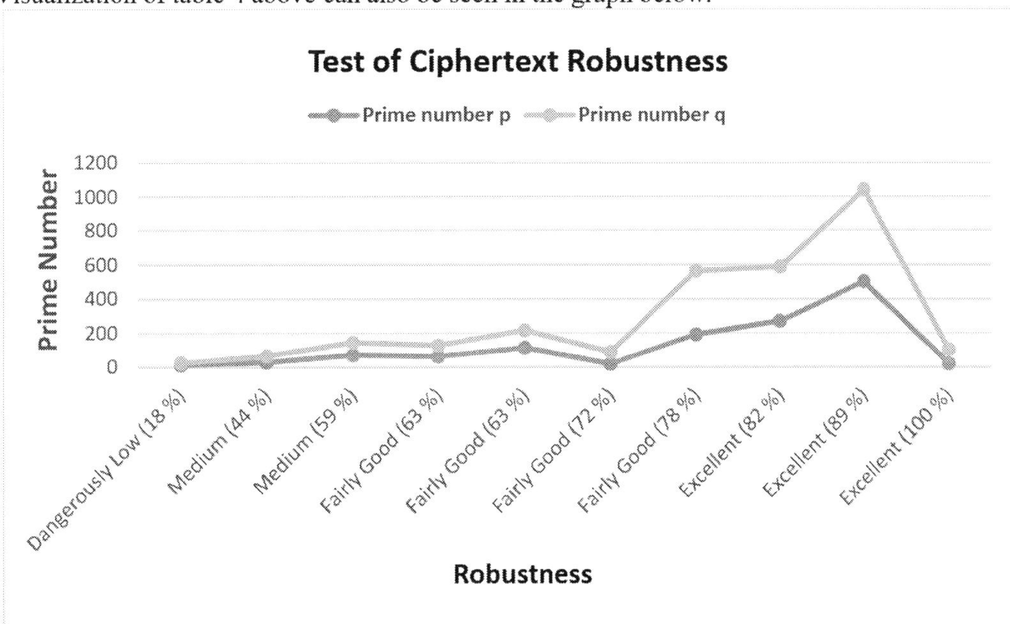

Figure 1. Visualization from Ciphertext Robustness

In Table. 4 and Figure. 1 above, the large prime number of p and q has a high percentage of ciphertext robustness. In the ciphertext robustness test above, 9 out of 10 data show a significant percentage change along with the increasing prime number p and prime number q. However, there is one data with a combination of a small prime number ($p = 19$) and the large prime number ($q = 83$) that achieve ciphertext robustness reaches 100%. Through these 10 test data, it can be said that the success rate of ciphertext robustness testing reaches ± 90%.

4 Conclusion

The conclusions of this study are :

1. The selection of prime number p and prime number q in generating RSA keys can affect the ciphertext robustness.
2. From the tests conducted, 90% of the test results showed a greater prime number p and prime number q will achieve a higher percentage of ciphertext robustness.

3. From the tests conducted, 10 % of the test result showed the combination of small prime number p with large prime number q is also able to achieve the highest percentage of ciphertext robustness.

5. Acknowledgment

Thanks to the Ministry of Research, Technology and Higher Education for funding this research in the Beginner Lecturer Research grant scheme in the fiscal year 2019.

References

[1] M. H. Dehkordi, A. Asgari, and A. Moradian, "A New Scheme to Secure Communication and Data, based on the Integration of Cryptography and Steganography," *Int. J. Cyber-Security Digit. Forensics*, vol. 7, no. 2, pp. 183–188, 2018.

[2] H. Al-Mahdi, M. Alruily, O. R.Shahin, and K. Alkhaldi, "Design and Analysis of DNA Encryption and Decryption Technique based on Asymmetric Cryptography System," *Int. J. Adv. Comput. Sci. Appl.*, vol. 10, no. 2, pp. 499–506, 2019.

[3] S. K, "An Optimal RSA Encryption Algorithm for Secret Images," *Int. J. Pure Appl. Math.*, vol. 118, no. 20, pp. 2491–2500, 2018.

[4] R. Prakash, P. Chithaluru, D. Sharma, and P. Srikanth, *Implementation of Trapdoor Functionality to Two-Layer Encryption and Decryption by Using RSA-AES Cryptography Algorithms*. Springer Singapore, 2018.

[5] M. I. Nazeer, G. A. Mallah, R. Bhatra, and R. A. Memon, "Implication of Genetic Algorithm in Cryptography to Enhance Security," *Int. J. Adv. Comput. Sci. Appl.*, vol. 9, no. 6, pp. 375–379, 2018.

[6] A. Gambhir and R. Arya, "Performance Analysis and Implementation of DES Algorithm and RSA Algorithm with Image and Audio Steganography Techniques," in *Computing, Communication and Signal Processing*, 2018, no. September, pp. 1021–1028.

[7] D. Rachmawati, M. A. Budiman, and A. F, "PDF File Encryption on Mobile Phone using Super-Encryption of Variably Modified Permutation Composition (VMPC) and Two Square Cipher Algorithm PDF File Encryption on Mobile Phone using Super-Encryption of Variably Modified Permutation Composition (VMPC," in *2nd International Conference on Computing and Applied Informatics (ICCAI)*, 2018, pp. 1–4.

[8] D. Apdillah, M. K. Harahap, and N. Khairina, "Generating Mersenne Prime Number Using Rabin Miller Primality Probability Test to Get Big Prime Number in RSA Cryptography," *Int. J. Inf. Syst. Technol.*, vol. 1, no. 1, pp. 1–7, 2017.

[9] M. K. Harahap and Rina, "Kombinasi Kriptografi RSA dengan Linear Congruential Generator," *SinkrOn*, vol. 3, no. 1, pp. 267–271, 2018.

[10] V. A. L. Muthuraman, "An Efficient Technique for Cryptography with Enhanced Key Security," *Int. J. Pure Appl. Math.*, vol. 118, no. 8, pp. 479–484, 2018.

[11] M. A. Islam, A. Islam, N. Islam, and B. Shabnam, "A Modified and Secured RSA Public Key Cryptosystem Based on ' n ' Prime Numbers," *J. Comput. Commun.*, vol. 6, pp. 78–90, 2018.

[12] D. Mahto and D. K. Yadav, "Performance Analysis of RSA and Elliptic Curve Cryptography," *Int. J. Netw. Secur.*, vol. 20, no. 4, pp. 625–635, 2018.

[13] S. Bhat and V. Kapoor, "Secure and Efficient Data Privacy, Authentication and Integrity Schemes Using Hybrid Cryptography," in *International Conference on Advanced Computing Networking and Informatics*, 2018, pp. 279–285.

[14] S. Keerthi, M. Mahita, J. N. Devi, and M. Nivedhitha, "Online User's Behavior Data Preserving Using RSA Based Selective Aggregation," *Int. J. Pure Appl. Math.*, vol. 118, no. 20, pp. 2347–2352, 2018.

[15] M. Jannah, B. Surarso, and Sutimin, "A Combination of Rivest Shamir Adleman (RSA) and Affine Cipher Method on Improvement of The Effectiveness and Security of Text Message," in *ISNPINSA*, 2019, pp. 1–7.

[16] C. Intila, B. Gerardo, and R. Medina, "A Study of Public Key ' e ' in RSA Algorithm," in *The International Conference on Information Technology and Digital Applications*, 2019, pp. 1–10.
[17] B. Schneier, *Applied Cryptography Second Edition : Protocols, Algorithms, and Source Code in C*. 1996.

Real Time Catalog of Uniqueness Face Using the CAMShift and Gabor Wavelet Algorithims

Anita Sindar Ros Maryana Sinaga[1], Arjon Samuel Sitio[2]

[1,2]Engineering Informatics, STMIK Pelita Nusantara

*haito_ita@yahoo.com

Abstract Moving image processing uses continuous moving images such as recording live video as a data set. The process of developing a unique database catalog application of student face recognition begins with storing the results of real time recordings of each face. Face tracking uses the CamShift algorithm to identify skin color based on the Hue histogram value. To bring out the special features of a face image that has been convoluted to the kernel, the Gabor Wavelet method is used. 2D Gabor Wavelet kernel which is obtained by modulating 2D sine waves at certain frequencies and orientations with a gaussian envelope. Training data taken from the live video of student activities consists of 30 data sets. Stages start from file conversion - tracking video to multiple frames - RGB-HSV conversion - Calculate Histogram - Face detection and finally face recognition process. If there are similarities with the largest value between querying the face image in the database, the face will be detected. Looping is done to take the face of other students. Video .mp4 is converted to .avi. The results of video conversion are tracked into several frames according to facial movements followed by RGB - HSV conversion, 1-2 minute duration videos produce 40-250 frames. Face recognition successfully identifies moving faces around 80%-90% with skin color detection and 80%-85% with points position detection.

Keywords: Moving Faces, Face Detection, Chamsift, Gabor Wavelet, Face Recognition

1. Introduction

A student is more or less educated for 5 years at a college. Many physical changes since entering college. To get a real face, a live video is recorded, from the results of this recording, it will be used to identify certain features that a student has stored in a digital face database. Moving image processing uses images moving in succession like live video recordings as input [1]. The process of developing a unique database catalog application of student face recognition begins with storing the results of real time recordings of each face. From the results of facial movements, face traking was obtained to find out the facial position and face recognition to recognize the face. Object detection and tracking are important in many computer vision applications including activity recognition, automotive safety, and surveillance [2]. Face tracking proces system: detect a face to track, identify facial features to track and track the face. Before begin tracking a face, need to first detect it. By default, the detector is configured to detect faces, but it can be configured for other object types [3]. Once the face is located in the video, the next step is to identify a feature that will help track the face. Use the shape, texture,

Journal of Physics: Conference Series **1424** (2019) 012012 doi:10.1088/1742-6596/1424/1/012012

or color. Choose a feature that is unique to the object and remains invariant even when the object moves. Get the skin tone information by extracting the Hue from the video frame converted to the HSV color space. HSV (H : Hue, S : Saturation, V : Value) and HSL (H: Hue, S : Saturation, L : Luminance, that represent colors space as seen by the human eye. The Chamsift algorithm is used for color detection of facial skin. The color captured by the human eye comes from the reflection of the color of the object consisting of Hue, Saturation and Brightness. Color space (a color model) as a coordinate system specification and a sub space in a system with each color expressed with one point in itThe histogram based tracker uses the CAMShift algorithm, which provides the capability to track an object using a histogram of pixel values [4]. These pixels are used to initialize the histogram for the tracker. Mean-shift is another technique of video object tracking that is based on primitive geometric shapes [5]. Since face recognition is not a difficult task for human beings, selection of biologically motivated Gabor filters is well suited to this problem. Gabor filters, modeling the responses of simple cells in the primary visual cortex, are simply plane waves restricted by a Gaussian envelope function [6].

2. Methodology

Digital images are expressed in a two-dimensional array or a matrix whose elements state the gray level of the color of each pixel. Pixel is the smallest element of an image, which is in the form of color the image [7]. The beginning Mean-shift (MS), it is defined a region of interest around the object to be tracked in the current frame, and then it is started an iterative process based on comparing the histogram of the region of interest in the current frame with the histograms obtained from candidate regions in the next frame where there exist the chances of finding the object being tracked. Finally, the location of the object being tracked in the next frame is defined by the candidate region whose histogram presents the greatest similarity with the histogram from the region of interest which surrounds the object in the current frame [8]. CAMshift stands for Continuously Adaptive Mean Shift, the development of the Mean Shift algorithm that is done continuously to adapt or adjust the ever-changing color probability distribution of each frame change from the video sequence [9]. The second stage: selection of meanshift location, then the value of the histogram Hue is calculated which is the probability for each pixel image. There is basic characteristics in the image, namely color, shape and texture. Color can distinguish objects in the image. The color characteristics of the image are indicated by color distribution and dominant color in the image. The color distribution states the frequency of appearance of colors or color histogram. H, S, V values based on R, G, and B, there are several ways:

$H = \tan(\frac{3(G-B)}{(R-B)+(R-G)} S = 1 - \frac{\min(R,G,B)}{V}$; $V = \frac{\min(R,G,B)}{V}$. ....(1)

If S = 0 then $r = \frac{R}{R+G+B}$, $g = \frac{G}{R+G+B}$, $b = \frac{B}{R+G+B}$; $V = \max(R,G,B)$ ....(2)

$S = \begin{cases} 0, & jika\ V = 0 \\ 1 - \frac{\min(r+g+b)}{3}, & V > 0 \end{cases}$; $H = \begin{cases} 0, & S = 0 \\ \frac{60*(g-b)}{S*V}, & V = r \\ 60 * [2 + b S - * V r], & V = g \\ 60 * [4 + r S - * V g], & V = b \end{cases}$ ....(3) ; $H = H + 360$ if $H < 0$................(4)

The Hue Histogram index uses the index of the value hue. To generate a hue index histogram: Read image - Convert RGB value to Hue - Calculate histogram. Using index from the value of Hue, saturation and value. The value of Hue is in the range 0 to 360, the saturation value is in the range 0 to 1, the value of the value is in the range 0/255. Color features can be obtained through statistical calculations such as mean, standard deviation, skewness, and kurtosis. YIQ color space, Y: luma (which states luminance); I and Q are called chroma. YIQ conversion based on RGB:

$$\begin{bmatrix} Y \\ I \\ Q \end{bmatrix} = \begin{bmatrix} 0.299 & 0.587 & 0.114 \\ 0.596 & -0.274 & -0.322 \\ 0.211 & -0.523 & 0.312 \end{bmatrix} \begin{bmatrix} R \\ G \\ B \end{bmatrix} \text{........(5)}$$

Object recognition methods can be defined as the process of determining object identification based on existing image databases. Gabor Wavelet is able to emerge special features of images that have been convoluted to the kernel. As a filter, Gabor Wavelet 2D kernels are used by modulating 2D sine waves at certain frequencies and orientations with the Gaussian envelope. An image can be represented by the Gabor wavelet transform allowing the description of both the spatial frequency structure and spatial relations [10]. Image capture is done frame by frame (sequence) with video input in AVI format. This image acquisition process is done by opening AVI format video files that have been previously stored in the storage media. The system breaks video input files based on the number of frames starting from the first frame to the last frame. The process of facial image recognition is done by tracking (position) and recognitition. To produce a digital image, input data is in the form of capture results from real time video recordings processed by facial extraction, Figure 1.

The steps of the ChamShift algorithm:

1. The video image frame is converted into the HSV color system, then creates an image histogram to find out the color probability.
2. Perform the mean shift algorithm above (one or many iterations) with the input in the form of the size and location of the search window and the color probability image and stored.
3. Set the value of x, y is used to determine the middle value of the search window. Repeat the previous step for each change in the video image frame.

The features of Gabor Wavelet extraction are carried out:

1. Image database, segmentation results are used as input for Gabor Wavelet. Perform kernel generator calculations based on frequency input and orientation angles that produce real and imaginary kernels. Every matrix element real image and imaginary image matrix are normalized by dividing it by the matrix element real image and imaginary image matrix.
2. The real image matrix and the normalized image matrix that have been normalized are then magnified to get the absolute matrix. The absolute matrix is extracted for its characteristics. The extraction output is a feature vector.
3. Characteristics of the texture are then stored in the database.

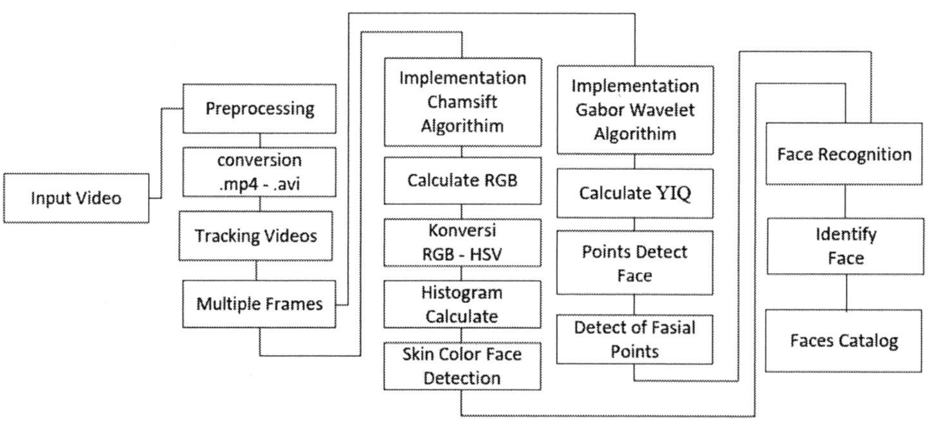

Figure 1. Faces Recognition Steps

3. Result and Discussion

Face detection and face tracking are important components in face recognition systems, which automatically detect or locate the face region in the input frames. The process of determining object identification is based on the image database that was previously stored. The way face recognition works is to convert the video into a series of digital images and then compare it with familiar faces.

Several stages of facial image recognition by the system are detection, position recognition, normalization, coding and comparison. The mean gray level informs the preprocessing results, conversion of image tracking to grayscale, filtering process to get the mean filter and then segmenting it by determining the threshold value to separate the object to be examined in the background. Adaptive background aims to separate the background. The dominant distribution is used as the background. The binarized difference informs the state of the image as having only two colors, black or white, Figure 2.

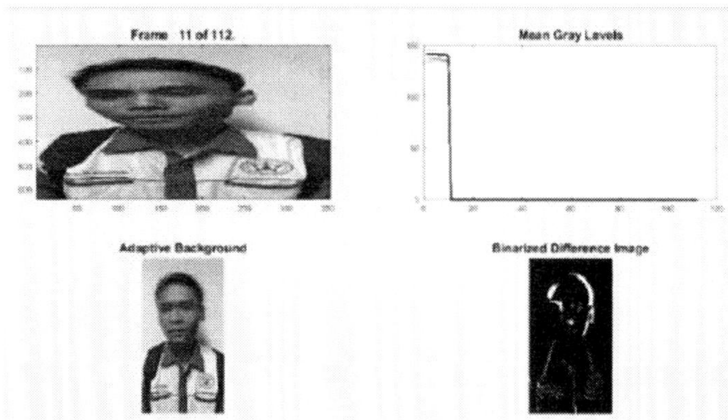

Figure 2. Tracking Video Process

Set data is collected from study consisted of collecting 30 video images of student facial images taken live. To facilitate tracking, video footage in the form of .mp4 is converted to .avi. forma. For taken 1-2 second video produces 43-250 frames, the image result is .png format. Take video to extract frames and get frame means from an avi movie and save individual frames to separate image files. Also computes the mean gray value of the color channels. Loop through the movie, writing all frames out. Calculate the mean gray level. Calculate the mean R, G, and B levels, Figure 3.

Table 1. Total Frame of Data Set

Video	Data Set	Frame Rate	Total Frame
1	Enggel	29.19/s	43
2	Ariguna	29.94/s	47
3	Ibnu	30.58/s	112
4	Fita	25.00/s	120
5	Agnes	29.98/s	157
6	Yudi	30/67/s	111
7	Luter	32.10/s	251
8	Leo	30.69/s	108
9	Lia	25.68/s	154
10	Yumna	28.20/s	80
11	Lasmian	33.20/s	144
12	Dilla	24.91/s	79
13	Fangalulu	29.00/s	192
14	Nurwindi	30.44/s	197
15	Riska	09.98/s	98

Figure 3. Tracking Video Result

Visual tracking of objects has been intensively studied in computer vision, and different approaches have been introduced, for example, particle filtering and mean shift. Read a video frame and run the detector. to detect the location of a face in a video frame uses the Chamsift algorithm and a trained classification model for detection. Draw the returned bounding box around the detected face. use the cascade object detector to track a face across successive video frames. performing face detection for every video frame is computationally intensive. Choose a feature that is unique to the object and remains invariant even when the object moves. use skin tone as the feature to track. The skin tone provides a good deal of contrast between the face and the background and does not change as the face rotates or moves. Get the skin tone information by extracting the Hue from the video frame converted to the HSV color space, Figure 4.

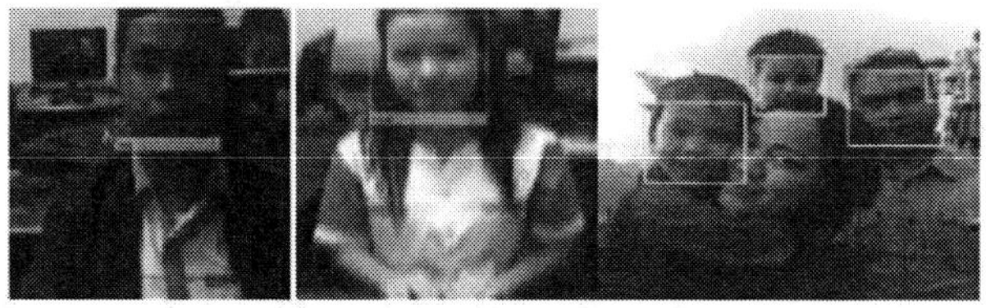

Figure 4. Face Detection Result

Face recognition could be used for video content description, indexing, and retrieval. Finding a specific person in videos is essential to understand and retrieve videos. The identity of the target face is determined by comparing it with the rendered views. Face information has been combined with text information for face or person retrieval in video.

Color plays a role in detecting faces. The HSV component because it is closer to the color of human skin. Image colors generally contain RGB, need to be extracted to produce more special colors, called the segmentation process. In this study, the RGB value of the face was first searched and then the HSV value was calculated to obtain a Histogram. The face image is convolved with a set of Gabor wavelets and the resulting images are further processed for recognition purpose. There have been a great amount of researches on face recognition recently, and various proposed approaches could be roughly classified into analytic and holistic approaches. Face detection using Gabor Wavelet produces dots around the face. There are 5 standards for face detection, namely the face, right eye, left eye, nose and mouth. First it will detect the identified points and then extract facial features, Figure 5.

Figure 5. Points Detection Result

The results of detection of points in pixel, positions: 'left eye-right eye'; 'right eye-nose'; 'left eye-nose'; 'right eye-mouth'; 'left eye-mouth'; 'nose-mouth' from several data sets, Table 2.

Table 2. Results of Position Detection Points on the Faces

data set	right eye-left eye	right eye-nose	left eye-nose	right eye-mouth	left eye-mouth	nose-mouth
1	61.5081	50.2121	49.1172	76.3217	77.6032	31.6425
2	216.1134	148.1773	120.9897	148.8027	233.8017	113.1028
3	161.0093	139.1869	139.517	207.8088	205.6271	76.6975
4	127.8534	92.3864	119.7174	167.6164	183.8049	79.69
5	52.5024	43.1393	38.7105	65.5229	64.158	28.0401
6	129.0591	106.0684	101.4458	160.2623	160.1694	65.5477
7	160.0031	124.1582	128.1025	200.3104	202.8158	87.5014
8	165.5755	132.9003	140.8128	198.0467	209.3161	77.1622
9	28.6575	25.1098	22.4109	32.1636	36.0035	13.6015
10	21.8689	31.8159	24.135	32.1403	33.5596	13.7568

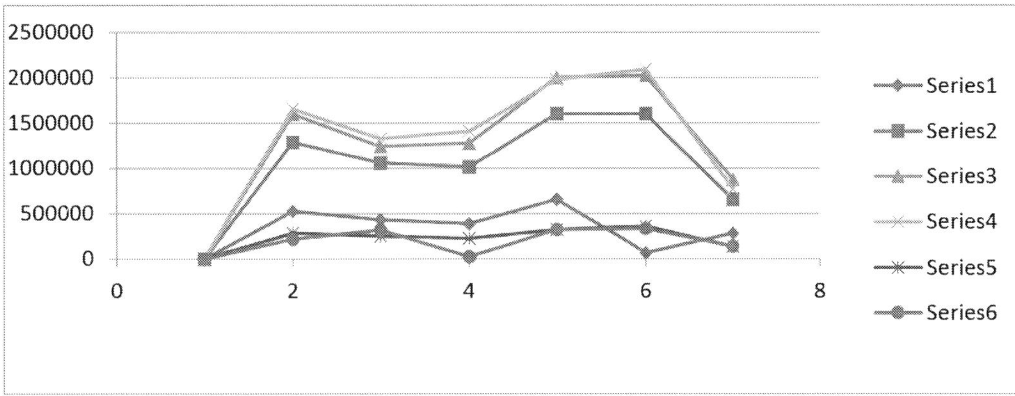

Figure 6. Position Detection Points on the Faces Chart

4. Conclusion

The process of color segmentation, separation of objects with background is an important point in detecting facial skin, the position of the face that does not match the rectangular selection makes the face undetectable and the color similar to skin color is easily detected. Tracking the position of the eyes, nose, mouth on the face is the main point as a reference for facial identification, but it will be a little difficult when detecting a moving image. Conclusions in face recognition and detection:

1. Detection of the 30 data sets is based on facial skin color if detected per person: detection reaches 80% -90%, if detection is in a number of people (collection), the detection is less than 70% -80%. For detection based on the distance between the nose and nose, the results: detection per person: detection reaches 85% -90%, the detection is less than 70% -80%.
2. The level of difficulty in identifying and identifying faces in moving images is more complicated than facial recognition in the form of images.
3. The design of digital face databases in real time can be documented in a dialog so that it is easy to recognize and identify faces, digital face information can be used to facilitate identity recognition.

Acknowledgement
This research was full supported by DIKTI.

References

[1] B. Dhivakar, C. Sridevi, S. Selvakumar, and P. Guhan, "Face detection and recognition using skin color," in 3rd International Conference on Signal Processing, Communication and Networking, pp. 1–7, March 2015.
[2] D. T. Nguyen, W. Li, and P. O. Ogunbona, "Human detection from images and videos: A survey," Pattern Recognition, vol. 51, pp. 148–175, March 2016.
[3] Chandrajit M, Girisha R and Vasudev T, "Multiple Objects Tracking In Surveillance Video Using Color And Hu Moments", Signal & Image Processing : An International Journal (SIPIJ) Vol.7, No.3, pp. 16-27, June 2016.
[4] S. N. Endah, R. Kusumaningrum, and H. A. Wibawa, "Color Space to Detect Skin Image: The Procedure and Implication," Sci. J. Informatics, vol. 4, no. 2, pp. 143–149, 2018.
[5] S. R. Borra, G. J. Reddy, and E. S. Reddy, "A broad survey on fingerprint recognition systems," Proc. 2016 IEEE Int. Conf. Wirel. Commun. Signal Process. Networking, WiSPNET 2016, no. May, pp. 1428–1434, 2016.
[6] Gopika, B, Srilaxmi, K Alekhya, D Rao, B Bhaskar Rama Mohan, "Face Recogition by Using Gabor Feature Extraction and Neural Networks", IOSR Journal of Electronics and Communication Engineering Ver. II, Vol.10, pp 2278-2834, 2015.
[7] J. R. Tang and N. A. Mat Isa, "Bi-histogram equalization using modified histogram bins," Appl. Soft Comput. J., 2017.

[8] Sinaga ASRM, "Optical Systems and Digital Image Acquisition," J. Insitusi Politek. Ganesha Medan Juripol, vol. 1, no. 1, pp. 61–67, 2018.
[9] K. S. Gautam, "Parallel Histogram Calculation for FPGA: Histogram Calculation," in Proceedings - 6th International Advanced Computing Conference, IACC 2016, 2016.
[10] S. V. Tathe, A. S. Narote, and S. P. Narote, "Face Recognition and Tracking in Videos," Adv. Sci. Technol. Eng. Syst. J., vol. 2, no. 3, pp. 1238–1244, 2017.
[11] B. Baby, "Identify the surgically altered face images using granular-PCA approach," IOSR J. Comput. Eng., vol. 16, no. 5, pp. 08–14, 2014.

Electromyography (EMG) signal classification for wrist movement using naïve bayes classifier

D S Putra[1], M A Ihsan[2], A D Kuraesin[3], Mustakim[4], Achmad Daengs, G. S.[5], and I B A I Iswara[6]

[1,2] Informatics, Politeknik Aceh Selatan, Tapaktuan, 23711, Indonesia
[3] Information System, STIE Muhammadiyah Jakarta, Jakarta, 12970, Indonesia
[4] Information System, Universitas Islam Negeri Sultan Syarif Kasim Riau, Pekanbaru, 28293, Indonesia
[5] Management, Universitas 45 Surabaya, Surabaya, 60256, Indonesia
[6] Informatics, STMIK STIKOM Indonesia, Denpasar, 80225, Indonesia

E-mail : darma.poltas@gmail.com

Abstract. Electromyography (EMG) signal is an myoelectric signal in the muscle layer. It occurs caused by contraction and relaxation muscle activity. This article provide numerical study of the classifying the electromyography signal for wrist movement combined with open and grasping finger flexor. The EMG signal has recorded using a device called electromyography. It has acquired by attaching an surface electrode in the skin then the electrode was capturing the raw signal. The volunteer involved were six where each volunteer has ten datasets the EMG signal. The surface electrode are sticked in the lower arm muscle. The EMG raw signal was processed using zero-mean normalization. The feature extraction method is root mean square (rms), mean absolute value (mav), variance (var), and standard deviation (std). This EMG signal has been classified by naïve bayes classifier. Training and testing data was using 5-cross validation. The result indicates that the classification accuracy for classifying the EMG signal for wrist movement combined open finger flexor (OFF) and grasping finger flexor (GFF) is 70% and 75% respectively. Therefore, the EMG signal can be applied for identificating of muscle disorder, prostheses hand and biometric system.

1. Introduction

The development of science and technology in the world has impacted in security system, health and human activity field. The study in electromyography (EMG) signal is a study that was developed rapidly to get solution for human rehabilitation, prostheses hand control and biometric system [1]–[3].

Every human has different muscle size. The EMG signal can help humans to identify the abnormalities muscle. In Previous research, we have carried out studies on the movements of opening and grasping fingers extensor using learning vector quantization classifier [4]. Any relevant research about EMG signal has been published. The artificial hand has used the EMG signal for controlling the hand prostheses. Two surface electrode are attached in the lower arm muscles to record the EMG signal. The EMG signal has been extracted and classified using different type of classification [5]–[9]. Flexion movement in a single finger be able to identified using a single channel surface electrode device. This surface EMG device has recorded the EMG signal during flexor movement in different fingers [9]–[11]. Multi channel of EMG electrode surface has used to classify a single and combination movement for artificial hand control. The feature extraction method has used time domain auto regression, fuzzy neighborhood discriminant analysis and classification has used linear discriminant analysis [7], [12].

Hence this study aims to extract the feature EMG signal using time domain feature and classify the open and grasping finger flexion movement using naïve bayes classifier.

2. Methods

The methodology of the research is explained by flow diagram in Figure 1.

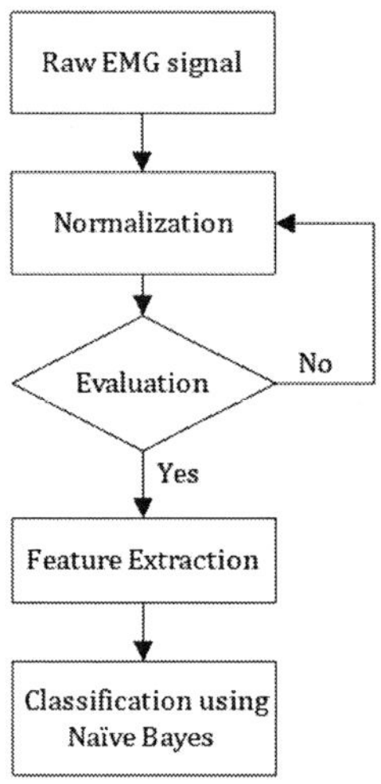

Figure 1. The methodology of the research

2.1. The object and location

This study was involving six participants with male gender. The average of participant's height and weight is 162.5 cm and 60.5kg respectively. The location of this study was at Politeknik Aceh Selatan.

2.2. Data collection

2.2.1. Preparation. The devices needed for EMG signal data recording are determined. The tools and devices are the EMG myoware sensor, the surface electrodes, arduino, data cable, laptop, coolterm software. The EMG myoware sensor and the surface electrode as shown in Figure 2 and Figure 3. All devices are installed to be a single unit.

Figure 2. An EMG myoware sensor

Figure 3. The surface electrode

2.2.2. Data recording. The data recording was performed for the data retrieval process for each participant. Before data recording, the subject must fill the inform consent to become the subjects. The subjects have no injuries and abnormalities in the muscles tissue. The subjects were attached three electrodes in the lower right arm muscles. The ilustration of placement electrodes is as shown in Figure 4.

Figure 4. Surface electrode placement in the lower arm muscle

The two types of movement for recording the EMG signal for each participant are *open finger flexor* (OFF) and *grasping finger flexor* (GFF) as shown in Figure 5 and Figure 6 respectively.

Figure 5. Open finger flexor movement

Figure 6. Grasping finger flexor movement

The CoolTerm software was used for recording the EMG signal. Every movement was started on the 5th and ends on the 15th second. The 1st to 5th second was named the initial phase of the EMG signal and the 16th to 20th second are named the end phase of the EMG signal. The plotting of EMG signals for OFF dan GFF movement to one participant is shown in Figure 7.

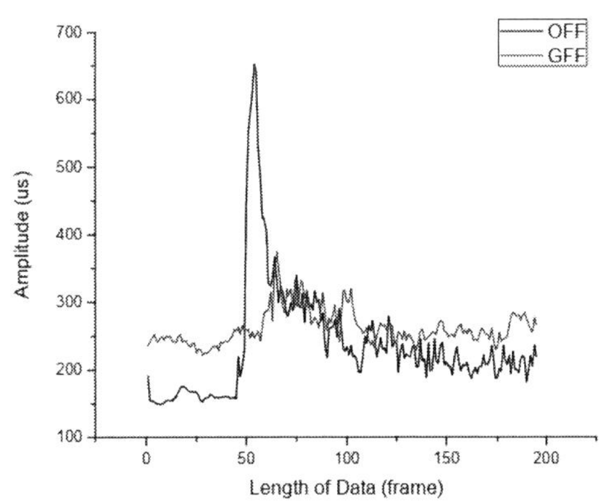

Figure 7. The plotting of EMG signal for OFF and GFF movement

2.2.3. Data processing. The raw EMG signal that has been recorded from each participant was processed to ensure not interference signal. All EMG raw signals have normalized using Equation (1).

$$zeromean = a - \bar{x} \quad (1)$$

Where a is the original EMG signal and $\bar{x}$ is mean value of a.

Mean absolute value (MAV). This method was one of the feature extraction that popularly used in the EMG signal analysis. The MAV equation is using Equation (2).

$$MAV = \frac{\sum_{i=1}^{N} |x_i|}{N} \tag{2}$$

where $|x_i|$ is absolute value and N is the number of data from the EMG signal.

Root mean square (RMS). Root mean square is the root of the mean of the squared the EMG signal. This method aims to obtain the effective value of the EMG signal. The RMS is using Equation (3).

$$RMS = \sqrt{\frac{\sum_{i=1}^{N} x_i^2}{N}} \tag{3}$$

where x_i^2 the square of the EMG signal amplitude.

Variance (*Var*). The variance is the mean of the two degree value of deviation variables. Variance is using Equation (4).

$$Var = \frac{\sum_{i=1}^{N} x_i^2}{N-1} \tag{4}$$

Standard Deviation (*STD*). The standard deviation is the distribution of the EMG signal value. The greater the standard deviation value, the greater the distance of each data to average value. The standard deviation equation is shown in Equation (5).

$$s = \sqrt{\frac{\sum_{i=1}^{n} (x_i - \bar{x})^2}{n-1}} \tag{5}$$

where s is standard deviation, x_i is EMG signal data, $\bar{x}$ is the mean of EMG signal, and n is the number of data.

2.2.4. The EMG signal classification. This classification is to get differences in EMG signal patterns for each subjects. The classification method is naïve bayes classifier. The classification performance is evaluated using accuracy, precision, and recall as shown in Equations (6), (7) and (8).

$$accuracy = \frac{TP + TN}{TP + TN + FP + FN} \tag{6}$$

$$precision = \frac{TP}{TP + FP} \tag{7}$$

$$recall = sensitivity = TPR = \frac{TP}{TP + FN} \tag{8}$$

3. Results and Discussion

Based on the research methods that have been explained in section 2, the results of EMG signal classification can be obtained as shown in Table 1 and Table 2. The classification performance evaluation of the naïve bayes classifier as shown in Table 3 and Table 4.

Table 1. Result of naïve bayes classifier for OFF movement

		Prediction (%)					
		Subject1	**Subject2**	**Subject3**	**Subject4**	**Subject5**	**Subject6**
Actual (%)	**Subject1**	70	10	0	10	0	10
	Subject2	30	30	0	10	0	30
	Subject3	0	0	100	0	0	0
	Subject4	20	10	0	70	0	0
	Subject5	0	0	0	10	90	0
	Subject6	20	20	0	0	0	60

Table 2. Result of naïve bayes classifier for GFF movement

		Prediction (%)					
		Subject1	**Subject2**	**Subject3**	**Subject4**	**Subject5**	**Subject6**
Actual (%)	**Subject1**	100	0	0	0	0	0
	Subject2	0	80	0	10	0	10
	Subject3	0	0	100	0	0	0
	Subject4	0	60	10	10	0	20
	Subject5	0	0	0	0	100	0
	Subject6	0	40	0	0	0	60

Table 3. Performance of naïve bayes classifier test for OFF movement

Subject	TP (%)	FP (%)	FN (%)	Precision (%)	Recall (%)
1	70	70	30	50	70
2	30	40	70	42.85	30
3	100	0	0	100	100
4	70	30	30	70	70
5	90	0	10	100	90
6	60	40	40	60	60

Table 4. Performance of naïve bayes classifier test for GFF movement

Subject	TP (%)	FP (%)	FN (%)	Precision (%)	Recall (%)
1	100	0	0	100	100
2	80	100	20	44.44	80
3	100	10	0	90.90	100
4	10	10	90	50	10
5	100	0	0	100	100
6	60	30	40	66.66	60

Using the Equation (8), the result of accuracy classification overall is 70% for OFF movement and 75% for GFF movement.

From the Table 1, it shows that the naïve bayes method classifying the EMG signal for OFF movement with true positive (TP) for subject3. It indicates that the EMG signal in open finger flexor combining wrist movement is unique movement. Meanwhile, the naïve bayes classifying the EMG signal with false negative (FN) for subject5 is middle unique movement and for subject1, subject2, subject4 and subject6 is not unique movement. From the Table 2, it shows that the naïve bayes method classifying the EMG signal for GFF movement with true positive (TP) for subject1, subject3 and subject5. It indicates that the EMG signal in grasping finger flexor combining wrist movement is unique movement. Meanwhile, the naïve bayes classifying the EMG signal with false negative (FN) for subject2 is middle unique movement and for subject4 and subject6 is not unique.

From the Table 3, it explains that subject3 and subject5 have the best precission and recall value. It indicates that the EMG signal for this subjects have the best classifier testing performance criteria. Meanwhile subject1, subject4, and subject6 have a good classifier testing performance criteria and subject2 has a bad classifier testing performance criteria.

From the Table 4, it shows that subject1, subject3 and subject5 have the best precission and recall value. It indicates that the EMG signal for this subjects have the best classifier testing performance criteria. Meanwhile subject6 has a good classifier testing performance criteria. Subject2 and subject4 have a bad classifier testing performance criteria.

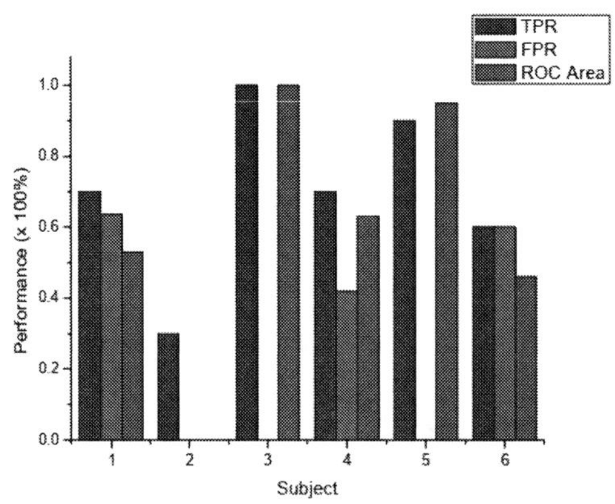

Figure 8. The naïve bayes classifier performance evaluation for OFF movement

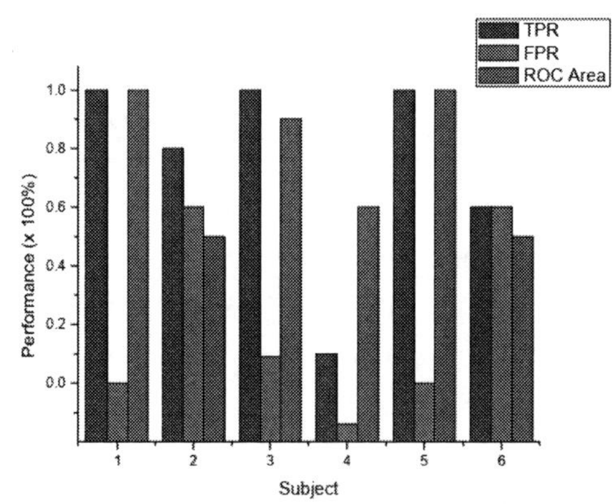

Figure 9. The naïve bayes classifier performance evaluation for GFF movement

Figure 8 and 9 show ROC Area, TPR, and FPR for naïve bayes classifier performance evaluation. The total area under ROC curve (AUC) is a single index for measuring the performance evaluation. The larger the AUC, the better is overall performance of naïve bayes classifier.

4. Conclussion

In this investigation has shown that the EMG signal in the lower arm muscle be able to used to identify people's wrist movement. The EMG signal be able to specify the pattern recognition from one to another applied to artificial hand. The overall classification accuracy is 70% for OFF movement and 75% for GFF movement. The finding of this study suggest that the feature used in this study and naïve bayes classifier was appropriated method for studying the EMG signal classification. A further study could assess reducing the number of instances in each subjects to achieve better accuracy.

Acknowledgement

This research was donated by Republic of Indonesia Ministry of Research, Technology and Higher Education (No. 117/SP2H/LT/DRPM/2018).

References

[1] W.-T. Shi, Z.-J. Lyu, S.-T. Tang, T.-L. Chia, and C.-Y. Yang, "A bionic hand controlled by hand gesture recognition based on surface EMG signals: A preliminary study," *Biocybern. Biomed. Eng.*, vol. 38, no. 1, pp. 126–135, 2018.

[2] R. N. Khushaba, S. Kodagoda, M. Takruri, and G. Dissanayake, "Toward improved control of prosthetic fingers using surface electromyogram (EMG) signals," *Expert Syst. Appl.*, vol. Vol. 39, no. 12, pp. 10731–10738, 2012.

[3] D. Karabulut, F. Ortes, Y. Z. Arslan, and M. A. Adli, "Comparative evaluation of EMG signal features for myoelectric controlled human arm prosthetics," *Biocybern. Biomed. Eng.*, vol. 37, no. 2, pp. 326–335, 2017.

[4] D. S. Putra, Y. U. W. Weru, and Fitriady, "Pattern recognition of electromyography (EMG) signal for wrist movement using learning vector quantization (LVQ)," *IOP Conf. Ser. Mater. Sci. Eng.*, vol. 506, p. 12020, 2019.

[5] O. S. Powar, K. Chemmangat, and S. Figarado, "A novel pre-processing procedure for enhanced feature extraction and characterization of electromyogram signals," *Biomed. Signal Process. Control*, vol. 42, pp. 277–286, 2018.

[6] O. W. Samuel *et al.*, "Pattern recognition of electromyography signals based on novel time domain features for amputees' limb motion classification," *Comput. Electr. Eng.*, vol. 67, pp. 646–655, 2018.

[7] A. H. Al-Timemy, G. Bugmann, J. Escudero, and N. Outram, "Classification of Finger Movements for the Dexterous Hand Prosthesis Control With Surface Electromyography," *J. Biomed. Heal. Informatics*, vol. Vol. 17, no. 3, pp. 608–618, 2013.

[8] D. Peleg, E. Braiman, E. Yom-tov, and G. F. Inbar, "Classification of Finger Activation for Use in a Robotic Prosthesis Arm," *IEEE Trans. Neural Syst. Rehabil. Eng.*, vol. Vol. 10, no. 4, pp. 290–293, 2002.

[9] N. Duan, L.-Z. Liu, X.-J. Yu, Q. Li, and S.-C. Yeh, "Classification of multichannel surface-electromyography signals based on convolutional neural networks," *J. Ind. Inf. Integr.*, 2018.

[10] D. K. Kumar, S. P. Arjunan, and V. P. Singh, "Towards identification of finger flexions using single channel surface electromyography – able bodied and amputee subjects," *J. Neuroeng. Rehabil.*, pp. 1–7, 2013.

[11] C. L. Lipinski, L. Donovan, T. J. McLoughlin, C. W. Armstrong, and G. E. Norte, "Surface electromyography of the forearm musculature during an overhead throwing rehabilitation progression program," *Phys. Ther. Sport*, vol. 33, pp. 109–116, 2018.

[12] N. van Beek, D. F. Stegeman, J. C. van den Noort, D. (H.E.J.) Veeger, and H. Maas, "Activity patterns of extrinsic finger flexors and extensors during movements of instructed and non-instructed fingers," *J. Electromyogr. Kinesiol.*, vol. 38, pp. 187–196, 2018.

Regression Tree Role for Interpret Monetizing of Game Live Streaming

Erlin Windia Ambarsari[1], Reza Avrizal[1], Erlando Doni Sirait[1], Saruni Dwiasnati[2] and Robbi Rahim[3,*]

[1]Informatics Department, Universitas Indraprasta PGRI
[2]Informatics Department, Universitas Mercu Buana
[3]Management Department, Sekolah Tinggi Ilmu Manajemen Sukma

*usurobbi85@zoho.com

Abstract. Game Live Streaming is streamers activity who play games with a purpose to monetizing. The running game recorded into a streaming statistic. Therefore, streamers can monitor it. The company of Game Live Streaming evaluated whether they are eligible for compensation to make it easier to interpret monetizing, a decision tree is needed. However, the dataset obtained from statistical streaming is numeral; the calculation uses Standard Deviation Reduction for construct Regression Tree which the concept is similar to ID3 Algorithm. The Regression stop constructs Tree when the CV is zero, or the number of instances is the least. Also, the mean reached dataset homogeneously.

1. Introduction

The live broadcast is broadcasting in real time, and it can reach an audience at the same time without delay. One of the Live shows has been carried out by Television stations since the advent of the cathode-ray tube, which the functions to broadcast sports tournament, entertainment, and news. Also, significant internet network deployment has influenced viewers to watch live broadcasts extensively through websites and Android platform, as known as live streaming.

Live streaming become children and young adults favorite. Therefore, they have the opportunity to become a creator, a presenter, and noticed by viewers. However, live streaming is uncensored and unedited. Thus, each of streamers [1] has responsibility for their content, although their actions caused controversy for the community. Nevertheless, On the feasible side from live streaming is that brands endorse increased for marketing [2].

One of the most prominent endorse brands is electronic sports (eSports) live streaming. eSports is a competitive digital game such as real-time strategy (RTS), first-person shooters, multiplayer online battle arena (MOBA) and arcade-style fighting games; which one plays is professional gamers (pro-gamers) [3]. The tournament is similar to a traditional sport, which players have intense training to improve their skill. Also, they got the champion prize from sponsorships and advertising on their live streams as their income. Although pro-gamers does a live stream on their tournament, in general, they still monetize by attracting viewers a daily using game live streaming platform. In Indonesia, a popular platform using by gamers is Cube TV which not only pro-gamer, the amateur plays the game to get paid from monetizing which follow protocol. The protocol for monetizing has considered such as duration playing games, several viewers, reward, and violation; which the policy determined by the company.

The data from game live streaming has recorded in real time every day; streamers could monitoring their record on Streaming Statistics. Their streaming statistics had evaluated by the company whether they deserve compensation every month — their overall Live stream report sent to the admin or agent who is in charge.

In live streaming case, various conditions must fulfill to monetizing by streamers, for the example: duration games which they play are three hours for a day, they have 500 followers who give reward to them, and no verbal violence. Therefore, the company gave a compensation category as a target. If they fulfill a goal more from limit specified, they get paid the same. However, when they could not achieve it, they still get paid base on condition determined. Unfortunately, the company did not explain how the flow of compensation is, especially for streamer who does not achieve the requirement.

According to the problem case about streamers compensation flow, we interpret monetizing of game live streaming which we obtain the dataset using Decision Tree. The dataset that we examine as the example is Cube TV's dataset. Several methods of Decision Tree to solve the problem, such as ID3, CART, and C4.5. ID3 is a classification algorithm using Information Gain to select the best attribute, which proposed by J. Ross in 1979 [4]. CART handle a classification and regression tree; if the variable has a category class, CART constructs classification tree based on the binary attributes splitting. Also, CART builds regression tree when the variable was continuous or numeric. C4.5 improved of ID3 algorithm which used for the statistical classifier. Similar to ID3, C4.5 uses information gain for splitting criteria [5].

Live stream dataset is numerical data obtained from the streaming statistic. The method that we use is Regression Tree which considered the streamer's compensation from the company. The concept which applies to this case is ID3. However, ID3 is a classification method that handles category data which use information gain. Therefore, the solve problem to construct a Decision Tree for numeric data on the streaming statistic is to use the coefficient of variation [6]. It is the ratio of standard deviation and the mean (average); the standard deviation expressed units as the variable under study that enable to interpret [7] monetizing game live streaming.

2. Methodology

Decision Tree is the method to classify homogeneous instances dataset. The way the Decision Tree work has split a dataset into the smallest subsets for construct decision nodes and leaf nodes. Also, at the same time, an associated Decision Tree is incrementally developed [8]. As mention before, the Decision Tree method for this case use is ID3 algorithm which the information gain has to replace Standard Deviation Reduction, by a decrease standard deviation after a dataset split on an attribute. Constructing a Decision Tree is to discover an attribute that returns the highest standard deviation reduction.

Standard deviation is useful to compute a numerical sample. If the example is entirely homogeneous, the standard deviation has zero value. The formula of standard deviation (σ) as follows [9]:

$$\sigma = \sqrt{\frac{1}{n}\sum_{i=1}^{n}(u_i - \mu)^2} \quad (1)$$

Standard deviation is functionated to measure spread data while the mean (μ) used to determined central tendency. Standard deviation is profoundly affected when the mean gives a poor measuring.

$$\mu = \frac{1}{n}\sum_{i=1}^{n}u_i \quad (2)$$

Therefore, the order to gain regression tree for the decision does as follows :

Journal of Physics: Conference Series **1424** (2019) 012014 doi:10.1088/1742-6596/1424/1/012014

a. Calculate the standard deviation of the target as a class (equation 1). The target obtains from streamer's compensation or honorarium while they do live stream from the streaming statistic.
b. Split the streaming statistic dataset into the different attributes. Separation is using a scatter plot to examine the relationship two attributes variable, namely target and predictor. The predictor predicts the class value for the new pattern. It able used to determine the root node and sub-node (decision node and leaf node).
c. Calculate the standard deviation for each branch. Also, the standard deviation reduction (SDR) calculated from subtracting the standard deviation for two attributes (target and predictor) and the standard deviation before the split (step number 1).

$$SDR = \sigma - \sigma\ (T, P) \quad (3)$$

where

$$\sigma(T, P) = \left(\frac{n_1}{n}\ x\ \sigma(P)_1\right) + \left(\frac{n_2}{n}\ x\ \sigma(P)_2\right) \quad (4)$$

The standard deviation for two attributes to be done with separate instances based on binary attribute splitting which total number of split (n_1 and n_2) divided the full amount instances before the split (n), and it multiple by standard deviation after the split ($\sigma(P)_1$ and $\sigma(P)_2$).
d. Select the attribute with the most substantial of standard deviation reduction for the decision node.
e. Calculate the Coefficient of Variant (CV) to stop the tree process based on a branch become smaller than the threshold (10% for example), or the number of instances (n) is the least.

$$CV = \frac{\sigma(P)}{\mu} x\ 100\% \quad (5)$$

3. Result and Discussion

Cube TV's streaming statistic which obtains 114 streamers instances (n) of live game streaming for one month. Attributes considered such as ID HOST, APCU Awal, Rate, APCU Akhir, Followers, Jam Siaran, Hari Siaran, Energy, Pelanggaran, and Komisi. The explanation is as follows:
a. ID HOST: streamers who register on game live streaming.
b. APCU (Average Peak Concurrent Users) Awal: the highest average viewers value which not audit yet.
c. Rate: The rate of the live stream which determined by Cube TV.
d. APCU Akhir: APCU Awal multiply by Rate which determined by Cube TV.
e. Followers: viewers who follow the streamers.
f. Jam Siaran: duration of games live streaming per second.
g. Hari Siaran: streamers play games for 30 days.
h. Energy: coin which gives by followers.
i. Pelanggaran: offense, such as verbal violation, pornography, illegal ads, and game cheating.
j. Komisi: compensation obtained by streamers.

Data Table

Data instances: 114
Features: 8
Meta attributes: 1
Target: Numeric variable 'Komisi'

	Komisi	ID HOST	APCU AWAL	Rate	APCU AKHIR	Followers	Jam Siaran(Detik)	Hari Siaran	Energy	Pelanggaran
1	31500.0	23324139.0	910.7272727	0.985673	897.6794786	812.0	341718.0	20.0	1530.0	0.0
2	28350.0	24233764.0	651.3478261	0.952585	620.4644313	861.0	275170.0	22.0	245.0	0.0
3	25200.0	23479587.0	651.3600000	0.948748	617.9766224	848.0	288491.0	21.0	836.0	1.0
4	28350.0	23375298.0	627.1739130	0.964798	605.0962394	826.0	256672.0	20.0	501.0	0.0
5	21000.0	23716622.0	644.7600000	0.925402	596.6619280	937.0	253492.0	20.0	1017.0	0.0
6	18900.0	24239514.0	602.6250000	0.963987	580.9226305	1013.0	305793.0	22.0	460.0	0.0
7	0.0	23744665.0	639.4285714	0.776459	496.4901837	835.0	288154.0	21.0	2101.0	0.0
8	18900.0	24010856.0	463.7619048	0.980644	454.7855166	770.0	247810.0	21.0	470.0	0.0
9	21000.0	24786201.0	444.9047619	0.993264	441.9078009	767.0	261464.0	20.0	1917.0	0.0
10	10500.0	17192578.0	433.5500000	0.993942	430.9235405	273.0	264965.0	20.0	510.0	0.0

Figure 1. Cube TV Dataset

Calculate standard deviation (σ) for specifying the root node. The result is σ = 6955.229709, μ = 6138.815789, and CV = 113.2992086%. The CV result indicates greater value than threshold = 10%; it is necessary to split instances on the attribute. Separate instances using a scatter plot as Figure 2 with obtained two attributes (Komisi and APCU Akhir for example). The separation can divide by taking midpoint value or the value that approaches. Also, it uses the highest SDR by iterations.

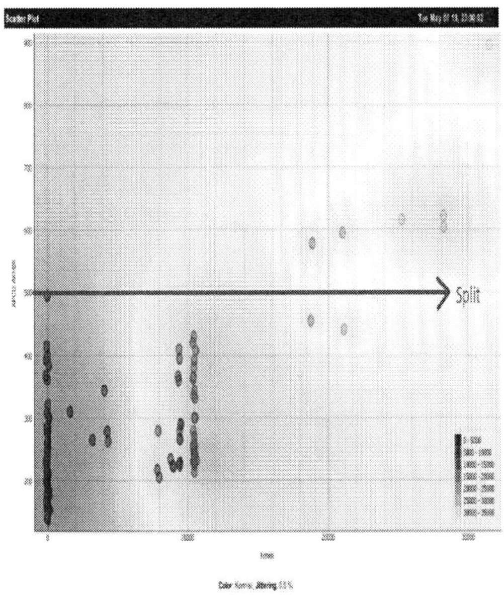

Figure 2. a Scatter Plot for two attributes

Table 1. The Example Dataset of APCU Akhir to Komisi for 17 instances

APCU AKHIR	Komisi
897.6794786	**31500**
620.4644313	**28350**
617.9766224	**25200**
605.0962394	**28350**
596.661928	**21000**
580.9226305	**18900**
496.4901837	0
454.7855166	18900
441.9078009	21000
430.9235405	10500
418.6256027	10500
415.5338098	0
407.2073429	10500
405.9665422	9450
399.1413763	0
394.3399995	9450
392.0473117	10500

Based on Figure 2, value that obtained for split is 496.4901837. Cluster the dataset by separating values below 496.4901837, and values above or equal to 496.4901837 (shown on Table 1). For the example, APCU Akhir > 496.4901837 then Komisi dataset is 31500, 28350, 25800, 28350, 21000, and 18900 which is marked by bold. Similary to APCU Akhir <= 496.4901837. The number of dataset when calculate: APCU Akhir <= 496.4901837 is 108 instances (n), σ = 5281.24096, μ = 5060.416667, and CV = 104.3637571%; APCU > 496.4901837 is 6 instances (n), σ = 4399.431781, μ = 25550, and CV = 17.21891108%.

Determination of root node is done by using SDR; which σ (Komisi, APCU Akhir) = ((108/114) x 5281.24096) + ((6/114) x 4399.431781) = 5234.82995 and SDR = 6955.229709 - 5234.82995 = 1720.4. The selection of attributes observed shown on the table below :

Table 2. The Split Attribute for 114 instances

Attributes	Split	n	Σ	μ	σ (T,P)	SDR	CV
APCU Awal	<=500	105	5285.438	5205			
	>500	9	12568.63	17033	5860.43	1094.8	
Rate	<=0.9	26	4872.5	4038			
	>0.9	88	7346.084	6759	6781.93	173.296	
APCU Akhir	**<=496.4901837**	**108**	**5281.241**	**5060**			**104.36%**
	>496.4901837	**6**	**4399.432**	**25550**	**5234.83**	**1720.4**	**17.22%**
Followers	<=755	94	4902.146	4759			
	>755	20	10571.26	12626	5896.73	1058.5	
Jam Siaran	<=344286	108	7079.492	6072			
	>344286	6	3928.74	7350	6913.66	41.5668	
Hari Siaran	<=20	80	6672.017	4692			
	>20	34	6398.91	9543	6590.56	364.666	
Energy	<=1388	101	6531.506	5993			
	>1388	13	9556.744	7269	6876.49	78.7406	
Pelanggaran	<=0.5	99	6924.145	6639			
	>0.5	15	6217.206	2835	6831.13	124.103	

Based on Table 2, the highest SDR is 1720.4. Therefore, the attribute used for the root node is APCU Akhir. However, CV for APCU Akhir is not the least than 10%. It need split for decision nodes and leaf nodes which the similar steps as the root node.

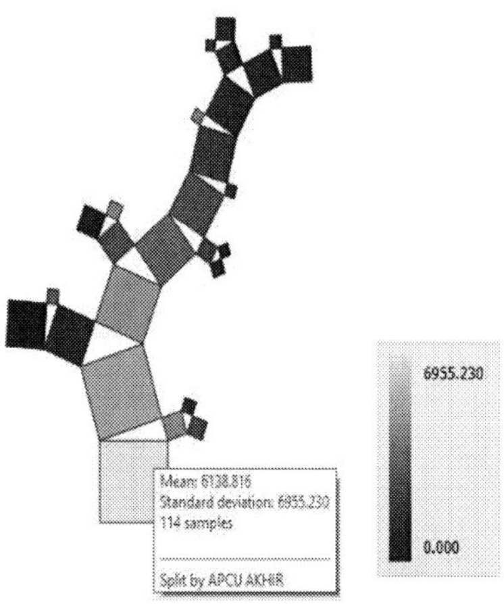

Figure 3. Construct Regression Tree

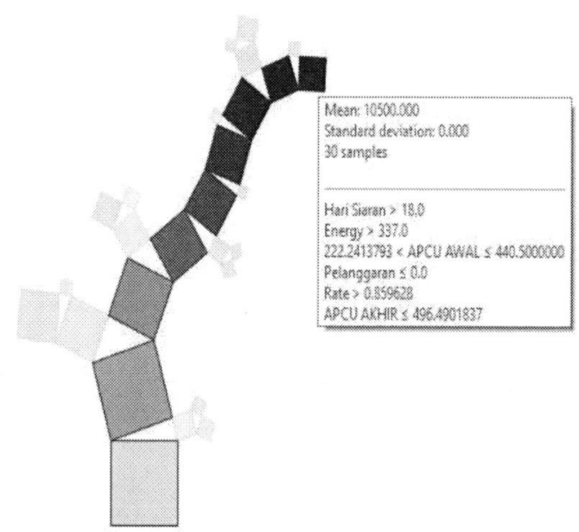

Figure 4. Compensation Decision Tree

The Regression has to be done to construct Tree which the step has stopped when CV has 0, or the number instances is the least (as shown in Figure 3). The mean evidence by reached homogeneous instances, and the longest path on a tree is the best decision (Figure 4). Therefore, streamers have compensation by monetizing in normally which the average is 10500 with the condition is Hari Saran

> 18, Energy > 337, $222.2413793 <$ APCU Awal $<= 440.5$, Pelanggaran $<= 0$, Rate > 0.859628, and APCU Akhir $<= 496.4901837$.

4. Conclusion

The activity has done by the streamer as a gamer in live stream platform was recorded on the streaming statistic. Dataset of streaming statistics are numeral has challenging to examine by Decision tree Classification. Therefore, this is the Regression Role to calculate the numeral dataset by using Standard Deviation Reduction. The Regression Tree stops construction when the CV is zero, or the number of instances is the least. Also, the mean reached homogeneous instances. Therefore, streamers have compensation by monetizing in normally which the average is 10500 with the condition is Hari Saran > 18, Energy > 337, $222.2413793 <$ APCU Awal $<= 440.5$, Pelanggaran $<= 0$, Rate > 0.859628, and APCU Akhir $<= 496.4901837$.

References

[1] A. N. Wendt, "The empirical potential of Live Streaming beyond cognitive psychology," *J. Dyn. Decis. Mak.*, vol. 3, no. 1, pp. 1–9, 2017.

[2] P. R. Todd and J. Melancon, "Gender and live-streaming: source credibility and motivation," *J. Res. Interact. Mark.*, vol. 12, no. 1, pp. 79–93, Mar. 2018.

[3] N. Edge, "Evolution of the Gaming Experience : Live Video Streaming and the Emergence of a New Web Community," *Elon J. Undergrad. Res. Commun.*, vol. 4, no. 2, pp. 33–39, 2013.

[4] S. Rai, P. Saini, and A. K. Jain, "Model for Prediction of Dropout Student Using ID3 Decision Tree Algorithm," *Int. J. Adv. Res. Comput. Sci. Technol.*, vol. 2, no. 1, pp. 142–149, 2014.

[5] H. Sharma and S. Kumar, "A Survey on Decision Tree Algorithms of Classification in Data Mining," *Int. J. Sci. Res.*, vol. 5, no. 4, pp. 2094–2097, 2016.

[6] H. Bindu K, S. Rani K, and R. Rao C, "Coefficient of Variation based Decision Tree (CvDT)," *Int. J. Innov. Technol. Creat. Eng.*, vol. 1, no. 6, pp. 1–6, 2011.

[7] J. D. Curto and J. C. Pinto, "The coefficient of variation asymptotic distribution in the case of non-iid random variables," *J. Appl. Stat.*, vol. 36, no. 1, pp. 21–32, 2009.

[8] M. Ouahilal, M. El Mohajir, M. Chahhou, and B. E. El Mohajir, "A novel hybrid model based on Hodrick–Prescott filter and support vector regression algorithm for optimizing stock market price prediction," *J. Big Data*, vol. 4, no. 31, pp. 1–22, 2017.

[9] M. F. Al-Saleh and A. E. Yousif, "Properties of the Standard Deviation that are Rarely Mentioned in Classrooms," *AUSTRIAN J. Stat.*, vol. 38, pp. 193–202, 2009.

Morphological Analysis of Siak Sri Indrapura City in 2005-2018

Febby Asteriani[1*], Hanifatul Jannah[1], Idham Nugraha[1], Catur Cahyaningsih[2], Mira Hafizhah Tanjung[1]

[1]Urban and Regional Planning Department, [2]Geological Engineering Department, Faculty of Engineering, Universitas Islam Riau, Address: Jln. Kaharuddin Nasution No. 113 Perhentian Marpoyan, Pekanbaru, 28284

*Correspondent Email: febby.uir.fr@gmail.com

Abstract. Siak Sri Indrapura City is the one of a quite fast city in developing; it has increasingly seen since Tengku Agung Sultanah Latifah Bridge in 2007 which is the link between two areas that glide by the river for perfect accessibility. This study aims to examine the morphology Siak Sri Indrapura City in 2005-2018. The method of this study to use descriptive qualitative and quantitative analysis with GIS (*Geographic Information System*), which is analyzing and explaining phenomena in depth through in-depth data collection. The results of this study are known that the City of Siak Sri Indrapura experienced changed in land cover, land use, road and changed in building patterns and functions each year. The increase of the not built area is 15589 Ha or 83.45%, changed from the not built area into the built area is 2052 Ha or 10.99% and increase of land built area is 1039 Ha or 5.56%. The most dominant changed in land use in Siak Sri Indrapura City was increased of plantations, changed of plantations into settlements and increased of settlements. Road pattern is grid building pattern is heterogeneous and linear, a function of buildings dominated by residential buildings.

1. Introduction

The form of regional morphology is reflected in spatial patterns, architectural forms of buildings, and other physical elements of the city in the overall context of the development of the city [1]. A city always experiences development from time to time. The morphology of the city is formed in a long process; any changes in the shape of the region morphologically could provide meaning and benefits that are very valuable for handling the development of an urban area [2][3][4]. Definition of Morphology has three components regarding the physical condition of the area. These components are reviewed from land cover and regional land use that reflects regional activities, circulation patterns or road network patterns that connect between regions, and building patterns and their functions [5][6].

This research took place in the City of Siak Sri Indrapura because among the 12 Regencies in Riau Province, the City of Siak Sri Indrapura is one of the cities that is very fast in terms of development, both in terms of population growth and in the development of urban development. This aim of the study is classified the morphological nature of Siak Sri Indrapura between 2005 and 2018 [7][8].

2. Methodology

This study uses a technique overlay map used for map analysis, the technique overlay consists of two or more map layers (as needed). Analyze the mapping technique was overlaying in *ArcGIS 10.1*. The overlay maps will use satellite imagery of Siak Sri Indrapura City from 2005-2018 to see changes in land cover and changes in land use, as well as the pattern of road networks in the of Siak Sri Indrapura City from 2005-2018[9].

The analysis phases in this study consist of:

a. Identify changes in land cover and changes in land use in the City of Siak Sri Indrapura in terms of 2005-2018. To see the development of land cover changes and land-use changes using GIS with overlay methods and using satellite imagery of the City of Siak Sri Indrapura from 2005-2018, After getting the image of the City of Siak Sri Indrapura, data management is using GIS (Geographic Information System). Land cover is basic information in the study of geoscience and global change (Jia *et al* 2014 in Sampurno and Thoriq 2016), while land use can also be interpreted as a human activity on land directly related to land location and conditions. Land use can also be interpreted as a form or form of business activity, utilization of a plot of land at a time [10].
b. Identifying the pattern of the road network in Siak Sri Indrapura City in 2005-2018 using GIS (overlay) and observation. To be able to make a map of the road network pattern of Siak Sri Indrapura City in 2005-2018, it is essential to use the satellite image of Siak Sri Indrapura City in 2005-2018. After obtaining some image, the Geographic Information System could be managed. After finishing, we could overlay the map, so we know how the pattern of the road network changes in Siak Sri Indrapura City. To prove the correctness of the map, field observations should be carried out, such as testing map accuracy. If it has matched the results of the map and in the field, then it will become a map of the road network pattern of Siak Sri Indrapura City in 2005-2018 [11].
c. Identifying Building Patterns and functions in Siak Sri Indrapura City in 2005-2018 using qualitative descriptive, in identifying building patterns and functions can be divided into three patterns, namely: Homogeneous Patterns, Heterogeneous Patterns and Spread Patterns. The building is a physical form the results of construction work that are integrated with the place of residence both above and below the ground and merging with the place of residence in the water. To find out the functions of buildings could be distinguished, whether the building is a trade and services, settlements, industry, etc. So it will be seen how the different patterns of buildings and their functions from 2005-2018 [12].

Table 1. Morphological Variable

Variable	Indicator
Changes in land cover and land use	a. Geography, Topography b. Land cover c. Land use d. Land area e. Siak Sri Indrapura City imagery in 2005-2018
Pattern of Road	a. Geographical b. Land area c. Road network policy d. Road area e. Road network f. Siak Sri Indrapura City imagery in 2005-2018
Pattern of Buildings and Function of Buildings	a. Distribution of buildings b. Building density c. Building functions d. Land area e. Land cahnges f. Inhabitant g. Admission Map of Siak

Morphology has three components regarding the physical condition of the area. These components are reviewed from land cover and regional land use that reflects regional activities, circulation patterns or road network patterns that connect between regions, and building patterns and their functions [13].

3. Result and Discussion

Land Closure (Ministry of Forestry in Darkono, 2006) is a condition of the surface or the shape of the earth that describes the appearance of vegetation. Changes in land cover would changes that occur in the description of objects on the surface of the earth that are obtained from selected data sources and grouped into closure classes according to their needs (Forest Planning Agency, 2004).

Table 2. Land Built area and non Built Area

Year	Built Area (Ha)	(Hectare) %	Non Built area (Hectare)	%
2005	591	3,16	18.089	96,83
2012	1.039	5,56	17.641	94,44
2016	3.091	16,54	15.589	83,45

The land Built area in 2005-2016 continue to increase, this built-up land was dominated by residential land. Whereas the land that was not built in 2005-2016 has decreased every time because of the development carried out in of Siak Sri Imdrapura City. This non-built land is dominated by plantation land [7].

Table 3. Changes in land closure in 2005-2016

Year	Changes in land	Area (Ha)	Percentage
2005 - 2012	Non built area into built area	448	18
2012 - 2016	Non built area into built area	2052	82

Increased change in land is non into built up land due to the already many infrastructure facilities in Siak Sri Indrapura City. The area of land that has been built has experienced many improvements in Kampung Dalam Village, Kampung Rempak Village, Langkai Village and Mempura River Village. While the area of non-built land which has decreased is in Langkai Village, Sungai Mempura Village, Benteng Hulu Village and Merempan Hilir Village. Changes in land cover from 2016 to 2018 consist of changes a land not built into built-up land because there are still many Subdistrict/ Villages that are under construction. And the change of land built to land not built in 2016 to 2018 is in Kampung Dalam Village precisely located in the Old Siak Market or known as *China Town* which is a famous trade and service center in Siak [14].

Table 4. Classification of road in Siak Sri Indrapura City in 2005-2016

Classification of Road	Year 2005	Year 2012	Year 2016
Collector	36	66	75
Arterial	140	339	375
Total	176	405	450

Road network pattern is one of the elements of city morphology. In 2005 the road network was in the form of irregular roads, in 2012 the road network had changed to an angled road pattern. The road network of Siak Sri Indrapura City follows the building blocks. In 2016 and 2018 the road network is still the same, in the form of an angular road pattern, but the number of road segments is increasing [10][15].

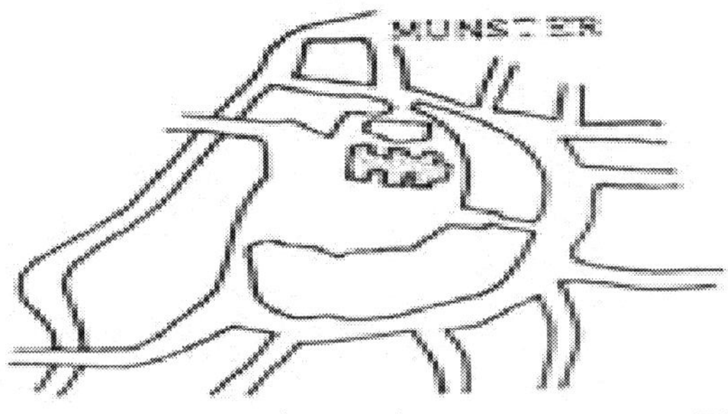

Figure 1. Irregular road pattern

Figure 2. Irregular road pattern imagery in 2005

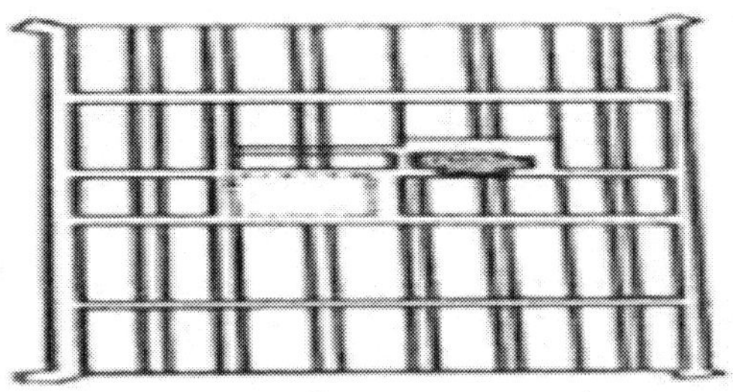

Figure 3. Elbow pattern road

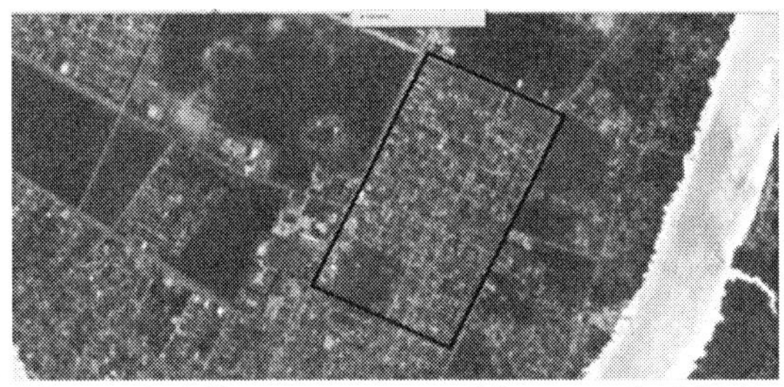

Figure 4. Elbow pattern road imagery in 2016

The pattern of building in this study explains the function of the building or referred to the allotment of the building. The function of allotment of buildings in the research location consists of settlements, trade and services, offices, industries, educational facilities, and health facilities, recreation. There are homogeneous, heterogeneous, linear, centralized and diffuse building patterns [15].

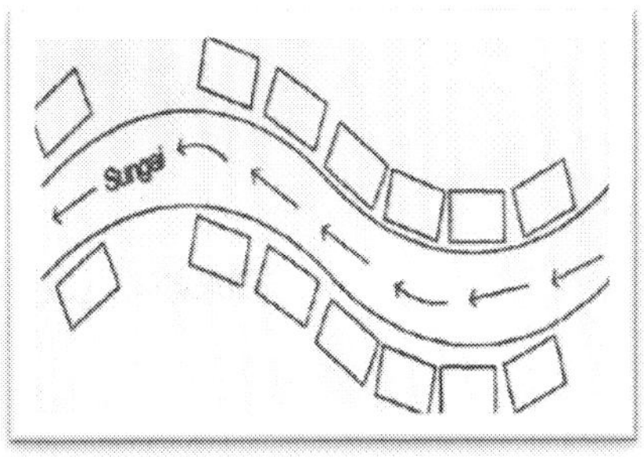

Figure 5. Linear Pattern road

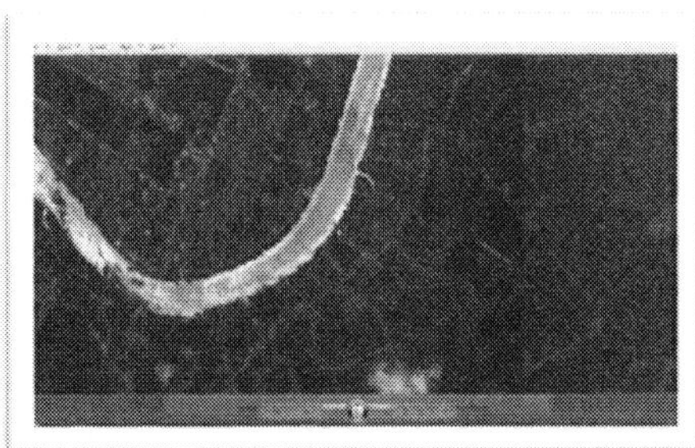

Figure 6. Linear pattern road imagery in 2005

In 2005 the pattern of buildings was dominated by linear patterns and homogeneous patterns due to the accessibility of the community still using ferries or boats to go to their destination so that people preferred to build buildings on river banks, and homogeneous patterns because the majority of the buildings in Siak Sri Indrapura City had architectural forms the same is a one-story building. The building patterns and functions in 2012, 2016 and 2018 in the City of Siak Sri Indrapura are heterogeneous because the forms of buildings in the City have begun to vary, there are one-story, two-storey with non-permanent, semi-permanent and permanent building types and patterns linear because the construction in Siak Sri Indrapura City in 2012 follows a lot of river flow and is already on the edge of the road [15].

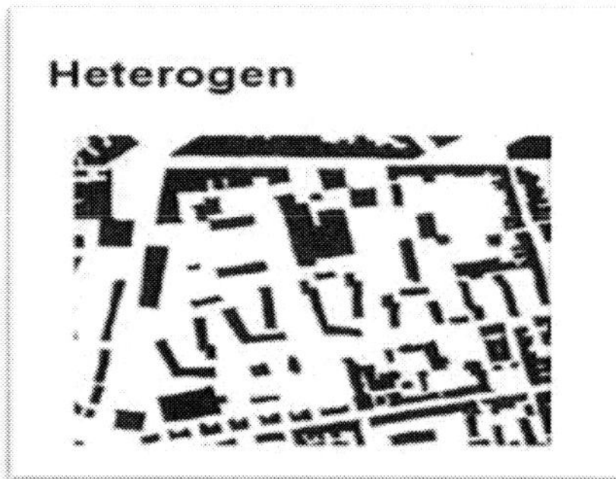

Figure 7. Heterogeneous road pattern

Figure 8. Heterogeneous road pattern in 2016

4. Conclusion

Siak Sri Indrapura City experienced rapid development. The city of Siak Sri Indrapura still has extensive land that is not built so that the land can be used for development either by the community or the government.

The form of the morphology of the City of Siak Sri Indrapura is the *ribbon-shaped cities* because the existing development in the City of Siak Sri Indrapura is more dominant than elongated. Morphology of the City of Siak Sri Indrapura tends to develop around the Siak River, this is due to the movement on the riverbank including fast, before 2007 the liaison between the City of Siak Sri Indrapura and other regions only used ferry crossing. To be closer to the transportation center, people prefer to build riverbanks, and now even though there is a bridge that connects Siak Sri Indrapura City with other regions, the community continues to build around the river to accelerate accessibility.

References

[1] F. Asteriani, "Preferensi Penghuni Perumahan Di Kota Pekanbaru Dalam Menentukan Lokasi Perumahan," *J. Ekon. Pembang. Kaji. Masal. Ekon. dan Pembang.*, vol. 12, no. 1, p. 77, 2011.

[2] C. Cahyaningsih *et al.*, "Geomorphology and structural geology characterization of landslide prone area in Riau-West of Sumatra Highway," *IOP Conf. Ser. Mater. Sci. Eng.*, vol. 536, no. 1, pp. 0–12, 2019.

[3] M. A. Putri, M. J. Rahayu, and R. A. Putri, "Bentuk Morfologi Kawasan Permukiman Urban Fringe Selatan Kota Surakarta," *J. Pengemb. Kota*, vol. 4, no. 2, p. 120, 2016.

[4] Catur Cahyaningsih; Puja Fransismik Crensonni., "Unicharacteristic of Geomorphological Landscape & Depositional Environment in Talawi Hilir : Geotourism Value of Sawahlunto," vol. 3, pp. 42–51, 2018.

[5] Z. Zaim and F. Asteriani, "The Spatial Analysis to Choose a Location to be Designated as a Landfill Site," *MATEC Web Conf.*, vol. 159, p. 01035, 2018.

[6] F. Asteriani, "Analisis peringkat faktor-faktor pemilihan lokasi Ruko dari sudut pandang pengguna dan pengembang Ruko di Kota Pekanbaru."

[7] I. Nugraha, F. Asteriani, and P. Astuti, "THE EFFECTS OF TENGKU AGUNG SULTANAH LATIFAH BRIDGE TOWARD PHYSICAL DEVELOPMENT IN SIAK SUB DISTRICTS," 2017, no. November, pp. 8–10.

[8] R. Fatriadi, F. Asteriani, and C. Cahyaningsih, "Effectiveness of the National Program for Community Empowerment (PNPM) for Infrastructure Development Accelerated and Geoplanology in District of Marpoyan Damai, Pekanbaru," *J. Geosci. Eng. Environ. Technol.*, vol. 2, no. 1, p. 53, 2017.

[9] R. Sampurno and A. Thoriq, "Klasifikasi Tutupan Lahan Menggunakan Citra Landsat 8

Operational Land Imager (Oli) Di Kabupaten Sumedang," *J. Teknotan*, vol. 10, no. 2, pp. 61–70, 2017.

[10] C. Cahyaningsih, "Safety Factor Characterization of Landslide in Riau-West of Sumatra Highway," *Int. J. GEOMATE*, vol. 17, no. 63, 2019.

[11] C. Cahyaningsih *et al.*, "Petrography, Geology Structure and Landslide Characterization of Sumatra Fault Deformation: Study Case In Km 10-15 Highway, Koto Baru Sub District, West of Sumatra," *J. Geosci. Eng. Environ. Technol.*, vol. 3, no. 4, p. 192, 2019.

[12] C. Cahyaningsih, "Hydrology Analysis and Rainwater Harversting Effectiveness as an Alternative to Face Water Crisis in Bantan Tua Village Bengkalis District-Riau," *J. Dyn.*, vol. 1, no. 1, pp. 27–30, 2016.

[13] R. H. Rio M Fauzi, Joko Nugroho R, "Analisa perubahan penutupan lahan pada kawasan hutan lindung gunung naning kabupaten sekadau provinsi kalimantan barat," vol. 4, pp. 520–526, 2016.

[14] C. Cahyaningsih, A. L. Ritonga, S. Aldila, and Z. Zulhikmah, "Lithofacies And Depositional Analysis Environment Of West Section Kolok Nan Tuo Village, Sawahlunto City, West Of Sumatera," *J. Geosci. Eng. Environ. Technol.*, vol. 3, no. 2, p. 128, 2018.

[15] C. A. B. Puji Astuti, Mardianto Manan, Febby Asteriani and and D. Akhyar, *The Prospective Analysis of Coastal Town Development Based on Waterfront City (Case Study: Bantan Sub-District, Bengkalis–Riau Province)*, no. August. 2014.

Using Spatial Analysis with weighted overlay on selecting area for fisheries agroindustry in Southeast Maluku, Indonesia

[1]Wellem Anselmus Teniwut, [1]Syahibul Kahfi Hamid, [2]Marvin Mario Makailipessy

[1]Fisheries Agribusiness Study Program, Tual State Fisheries Polytechnic, Langgur-Sathean Ringroad Km 6, Southeast Maluku District 97611, Indonesia
[2]Fishing Technology Study Program, Tual State Fisheries Polytechnic, Langgur-Sathean Ringroad Km 6, Southeast Maluku District 97611, Indonesia

Email: wateniwut@polikant.ac.id

Abstract. As a archipelagic region with rich marine biodiversity, the abundance of raw material of fishes and other marine resources are high. This advantage should be capitalized to enhance and spread the welfare of coastal community in the region. Fisheries agroindustry is way to boost sustainable economic of a coastal region. Although, to select the appropriate sub-district as the best suite location for either fishes-based agroindustry and marine culture-based agroindustry have to face a complicated barriers, therefore the purpose of this study is to select the suitable sub-district location as the center of fishery agro-industry in Southeast Maluku Regency based on several related factors such as resource potential, number of fishermen and farmers, number of production per year, supporting infrastructure and transportation. The method used in this study is spatial analysis with ArcGIS with utilizing weight overlay to process all the data based on experts perceptions. The results of this study are expected to be a comprehensive guidance for policy makers in this region in the context of sustainable regional development. The paper offers specific path based on empirical evidence for sustainable development in fisheries sector of Southeast Maluku District.

1. Introduction

Until now, research on the impact of agro-industry on improving the welfare of rural and coastal communities has not been too extensive and extensive. Research on agroindustry still focuses on technology, supply chains, and related government policies. Although there are also many criticisms of the impact of agro-industry on increasing value added and the lack of an economic multiplier effect when the production output is raw material [1]. Nevertheless, agroindustry can help increase the income of rural and coastal communities through improving technology, knowledge and productivity through the conversion of raw materials into intermediate and final products that are consistent and sustainable in quality and quantity. Current challenges that still become a cliché issue in rural and coastal areas are the ability related to knowledge and capabilities in entering markets and meeting global market standards, which is also a concern in Indonesia where [2] questions the ability of small-medium enterprises to serve the needs overseas market.

With high catch and cultivation fisheries resources, Southeast Maluku Regency which is part of the Kei Islands and is located in a coral triangle area with high marine biodiversity makes the availability of fishery raw materials very high [3]. On the other hand, the ability of human resources related to the management of raw materials and businesses in obtaining fisheries resources is still inadequate. This is due to the level of formal education of most of the coastal communities that carry out fisheries activities is still very low coupled with the absence of informal education related to fisheries business [4]; [5]; [6]; [7]. The abundant of raw materials are not yet accompanied by the support of human resources coupled with the distortion of the flow of information between regions, between coastal communities inside and outside areas due to geographical conditions in the Kei Islands region [8]; [9], making the challenges of developing fisheries agroindustry become increasingly complex. For this reason, it is very important to have research that combines all-natural resources, human resources and related infrastructure in determining the location of a suitable area to be the location of fisheries agroindustry in this region. For this reason, it is very important to have research that combines natural resources, human resources and related infrastructure in determining the location of a suitable area to be the location of fisheries agroindustry in this region.

The use of spatial analysis in the study of regional development, especially in the small islands region, such as in the Kei Islands presents more advantages compared to several other approaches. Many studies have well demonstrated the ability of spatial analysis to help facilitate the determination of suitable locations for waste handling locations [10], tours [11], agriculture and plantations [12], wind farms [13]; [14], healthcare [15], to regional development planning [16]. With the characteristics of the region of Southeast Maluku Regency which has difficult terrain and not extensive access to the region, the use of spatial analysis makes it very easy to analyze and map the area based on related factors in having a suitable sub-district area location to become the location of an agroindustry development center in this region. This study aims to determine the location of agroindustry centers in Southeast Maluku Regency with spatial analysis approach using weight overlay techniques with ArcGIS 10.5.

2. Methodology

This study uses field survey data collection techniques, where the type of data used in this study is primary data in the form of in situ field survey data in all districts in Southeast Maluku Regency. For secondary data in the form of Southeast Maluku satellite imagery from USGS for digitization, other data are obtained from the Southeast Maluku Regency Fisheries and Maritime Service and Statistics. The framework of the research flow can be seen in Fig 1. Data obtained from three sources, namely the results of the field survey in the form of identification of the potential for marine resource retention, human resource potential and socio-economic conditions, data from the local government and Statistic Indonesia in the form of data on the number of fishermen and farmers cultivating fisheries, the amount of capture fisheries production and aquaculture, then satellite imagery for digitizing the location of Southeast Maluku Regency.

The data obtained is then converted into shapefile form for further processing using the weighted overlay method (WOM), where classify data on each indicator is ranked based on input from experts and empirical conditions in the field (Table 1). Data processing using ArcGIS 10.5, the study location of the study is conducted in Southeast Maluku District located in Kei Islands, on Maluku province Indonesia. We use a rating scale of 1-5 with 1 being the best value and 5 representing the opposite. Furthermore, for the weight of each indicator, divided as follows, infrastructure and transportation have a weight of 25%; marine resource potential and HR 25%; the number of cultivated farmers and fishermen both have a weight of 10%; the amount of capture fisheries production is 20% and the amount of aquaculture production is 10%.

Table 1. Classification weighted factors for selecting center of agro-industry region

Indicator	Actual Value	Ranking	Weightage
Infrastructure and Transportation	Very good	1	
	Good	2	
	Fair	3	
	Bad	4	25%
	Very Bad	5	
Potency of each sub district	Fishing	1	
	Mariculture	2	25%
	Marine Ecotourism	3	
Number of fishers	>600	1	
	500-600	2	
	400-500	3	10%
	300-400	4	
	<300	5	
Number of marine farmers	>500	1	
	400-500	2	
	300-400	3	10%
	200-300	4	
	<200	5	
Production of fishing sub sector	>10.000 ton	1	
	8000-10.000 ton	2	
	6000-8000 ton	3	20$
	4000-6000 ton	4	
	<4000 ton	5	
Production of marine culture sub sector	>10.000 ton	1	
	8000-10.000 ton	2	10%
	6000-8000 ton	3	
	4000-6000 ton	4	
	<4000 ton	5	

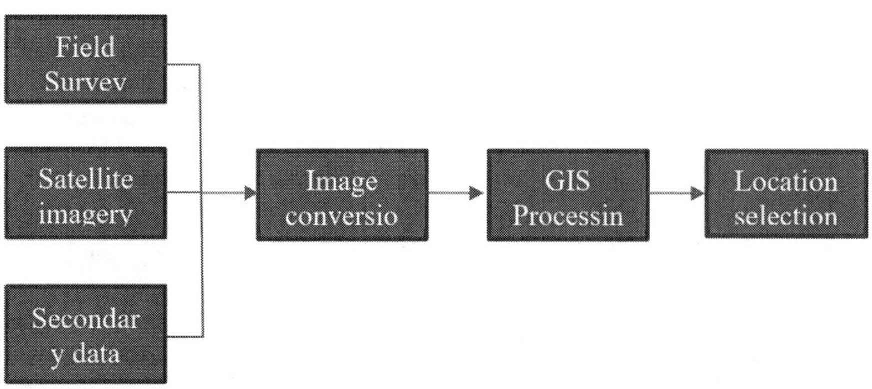

Figure 1. Research flowchart

Journal of Physics: Conference Series **1424** (2019) 012016 doi:10.1088/1742-6596/1424/1/012016

3. Result and Discussion

3.1. Indicators per sub-district in Southeast Maluku District

Based on the data obtained it can be seen that, Kei Kecil and Kei Besar Utara Timur Sub Districts are the districts with the highest number of fishermen in southeast Maluku regency, for marine farmers, Kei Kecil sub-district, Hoat Sorbay Subdistrict and Kei Besar Selatan Sub-district have the highest number of farmer for marine cultivation in this region. For fisheries potential, only Kei Kecil Timur Selatan sub-district is not suitable as an agroindustry location because it is more suitable for the development of marine tourism. For transportation and infrastructure indicators, all subdistricts on the island of Kei Kecil meet the conditions and only the kei besar sub-district meets the criteria for sub-districts located on the island of Kei Besar. For indicators of capture fisheries production, it can be seen that the Kei Kecil sub-district is the highest producer and followed by the big south-west kei sub-district, while for the fisheries production of Heat Sorbayt sub-district is the biggest producer and the Kei Besar Selatan Barat sub-district is the highest on the large Kei island (Fig 2).

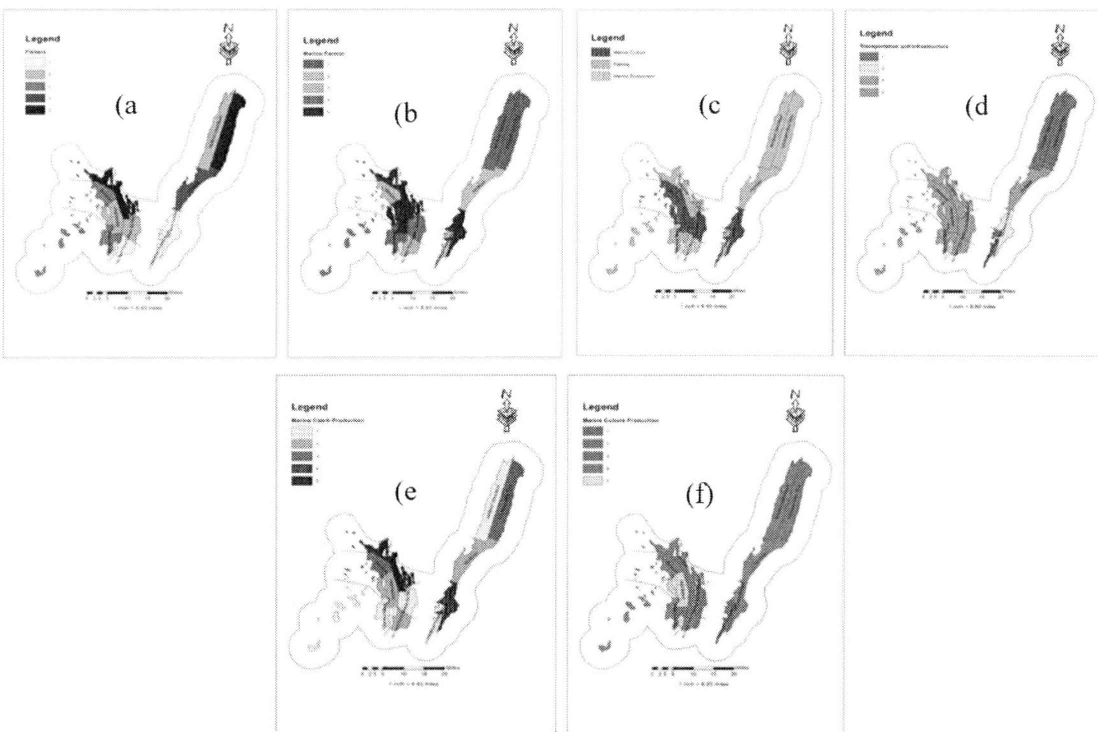

Figure 2. Indicators per sub-districts: (a) number of fishers; (b) number of marine farmers; (c) fisheries potential of each sub-district; (d) transportation and infrastructure; (e) marine catch production; (f) marine culture production

3.2. Suitable location for canter of agroindustry fisheries in Southeast Maluku District

The weighted overlay method (WOM) processing with ArcGIS showed that the most suitable location for the center of fisheries agro-industry development in Southeast Maluku Regency is Kei Kecil sub-district. With the potential of catch fisheries and aquaculture supported by infrastructure and transportation support either between regions and outside areas, makes the Kei Kecil sub-district is an obvious choice. However, by considering equitable development, location that also meets the criteria is Kei Besar Selatan Barat sub-district. This is because only the infrastructure and transportation are very limited in that sub-district but other indicators are very supportive. Thus, the local government' role is important in this regard in enhancing infrastructure and transportation development so that the acceleration of equitable development and improvement of the welfare of coastal communities on the island of Kei Besar in particular (Fig 3).

The combination of locations that tend to be isolated but which have very abundant marine resources makes it necessary to have a trigger which is not only increase the level of welfare of the people in Southeast Maluku Regency but also equitable development. With the existence of fisheries agro-industry, the unemployment rate can be reduced, such as research conducted by [17] in Brazil. In addition, increasing the level of income of coastal communities due to consistent production so the demand for fishery raw materials become stable so income of coastal communities can become consistent and increase. The existence of agro-industry also increases the competitiveness of fishery products from Southeast Maluku Regency which tends to be low in quality and less consistent in quantity. However, there is a need for heavy investment from the local government, investment is also expected to be right on target so that it helps accelerate the regional equitable development in Southeast Maluku Regency.

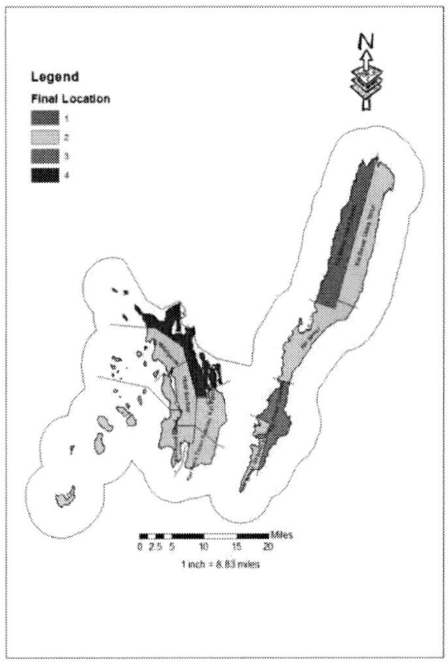

Figure 3. Final suitable location for center agroindustry fisheries in Southeast Maluku District

4. Conclusion

Southeast Maluku Regency has the characteristics of an archipelago that does not yet have easy access between regions coupled with a considerable distance from major cities in Indonesia, making Maluku Maluku Regency obliged to develop and develop fisheries agroindustry areas to improve equitable development. The use of spatial analysis can help solve complex problems by giving results that are easy to use. From the use of the weighted overlay method (WOM) with ArcGIS based on 6 existing indicators, obtained two sub-district locations that are suitable as fisheries agroindustry development sites in Southeast Maluku Regency, first Kei Kecil subdistrict and, the second, Kei Besar Selatan Barat subistrict. Therefore. it is hoped that the welfare of coastal communities can improve. Fishery processing industry is of the choice along with processing of semi-finished materials industry such as carrageenan, so that the added value of fishery products can increase in accordance with the selling price offered to the market.

Acknowledgement
The authors thank Directorate of Research and Community Service, Ministry of Research and Technology the Higher Education Republic of Indonesia for funding this research under PDUPT (Penelitian Dasar Unggulan Perguruan Tinggi) scheme 2019-2020.

References
[1] Ginting, G 2015 Open innovation model: empowering entrepreneurial orientation and utilizing network resources as determinant for internationalization performance of small medium agroindustry. *Agriculture and Agricultural Science Procedia*, 3, 56-61.
[2] Tambunan, T 2007 Entrepreneurship development: SMES in Indonesia. *Journal of Developmental Entrepreneurship*, 12(01), 95-118.
[3] Teniwut, W. A 2016 For sustainable revenue of fisheries sector in small islands: evidence of Maluku, Indonesia. *Aquaculture, Aquarium, Conservation & Legislation*, *9*(3), 722-732.
[4] Teniwut, W. A., Teniwut, Y. K., Teniwut, R. M., and Hasyim, C. L 2017 Family vs Village-Based: Intangible View on the Sustainable of Seaweed Farming. In *IOP Conference Series: Earth and Environmental Science* (Vol. 89, No. 1, p. 012021). IOP Publishing.
[5] Picaulima, S., Teniwut, W. A., Kahfi, S., Teniwut, R. M., Susanti, I. I. D. A. R., Hungan, M., ... and Ngabalin, A. M 2017 Mapping marine resources utilization based on seascapes area: a study on gender comparison. In *IOP Conference Series: Earth and Environmental Science* (Vol. 89, No. 1, p. 012027). IOP Publishing.
[6] Teniwut, R. M., Hasyim, C. L., and Teniwut, W. A 2017 Resource-Based Capability on Development Knowledge Management Capabilities of Coastal Community. In *IOP Conference Series: Earth and Environmental Science* (Vol. 89, No. 1, p. 012017). IOP Publishing.
[7] Teniwut, W. A., and Teniwut, R. M 2018) Minimizing the instability of seaweed cultivation productivity on rural coastal area: a case study from Indonesia. *Aquaculture, Aquarium, Conservation & Legislation*, *11*(1), 259-271.
[8] Teniwut, W. A., Betaubun, K. D., and Djatna, T 2017 A conceptual mitigation model for asymmetric information of supply chain in seaweed cultivation. In *IOP Conference Series: Earth and Environmental Science* (Vol. 89, No. 1, p. 012022). IOP Publishing.
[9] Teniwut, W., Marimin, M., and Djatna, T 2019 GIS-Based multi-criteria decision making model for site selection of seaweed farming information centre: A lesson from small islands, Indonesia. *Decision Science Letters*, *8*(2), 137-150.
[10] Shi, Q., Ren, H., Ma, X., and Xiao, Y 2019 Site selection of construction waste recycling plant. *Journal of Cleaner Production*, *227*, 532-542.
[11] Ayhan, Ç. K., Taşlı, T. C., Özkök, F., and Tatlı, H 2020 Land use suitability analysis of rural tourism activities: Yenice, Turkey. *Tourism Management*, *76*, 103949.
[12] Bodaghabadi, M. B., Faskhodi, A. A., Salehi, M. H., Hosseinifard, S. J., and Heydari, M 2019 Soil suitability analysis and evaluation of pistachio orchard farming, using canonical multivariate analysis. *Scientia horticulturae*, *246*, 528-534.
[13] Baseer, M. A., Rehman, S., Meyer, J. P., and Alam, M. M 2017 GIS-based site suitability analysis for wind farm development in Saudi Arabia. *Energy*, *141*, 1166-1176.
[14] Ostovari, Y., Honarbakhsh, A., Sangoony, H., Zolfaghari, F., Maleki, K., and Ingram, B 2019 GIS and multi-criteria decision-making analysis assessment of land suitability for rapeseed farming in calcareous soils of semi-arid regions. *Ecological Indicators*, *103*, 479-487.
[15] Mishra, S., Sahu, P. K., Sarkar, A. K., Mehran, B., and Sharma, S 2019 Geo-spatial site suitability analysis for development of health care units in rural India: Effects on habitation accessibility, facility utilization and zonal equity in facility distribution. *Journal of Transport Geography*, *78*, 135-149.

[16] Qiu, L., Zhu, J., Pan, Y., Hu, W., and Amable, G. S 2017 Multi-criteria land use suitability analysis for livestock development planning in Hangzhou metropolitan area, China. *Journal of Cleaner Production*, *161*, 1011-1019.
[17] Ometto, A. R., Ramos, P. A. R., and Lombardi, G 2007 The benefits of a Brazilian agro-industrial symbiosis system and the strategies to make it happen. *Journal of Cleaner Production*, *15*(13-14), 1253-1258.

Outliers Detection on Fisheries Commodity Transaction from Local Market in Tual City based on The x-means Clustering

[1]Syahibul Kahfi Hamid, [1]Wellem Anselmus Teniwut, [1]Roberto Mario Kabi Teniwut, [2]Maimuna Renhoran

[1]Fisheries Agribusiness Study Program, Tual State Fisheries Polytechnic, Langgur-Sathean Ringroad Km 6, Southeast Maluku District 97611, Indonesia

[2]Fishing Technology Study Program, Tual State Fisheries Polytechnic, Langgur-Sathean Ringroad Km 6, Southeast Maluku District 97611, Indonesia

Email: wateniwut@polikant.ac.id

Abstract. Correctly estimating factors that affect the income and profitability of coastal communities specifically catching fishermen is one way to increase the level of sustainable coastal community welfare. The purpose of this study is to detect outliers from fisheries commodity transactions in the fish market in Tual City, to have full view on which cases that may have certain traits that can improve the profitability of fishermen in the region. The research method used in this study is a survey method in which used as many as 1420 datasets obtained from interviews conducted starting from June 2017 to June 2018. The data mining approach is used in this study, in order to obtain the appropriate outliers then x- means clustering is used to cluster profitable transaction data, then data collected further being process with outlier detection. We found 25 high profitability transactions of 1420 datasets. The outlier analysis was carried out and found 1 outliers. We have figured that the species of fish and the location of the seller are the factors that influence the level of profit from the fishermen. This result is an input for policy makers and catch fishermen in conducting capture fisheries.

1. Introduction

Increasing the dynamics of development in an area is highly dependent on an effort to optimize the use of existing resources in the region for the prosperity of the people and for the benefit of all components in the region. Optimization of resource utilization in an area will depend on the knowledge of those who utilize the potential of natural resources In Maluku province there are three fisheries management areas (WPP), respectively: WPP-714 Banda Sea, WPP-715 Arafura Sea and WPP-716 Seram Sea, Maluku Sea and Tomini Bay. The Kei Islands is one of the regencies and cities in Maluku, which is between two WPPs that needs good capabilities and good knowledge. The potential of capture fisheries in the Kei Islands and Maluku Province in general according to [1] is still a source of income for fishermen in this region but refers to the existing conditions and trends related to behavior, fleets and techniques including illegal fishing so the tendency of the development of marine resources in particular fish will decrease in the next 8-10 years. The Kei Islands are included in the coral triangle waters region, so it is important to maintain the rich biodiversity of marine resources in the waters.

Based on studies by [2] showed that a total of over IDR 240 million was generated on average by each of the coastal villages in the Kei Islands, including the coastal villages in Tual City. This value is still below an optimal number based on the daily needs of coastal communities and the population of coastal communities added with given the characteristics of coastal communities in this region which has a very high number of family dependents with an average low level of education [3]; [4]; [5]; [6]. This condition is compounded by the financial and capital capability to procure better fishing gear. This causes income from capture fishermen in this region to be very inconsistent, one of the causes which also one cliché problem of coastal communities in the Kei Islands is a result of the use of technology that is still limited and limited knowledge [5]. Another factor that is an obstacle to improving the welfare of coastal communities is the potential for conflicts that arise both between villages, in villages between families in the village due to the use of marine resources [6]; [7]. For this reason, research is needed that not only improves the efficiency of fish catching activities, but also avoids potential conflicts and increases in income of coastal communities. A data mining approach can be used to help resolve this problem.

Data mining is process where an intelligent method are applied to extract knowledge and often associated with term knowledge discovery [8]. The data mining approach has been widely used by various fields including in the fields of fisheries and marine affairs [9]; [10]; [11]; [12]. Utilization of this technique is very helpful in breaking down information and knowledge from the data obtained. To increase fish catch efficiency by fishermen and increase fishermen profits, it can be obtained by identifying fish species, catch time and fishing location. Thus, the purpose of this study is to detect outliers from consumer transaction datasets on fisheries commodities in the local fish market in Tual City, Kei Islands, which provide the highest profit value to obtain the empirical characteristics of these transactions to obtain a fishing profile that benefits fishermen in the area.

2. Methodology

The stages of this research can be seen in Fig. 1, where the data mining approach used in this study consists of two parts, namely cluster analysis using the x means clustering approach which then clusters with the highest profits will be used to be analyzed for outliers to identify transactions that are outside trend and has a unique profile, namely transactions that provide high profits with fish species, catch time and unique fishing locations. The data used in this study were 1420 datasets obtained from a survey for one year in May 2017-June 2018 on fish commodity transactions in the local fish market in Tual City.

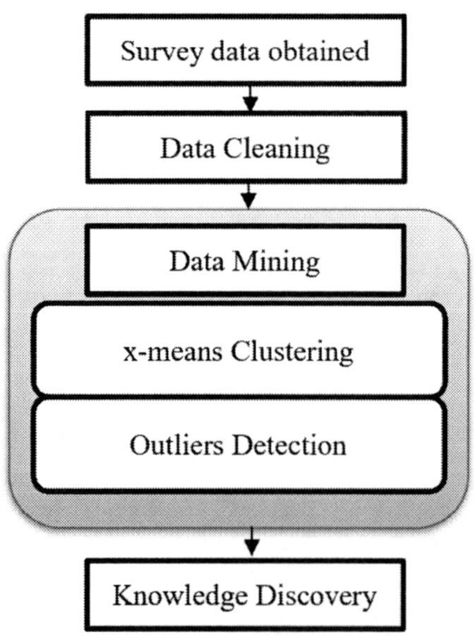

Figure 1. Research framework

2.1. X-means clustering

X-means clustering is a variation of K-means clustering, x-means clustering is a variation of K-means clustering. The x-means clustering algorithm is summarized as follows [13]:

1. Set initial number of clusters k_0, then applied k-means algorithm into all data ($k_0 = k$) and repeat all procedure (setting $i = 1, 2, \dots, k_0$).
2. Apply k-means cluster with setting $k = 2$ for cluster $C_1, C_2 \dots, C_{k0}$.
3. Assume p-dimensional distribute normal for data x in C_1. Calculate BIC as, $BIC = -2\log L(\hat{\theta}_i; x \in C_i) + q\ log\ n_i$ and BIC' as, $BIC' = -2\log L'\left(\hat{\theta}_i{}'; x \in C_i\right) + q\,'log\ n_i$. Where, $\hat{\theta}_i$ is the maximum likelihood estimation of p-dimensional, q is number of dimensions of the parameters and becomes $2p$, n_i is the number of elements contained C_1 and $L(.)$ is likelihood function, $q\,'$ is given by $2\ x\ 2p = 4p$.
4. When BIC > BIC' continue the division and if the vice versa, BIC < BIC', divide further clusters and stop.
5. Number the cluster output identification from the center of each cluster, and the number of elements in each cluster.

2.2. Outliers detections
Anomaly detection identifies single records in datasets which significantly deviate from the normal data, further it is process of finding patterns in a given dataset which deviate from the characteristics of the majority. We can use the maximum likelihood method to estimate the parameters μ and σ by maximize the log-likelihood function, as follows [8]:

$$\ln \mathcal{L}(\mu, \sigma^2) = \sum_{i=1}^{n} \ln f\left(x_i \middle| (\mu, \sigma^2)\right) = -\frac{n}{2}(2\pi) - \frac{n}{2}\ln \sigma^2 - \frac{1}{2\sigma^2}\sum_{i=1}^{n}(x_i - \mu)^2$$

Where n is the total number of samples, for derivatives μ and σ^2 solving the first order condition leads to maximum likelihood estimates:

$$\hat{\mu} = \bar{x} = \frac{1}{n}\sum_{i=1}^{n} x_i$$

$$\hat{\sigma}^2 = \frac{1}{n}\sum_{i=1}^{n}(x_i - \bar{x})^2$$

3. Result and Discussion

3.1. Clustering analysis

The results of data analysis using the x-means clustering approach found that there were 4 clusters, each of which was divided based on the main criteria, namely the level of income received from fishermen. Measurement of cluster distance using euchlidean distance with the average distance of each cluster is 0.950 with Davies-Bouldin Index of 0.752. In Fig 2, it appears that cluster 3 is a cluster that is used to the next stage for detecting outliers. Obtained 25 transaction data from 1420 datasets which have a price of only 5.97% higher than other datasets, but have an earning rate of 604.59% greater than other datasets and a higher sales amount of 517.73%.

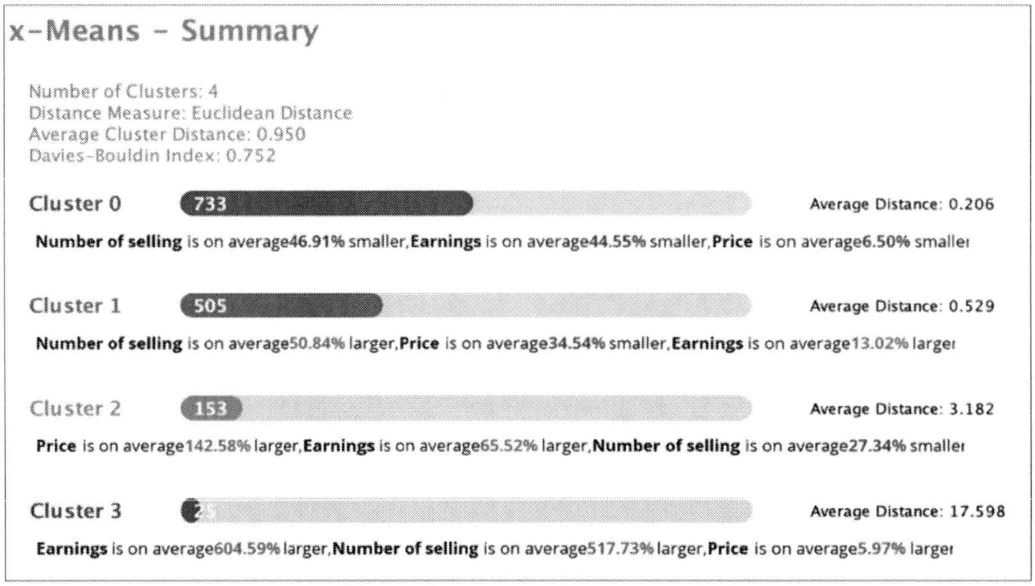

Figure 2. x-means clusters summary

Table 1. Cluster 3 datasets

No	Year	Period (month)	Seller Origin	Fish' Species	Number of selling	Price	Earnings	Fishing Ground
1	2017	June	Fair	Lethrinidae	143	50.000	7.150.000	Langgiar
2	2017	June	Warbal	Stolephorus	300	10.000	3.000.000	Debut
3	2017	July	Mangon	Sardinella maderensis	430	10.000	4.300.000	Kaimana
4	2017	July	Komp Banda Ely	Auxis rochei	270	30.000	8.100.000	Banda Ely
5	2017	July	Ngursit	Auxis rochei	145	50.000	7.250.000	Ngursit
6	2017	July	Sathean	Sardina pilchardus	290	20.000	5.800.000	Ohoililir
7	2017	July	Sathean	Sardina pilchardus	195	20.000	3.900.000	Ohoililir
8	2017	July	Banda Ely	Auxis rochei	150	25.000	3.750.000	Banda Ely
9	2017	July	Banda Ely	Auxis rochei	203	25.000	5.075.000	Banda Ely
10	2017	July	Banda Ely	Moolgarda seheli	165	25.000	4.125.000	Kaimana
11	2017	Jully	Wara	Auxis rochei	211	25.000	5.275.000	Ngursit
12	2017	July	Wara	Auxis rochei	213	25.000	5.325.000	Kei Besar
13	2017	July	Un	Sardina pilchardus	500	20.000	10.000.000	Pulau Ut
14	2017	July	Fair	Auxis rochei	39	30.000	1.170.000	Ngursit
15	2017	July	Banda Ely	Auxis rochei	150	25.000	3.750.000	Banda Ely
16	2017	August	Wara	Katsuwonus pelamis	70	50.000	3.500.000	Perairan Tan. Kei
17	2017	August	Fair	Auxis rochei	315	20.000	6.300.000	Ngursit
18	2017	August	Mangon	Rastrelliger faughni	150	25.000	3.750.000	Dobo
19	2017	August	Un	Stolephorus	60	10.000	600.000	Lupus
20	2017	August	Banda Eli	Auxis rochei	15	50.000	750.000	Ngursit
21	2017	August	Kiom	Parexocoetus brachypterus	150	20.000	3.000.000	Perairan Dullah Laut
22	2017	August	Fidabot	Ocyurus chrysurus	20	50.000	1.000.000	Tam
23	2017	August	Fidabot	Stolephorus	70	10.000	700.000	Lupus
24	2017	August	Fair	Auxis rochei	200	20.000	4.000.000	Ngursit
25	2018	April	Kompleks Banda	Selar crumenophthalmus	60	20.000	1.200.000	Sathean

3.2. Outliers detection

Based on the results of the x-means cluster analysis that has been done in the previous stage, then it has been used to detect outliers. Thus, it is expected to produce outliers that have unique characteristics and are out of trend to further serve as a reference for fishermen and fishery commodity sellers in Tual City in running related businesses. In Fig 3a, a comparison of fish species and fishing ground is seen that has unique characteristics. It is seen that on Langgiar fishing ground which is located in Kei Besar Selatan Barat sub-district, Southeast Maluku district with fish species that being capture was Lethrinidae. Then when seen in Fig. 3b, information on catch periods has been obtained in June which is included in The Indian summer monsoon. Thus, both fish species and period of catching information have been found, which it has the potential to provide more benefits to fishermen and sellers in terms of earning and profit.

Figure 3. Outliers detection: (a) combination between period of catching and fish species; (b) combination between fishing ground and fish species

4. Conclusion

Based on the results of research that has been done, x-means clustering gained 25 datasets out of 1420 datasets of fishery commodity transactions in the local fish market in Tual City which has high profit characteristics. Further results were obtained that, there were 1 data outliers from 25 datasets included in cluster 3. From the information of the outliers, it is obtained that the benefits obtained by the data are 57% higher than the average datasets in cluster 3, with a selling price of 53% higher than the average price in cluster 3. The results also found two factors that differentiate the fish species, Lethrinidae and the fishing period in June. Implication of research is that fishermen in this area should pay attention to fish species that are caught and sold because it is very influential on the income earned, this is related to taste of customer and distance of fish capture. In addition, fishers and seller also have to put more attention to period of fishing (monsoon), so it could be more efficient considering the natural and marine conditions in the Kei Islands, this is related to the demand and availability of fish in the market.

Acknowledgement

The authors thank Directorate of Research and Community Service, Ministry of Research and Technology the Higher Education Republic of Indonesia for funding this research under PTUPT (Penelitian Terapan Unggulan Perguruan Tinggi) scheme 2019-2020.

References

[1] Teniwut, W. A 2016 For sustainable revenue of fisheries sector in small islands: evidence of Maluku, Indonesia. Aquaculture, Aquarium, Conservation & Legislation, 9(3), 722-732.
[2] Picaulima, S., Teniwut, W. A., Kahfi, S., Teniwut, R. M., Susanti, I. I. D. A. R., Hungan, M., ... and Ngabalin, A. M 2017 Mapping marine resources utilization based on seascapes area: a study on gender comparison. In *IOP Conference Series: Earth and Environmental Science* (Vol. 89, No. 1, p. 012027). IOP Publishing.
[3] Hamid, S. K., Teniwut, W. A., Teniwut, R. M., Rahantoknam, M. A., Hasyim, C. L., and Hungan, M 2017 The Support of MPA (Marine Protected Area) in Coral Triangle Area: Evidence from Kei Islands, Indonesia. In *IOP Conference Series: Earth and Environmental Science* (Vol. 89, No. 1, p. 012025). IOP Publishing.
[4] Teniwut, R. M., Hasyim, C. L., and Teniwut, W. A 2017 Resource-Based Capability on Development Knowledge Management Capabilities of Coastal Community. In *IOP Conference Series: Earth and Environmental Science* (Vol. 89, No. 1, p. 012017). IOP Publishing.
[5] Teniwut, W. A., and Teniwut, R. M 2018 Minimizing the instability of seaweed cultivation productivity on rural coastal area: a case study from Indonesia. *Aquaculture, Aquarium, Conservation & Legislation*, *11*(1), 259-271.
[6] Teniwut, W. A., Teniwut, Y. K., Teniwut, R. M., and Hasyim, C. L 2017 Family vs Village-Based: Intangible View on the Sustainable of Seaweed Farming. In *IOP Conference Series: Earth and Environmental Science* (Vol. 89, No. 1, p. 012021). IOP Publishing.
[7] Yamazaki, S., Resosudarmo, B. P., Girsang, W., and Hoshino, E 2018 Intra-village and inter-village resource use conflict in Indonesia: The case of the Kei Islands. *Ocean & coastal management*, *155*, 50-59.
[8] Han, J., Pei, J., and Kamber, M 2011 *Data mining: concepts and techniques*. Elsevier.
[9] Tsai, W. P., Huang, S. P., Cheng, S. T., Shao, K. T., and Chang, F. J 2017 A data-mining framework for exploring the multi-relation between fish species and water quality through self-organizing map. *Science of the Total Environment*, *579*, 474-483.
[10] Su, F., Zhou, C., Lyne, V., Du, Y., and Shi, W 2004 A data-mining approach to determine the spatio-temporal relationship between environmental factors and fish distribution. *Ecological Modelling*, *174*(4), 421-431.
[11] Shen, W., Guo, X., Wu, C., and Wu, D 2011 Forecasting stock indices using radial basis function neural networks optimized by artificial fish swarm algorithm. *Knowledge-Based Systems*, *24*(3), 378-385.

[12] Cheng, Y., Jiang, M., and Yuan, D 2009 Novel clustering algorithms based on improved artificial fish swarm algorithm. In *2009 Sixth International Conference on Fuzzy Systems and Knowledge Discovery* (Vol. 3, pp. 141-145). IEEE.

[13] T. Ishioka,T 2005 An Expansion of X-means for Automat- ically Determining the Optimal Number of Clusters, Proc. of The 4th IASTED International Conference on Computational Intelligence, pp.91-96.

Implementation of Computer-Based Systems in Efficient Credit Acceptance Decisions Applying the Additive Ratio Assessment (ARAS) Method

Supriyanto[1], Mesran[2], Diding Kusnady[3], Weny[4], Murtopo[5]

[1]Politeknik LP3I Medan, Medan, Indonesia
[2]STMIK Budi Darma, Medan, Indonesia
[3]Politeknik Ganesha Medan, Medan, Indonesia
[4]Akademi Sekretari Manajemen Cendana, Medan, Indonesia
[5]Politeknik Negeri Media Kreatif, Medan, Indonesia

*mesran.skom.mkom@gmail.com

Abstract. The development of computer-based systems, increasingly helping the banking world, in doing work efficiency in a bank. Especially in selecting credit recipients at the bank. This is needed so that the resulting decision can be valued objectively and effectively for management. In processing credit receipt decisions, using an information system known as the Decision Support System (SPK). The selection of credit recipients requires alternatives and criteria so that the information retrieval process can run well. By applying the Additive Ratio Assessment (ARAS) method, decision-makers can produce better calculations. ARAS is a method for decision-based support.

1. Introduction

Banks have a very important role in the development of the people's economy. One form of bank service for the public is credit distribution. Lending in a bank plays an important role in helping to solve financial problems for the community. In granting credit the bank must conduct careful research and calculation of the customer. The ability of banks to make quick and careful decisions will be the key to success in banks to extend credit in the global competition in the future.

The existence of credit activities at a bank is certainly one of the benefits for the bank, including the customers who will be selected based on income, age, occupation, collateral, and the number of dependents. Problems that are often encountered by banks in lending include the determination of credit recipients. Every customer who is entitled to get credit from the bank is a customer who has predetermined criteria. Customers will be selected based on income, age, occupation, guarantee, and a number of dependents. But selection without using a computer-based information system, it will produce decisions that are not good, because decisions are made personally and tend to be manipulated. The use of assistive devices, in this case, the computer system in the form of a decision support system (SPK) is considered appropriate to avoid the fraud caused[1]–[4].

Decision Support System (SPK) is a computer-based system that is able to produce the best alternative decisions to support decisions taken by decision-makers. In its application, SPK uses methods in producing the best alternatives, such as VIKOR, ARAS, ELECTRE[5], [6]

In this study, the authors tried to conduct research for credit-granting decisions by implementing a computer-based information system known as a decision support system by applying the Additive Ratio Assessment (ARAS) method.

2. Methodology

Much research has been done by using DSS in generating effective decisions such as that of Syafrida in selling used laptops[7], Determination of work incentive recipients applying VIKOR[8]. In this study, the authors used the ARAS method in determining bank credit recipients.

The Additive Ratio Assessment (ARAS) method is a method used to rank criteria for decision making. In the ranking process, the ARAS method has several steps that must be carried out[9]–[11], namely:

Step-1: Formation of the Decision Making Matrix

$$X = \begin{bmatrix} X01 & X0j & \cdots & X0n \\ Xi1 & Xij & \cdots & Xin \\ \vdots & \vdots & \ddots & \vdots \\ Xn1 & Xmj & \cdots & Xmn \end{bmatrix} (i = 0, \text{m}; \ldots j = 1, \text{n}) \quad (1)$$

Information

m = number of alternatives

n = number of criteria

x_{ij} = performance value of alternative i to criterion j x0j = optimum value of criterion j

For the value of X0j obtained from the maximum/minimum value of the criteria contained in the criteria column according to the type of maximum or minimum criteria.

Step-2: Normalize the decision matrix

If the Beneficial criteria are normalized, follow:

$$Rij = \frac{Xij}{\sum_{i=0}^{m} Xij} \quad (2)$$

R_{ij} is a normalized matrix.

If the criteria are Non-Beneficial, then normalization follows:

Stage-1: $Xij^* = \frac{1}{Xij}$ (3)

Stage-2: $Rij = \frac{Xij^*}{\sum_{i=0}^{m} Xij^*}$ (4)

Step-3: Determine the normalized matrix weight

$$D = [dij]mxn = Rij.Wj \quad (5)$$

Step-4: Determine the value of the optimization function (S_i)

$$Si = \sum_{j=1}^{n} dij; (i = 1,2 \ldots, m; j = 1,2 \ldots, n) \quad (6)$$

Step-5: Determine the highest rank level

$$i = \frac{Si}{S0} \quad (7)$$

S_i and S_0 are the optimality criteria values, which are obtained from the equation. U_i values are calculated at intervals [0,1].

3. Result and Discussion

In the process of granting credit to customers, a customer must meet all the requirements or criteria set by management. The expected results of the selection of credit are expected to be objective so as to avoid the element of data manipulation. In computer-based systems, especially decision support systems require several conditions or criteria needed in granting credit, namely:

Table 1. The Criteria and Type of Criteria

Criteria	Weighted	Type
Business Turnover (C^1)	0.457	Benefit
Customer Revenue (C^2)	0.257	Benefit
Value of Guarantee (C3)	0.157	Benefit
Number of Dependents (C^4)	0.090	Cost
Age (C^5)	0.040	Cost

Weights against the criteria in table 1 are obtained by applying the Centroid Rank Order method. The alternatives will be processed as many as 5 alternatives, namely Heriansyah (A_1), Muhammad Doni (A_2), Siregar (A_3), Kiki Prasetia (A_4), and Karen (A_5). In table 2 are alternative data and criteria.

Table 2. the Alternatives

Alternative	C_1	C_2	C_3	C_4	C_5
A_1	43 Million	2.5 Million	51 Million	3	40
A_2	41 Million	3 Million	49 Million	5	45
A_3	43 Million	2.75 Million	48 Million	2	50
A_4	49 Million	3.25 Million	55 Million	3	48
A_5	44 Million	2.9 Million	45 Million	4	49

From table 2 above, the matching data is obtained in the form of a match rating, as shown in Table 3 as follows.

Table 3. Rating of alternative matches and criteria

Alternative	C_1	C_2	C_3	C_4	C_5
A_1	43	2.5	51	3	40
A_2	41	3	49	5	45
A_3	43	2.75	48	2	50
A_4	49	3.25	55	3	48
A_5	44	2.9	45	4	49

After the compatibility rating is known, it can be obtained by using a computer-based information system that data will be obtained by applying the ARAS method. The initial steps taken are determining the decision matrix (X_{ij}) as follows:

Table 4. Rating of alternative matches and criteria

Alternative	C_1	C_2	C_3	C_4	C_5
	Max	Max	Max	Min	Min
A_0	49	3.25	55	2	40
A_1	43	2.5	51	3	40
A_2	41	3	49	5	45
A_3	43	2.75	48	2	50
A_4	49	3.25	55	3	48

A_5	44	2.9	45	4	49

For the value of A_0, obtained from the maximum or minimum value according to the type of criteria used. Furthermore, by using equations 2, 3 and 4 the normalized matrix is obtained, the results are shown in Table 4.

Table 5. Normalized Matrix (R_{ij})

Alternative	C_1	C_2	C_3	C_4	C_5
	Max	Max	Max	Min	Min
A_0	0.1822	0.1841	0.1815	0.2362	0.1873
A_1	0.1599	0.1416	0.1683	0.1575	0.1873
A_2	0.1524	0.1700	0.1617	0.0945	0.1665
A_3	0.1599	0.1558	0.1584	0.2362	0.1499
A_4	0.1822	0.1841	0.1815	0.1575	0.1561
A_5	0.1636	0.1643	0.1485	0.1181	0.1529

After calculating the normalized matrix, the next step is to determine the weighted normalized matrix. This step uses equation 5. The results of the weighted normalized matrix are obtained in table 5.

Table 6. The Weighted of Normalization Matrix (D)

Alternative	C_1	C_2	C_3	C_4	C_5
	Max	Max	Max	Min	Min
A_0	0.0832	0.0473	0.0285	0.0213	0.0075
A_1	0.0731	0.0364	0.0264	0.0142	0.0075
A_2	0.0697	0.0437	0.0254	0.0085	0.0067
A_3	0.0731	0.0400	0.0249	0.0213	0.0060
A_4	0.0832	0.0473	0.0285	0.0142	0.0062
A_5	0.0748	0.0422	0.0233	0.0106	0.0061

The next process determines the optimal function with equation 6. The results of the calculations obtained optimal function (Si) seen in table 6.

Table 7. The Weighted Normalization Matrix (D)

Alternative	S_i
A_0	0.1878
A_1	0.1575
A_2	0.1539
A_3	0.1652
A_4	0.1795
A_5	0.1570

The final step determines the highest-ranking level (Ui) using equation 7.

$S_1 = 0.1575 / 0.1878 = 0.839$
$S_2 = 0.1539 / 0.1878 = 0.819$
$S_3 = 0.1652 / 0.1878 = 0.880$
$S_4 = 0.1795 / 0.1878 = 0.956$
$S_5 = 0.1570 / 0.1878 = 0.836$

Table 8. The highest-ranking level (Ui) and alternative rank

Alternative	U_i	Rank
A_1	0.839	3
A_2	0.819	5
A_3	0.880	2
A_4	0.956	1
A_5	0.836	4

From the calculation using the ARAS method, it appears that $A_4 > A_3 > A_1 > A_5 > A_2$, so it can be concluded that A_4 (Kiki Prasetia) is the highest-ranked customer and deserves to be given credit.

4. Conclusion

From the research conducted it can be concluded that the determination of the criteria in selecting credit recipients greatly affects the calculation that occurs. The application of the ARAS method is quite easy to apply to computer-based information systems, making it easier for banks to determine credit recipients. Decision support systems can overcome the problem of receiving a credit to be more systematic and appropriate for customers or the public who really need credit. Computer-based systems can reduce fraud caused by credit selectors.

References

[1] S. Kusumadewi, S. Hartati, A. Harjoko, and Retantyo Wardoyo, "Fuzzy Multi Attribute Decision Making (FUZZY MADM)," *Ed. Pertama Cetakan Pertama. Graha Ilmu. Yogyakarta.*, 2006.

[2] G.-H. Tzeng and J.-J. Huang, *Multiple Attribute Decision Making Method And Applications.* CRC Press, 2011.

[3] D. Nofriansyah and S. Defit, *Multi Criteria Decision Making (MCDM) pada Sistem Pendukung Keputusan.* 2018.

[4] D. Nofriansyah, *Konsep Data Mining Vs Sistem Pendukung Keputusan.* 2015.

[5] D. Siregar *et al.*, "Multi-Attribute Decision Making with VIKOR Method for Any Purpose Decision," *J. Phys. Conf. Ser.*, vol. 1019, no. 1, 2018.

[6] A. Yanie *et al.*, "Web Based Application for Decision Support System with ELECTRE Method," *J. Phys. Conf. Ser.*, vol. 1028, no. 1, 2018.

[7] S. H. Sahir, R. Rosmawati, and R. Rahim, "Fuzzy model tahani as a decision support system for selection computer tablet," *Int. J. Eng. Technol.*, vol. 7, no. 2.9, pp. 61–65, 2018.

[8] M. Mesran *et al.*, "The VIKOR Method to Support the Effectiveness of Decisions in Determining Work Incentive Recipients," *J. Phys. Conf. Ser.*, vol. 1175, p. 012043, 2019.

[9] S. Dharma Hardi *et al.*, "Implementation of Computer Based Systems for Effective Deci-sions in Acceptance of Vikar," *Int. J. Eng. Technol.*, vol. 7, no. 3, pp. 101–104, 2018.

[10] Esra; and AyÇegül, "AIR CONDITIONER SELECTION PROBLEM WITH COPRAS AND ARAS METHODS," *Manas J. Soc. Stud.*, vol. 5, no. 2, 2016.

[11] E. K. Zavadskas and Z. Turskis, "A new additive ratio assessment (ARAS) method in multicriteria decision - making," vol. 8619, 2011.

Journal of Physics: Conference Series **1424** (2019) 012019 doi:10.1088/1742-6596/1424/1/012019

Implementation of Multi-Objective Optimazation on the Base of Ratio Analysis (MOORA) in Improving Support for Decision on Sales Location Determination

S Sutarno[1], M Mesran[2], S Supriyanto[3], Y Yuliana[1], A Dewi[1]

[1]Akademi Sekretari Manajemen Cendana, Medan, Indonesia
[2]STMIK Budi Darma, Medan, Indonesia
[3]Politeknik LP3I Medan, Medan, Indonesia

*mesran.skom.mkom@gmail.com

Abstract. The location of sales is one of the most important factors, in the success of selling goods. At present, there are many business locations, in the form of new buildings being made which can later be used as sales locations. But in the selection of the sales location, sometimes the business actors do not consider whether the location will be strategic or not. So that the price is relatively cheap, the location of the business will be purchased without considering the position of the strategic sales location or not. To avoid losses for business actors, the application of a decision support system is absolutely necessary. It aims to provide assistance for business actors to produce effective, appropriate and efficient decisions. Decision support systems can use certain methods. In this study, the application of the Multi-Objective Optimization On The Base Of Ratio Analysis (MOORA) method is a simple method in ranking alternative business locations, so it is expected that the use of decision support systems in the selection of sales locations will provide large benefits for business actors.

1. Introduction

The success in carrying out sales activities is inseparable from the selection of sales locations. The rapid economic growth at this time encourages business people to compete in the selection of places, as a sales location. Many locations are now used as places aiming at, that place can be used as a business location by investors. So that even in new areas many vendors erect buildings to be resold. Cheap building prices make businesses to buy the building, with the aim of becoming a business location, without thinking about whether the place is crowded, close to the market, close to residential housing. Of course, this will make a loss for businesses because the merchandise will not sell, so that it will cause losses because the merchandise is only as a display.

In the selection of sales locations, there are many factors that are used as criteria. The goal is that the selection of the sales location can provide benefits for the business carried out. This is a complex problem for business actors in considering the selection of the sale location. So that the investment spent is worth more return, then we need an appropriate solution, so that the decisions made can be effective and efficient. For that, we need a computer-based decision-support information system. This information system is known as a decision support system (DSS) [1]–[4].

DSS is a tool in the form of computer software that is intended for users or management in producing effective decisions. In the process carried out, the decision support system (DSS) uses methods in its resolution [5], [6]. Among them are decision support system methods, namely Simple Additive Weighting (SAW), ELECTRE, VIKOR, Preference Selection Index, ARAS [7]–[10]. Besides these methods, there are also Multi-Objective Optimization On The Base Of Ratio Analysis (MOORA) methods, which are easier to use in producing the best ranking for the selection of sales locations.

From this explanation, the problem that will be discussed in this research is about the selection of sales locations to be more effective, and appropriate by using the Multi-Objective Optimization On The Base Of Ratio Analysis (MOORA) method, so that the benefits for business actors will not suffer losses in doing investment in sales location.

2. Methodology

Some research that has been done related to site selection, was conducted by Syafrida Hafni Sahir. The application of the Preference Selection Index method in DSS is able to provide good results in determining the location of used laptop sales [11], determining the location of facilities by applying Promethee II [12], determining the location of bank branches by applying MOORA [13].

The Multi-Objective Optimization method on the Basis of Ratio Analysis (MOORA), is a method of multi-objective optimization. The MOORA method is also known as the multi-criteria or multi-optimization method for attributes. The work process of the MOORA method is to carry out processes simultaneously to optimize two or more conflicting criteria, attributes (goals) are subject to certain restrictions[14].

The MOORA method was first introduced by Brauers (2004). The application is made in solving various types of complex decision-making problems in the manufacturing environment. The completion of implementing MOORA uses several steps[15], [16], namely:

Step 1: Create a Decision Matrix.

$$x = \begin{bmatrix} x_{11} & x_{12} & \cdot & x_{1n} \\ x_{21} & x_{22} & \cdot & x_{2n} \\ \cdot & \cdot & \cdot & \cdot \\ x_{m1} & x_{m2} & \cdot & x_{mn} \end{bmatrix} \quad (1)$$

Step 2: Normalize the Decision Matrix

$$x_{ij}^{*} = x_{ij} / \sqrt{\left[\sum_{i=1}^{m} x_{ij}^{2}\right]} \; (j = 1,2,\ldots,n) \quad (2)$$

Step 3: Optimize attributes.

$$y_i = \sum_{j=1}^{g} x_{ij}^{*} - \sum_{j=g+1}^{n} x_{ij}^{*} \quad (3)$$

If the results are more optimal, weight can be used to get the value of Yi, so equation 3 can be shaped as follows.

$$y_i = \sum_{j=1}^{g} w_j x_{ij}^{*} - \sum_{j=g+1}^{n} w_j x_{ij}^{*} \; (j = 1,2,\ldots,n) \quad (4)$$

W_j is the attribute weight j^{th}. The value of Y^i can be either a positive or negative value, according to which type of attribute affects the decision matrix. The value obtained at Y^i is the final preference. The best alternative has the highest Y^i value.

3. Result and Discussion

In the decision making process using computers as data processing aids, using the MOORA method in its completion. It is intended that the results achieved are more effective and objective for decision-makers. The implementation of a Decision Support System requires data, namely criteria, weights, and alternative business locations. From these data, it can be obtained the decisions needed for business owners. The following criteria are used.

Table 1. Criteria and Weighted

Criteria	Description	Weighted	Type
C^1	Land area	0,30	Benefit
C^2	Many visitors	0,15	Benefit
C^3	Number of Competitors	0.15	Benefit
C^4	Distance from Nearest City	0,10	Cost
C^5	Rental price	0,30	Cost

The following are alternative data and the value of the compatibility rating for alternatives and criteria.

Table 2. Alternative match ratings and criteria

Alternative	C^1	C^2	C^3	C^4	C^5
Suzuya Katamso (A_1)	20 m^2	4000 People/Week	2	3 Km	15 Million
Maju Bersama Simp SM Raja (A_2)	25 m^2	3500 People/Week	2	5 Km	15 Million
Medan Fair Plaza (A_3)	15 m^2	7000 People/Week	4	0.5 Km	21 Million
Suzuya Marelan (A_4)	19 m^2	4000 People/Week	2	0.6 Km	10 Million
Suzuya Tj Morawa (A_5)	22 m^2	2000 People/Week	1	3 Km	10 Million
Ramayana SM Raja (A_6)	24 m^2	6000 People/Week	2	4 Km	18 Million

Table 2 above, the rating data is simplified, as shown in table 3.

Table 3. Simplified Match Rating

Alternative	C^1	C^2	C^3	C^4	C^5
A_1	20	4000	2	3	15
A_2	25	3500	2	5	15
A_3	15	7000	4	0.5	21
A_4	19	4000	2	0.6	10
A_5	22	2000	1	3	10
A_6	24	6000	2	4	18

Next, the calculation process is carried out by applying the MOORA method. The initial step provides the decision matrix (Xij) obtained from table 3, as follows:

$$X_{ij} \begin{vmatrix} 20 & 4000 & 2 & 3 & 15 \\ 25 & 3500 & 2 & 5 & 15 \\ 15 & 7000 & 4 & 0.5 & 21 \\ 19 & 4000 & 2 & 0.6 & 10 \\ 22 & 2000 & 1 & 3 & 10 \\ 24 & 6000 & 2 & 4 & 18 \end{vmatrix}$$

After getting the X_{ij} matrix, then calculate the X_{ij}^* normalized matrix using equation 2. The calculation results for the normalized matrix (X_{ij}^*) are obtained namely:

$$Xij^* \begin{vmatrix} 0.3870 & 0.3465 & 0.3482 & 0.3886 & 0.3988 \\ 0.4837 & 0.3032 & 0.3482 & 0.6476 & 0.3988 \\ 0.2902 & 0.6064 & 0.6963 & 0.0648 & 0.5583 \\ 0.4257 & 0.1733 & 0.1741 & 0.3886 & 0.2658 \\ 0.4644 & 0.5198 & 0.3482 & 0.5181 & 0.4785 \end{vmatrix}$$

$$\begin{vmatrix} 0.3870 & 0.3465 & 0.3482 & 0.3886 & 0.3988 \end{vmatrix}$$

The final step is to optimize the X_{ij}^* matrix by using equation 4. This aims to make the final result better by including the processing of the weights of each criterion. The results of the optimization of the attributes can be seen by the value of Yi in table 4.

Table 4. Y_i List

Alternative	Maximum ($C^1+C^2+C^3$)	Minimum (C^4+C^5)	Y_i (Max – Min)
A_1	0.324	0.236	0.088
A_2	0.341	0.314	0.027
A_3	0.478	0.187	0.291
A_4	0.319	0.103	0.216
A_5	0.232	0.196	0.036
A_6	0.400	0.299	0.101

From the results of the calculation of the value of Yi, ranking the value obtained. Table 5 ranks the alternatives with the best to the worst scores.

Table 5. Rank Alternative

Alternative	Y_i	Rank
A_3	0.291	1
A_4	0.216	2
A_6	0.101	3
A_1	0.088	4
A_5	0.036	5
A_2	0.027	6

The results are shown in table 5, it can be seen that the value of $A_3>A_4>A_6>A_1>A_5>A_2$, so that in processing the results that Medan Fair Plaza (A_3) is the best selling location compared to other sales alternatives.

4. Conclusion

From the discussion above, the application of the MOORA method in the selection of sales locations is quite easy, the decision-maker only includes the criteria and weights used as priorities for the criteria made. The application of computers, in this case, can provide more effective and objective decisions. The application of the MOORA method is quite simple and effective in producing the decisions desired by business owners.

References

[1] D. Nofriansyah and S. Defit, *Multi Criteria Decision Making (MCDM) pada Sistem Pendukung Keputusan*. 2018.
[2] D. Nofriansyah, *Konsep Data Mining Vs Sistem Pendukung Keputusan*. 2015.
[3] G.-H. Tzeng and J.-J. Huang, *Multiple Attribute Decision Making Method And Applications*. CRC Press, 2011.
[4] E. Turban, J. E. Aronson, and T. Liang, "Decision Support Systems and Intelligent Systems."
[5] Kusrini, *Konsep dan Aplikasi Sistem Pendukung Keputusan*. Yogyakarta: Andi, 2007.
[6] S. Kusumadewi, S. Hartati, A. Harjoko, and Retantyo Wardoyo, "Fuzzy Multi Attribute Decision Making (FUZZY MADM)," *Ed. Pertama Cetakan Pertama. Graha Ilmu. Yogyakarta.*, 2006.

[7] A. Yanie *et al.*, "Web Based Application for Decision Support System with ELECTRE Method," *J. Phys. Conf. Ser.*, vol. 1028, no. 1, 2018.
[8] S. H. Sahir, R. Rosmawati, and K. Minan, "Simple Additive Weighting Method to Determining Employee Salary Increase Rate," *Int. J. Sci. Res. Sci. Technol.*, vol. 3, no. 8, pp. 42–48, 2017.
[9] D. Siregar *et al.*, "Multi-Attribute Decision Making with VIKOR Method for Any Purpose Decision," in *Journal of Physics: Conference Series*, 2018, vol. 1019, no. 1.
[10] M. Mesran *et al.*, "The VIKOR Method to Support the Effectiveness of Decisions in Determining Work Incentive Recipients," *J. Phys. Conf. Ser.*, vol. 1175, p. 012043, 2019.
[11] S. H. Sahir *et al.*, "The Preference Selection Index method in determining the location of used laptop marketing," *Int. J. Eng. Technol.*, vol. 7, no. 3.4 Special Issue 4, 2018.
[12] A. V. Manikrao and C. Shankar, "Facility Location Selection using PROMETHEE II Method," *Int. Conf. Ind. Eng. Oper. Manag. Dhaka*, pp. 59–64, 2010.
[13] P. Karande and S. Chakraborty, "Application of multi-objective optimization on the basis of ratio analysis (MOORA) method for materials selection," *Mater. Des.*, vol. 37, no. 2, pp. 317–324, 2012.
[14] Mesran, R. K. Hondro, M. Syahrizal, A. P. U. Siahaan, R. Rahim, and Suginam, "Student Admission Assessment using Multi-Objective Optimization on the Basis of Ratio Analysis (MOORA)," *J. Online Jar. COT POLIPT*, vol. 10, no. 7, pp. 1–6, 2017.
[15] S. Chakraborty, "Applications of the MOORA method for decision making in manufacturing environment," *Int. J. Adv. Manuf. Technol.*, vol. 54, no. 9–12, pp. 1155–1166, 2011.
[16] Onur Önay and B. F. Yıldırım, "Evaluation of NUTS Level 2 Regions of Turkey by TOPSIS , MOORA and VIKOR 1," *Int. J. Humanit. Soc. Sci.*, vol. 6, no. 1, pp. 212–221, 2016.

Intelligent LPG Gas Leak Detection Tool with SMS Notification

Muhammad Siddik Hasibuan[1*], Syafriwel[2], Iswandi Idris[3]

[1,2]Computer Technology, Politeknik LP3I Medan, Indonesia
[3]Industrial Engineering, Politeknik LP3I Medan, Indonesia

*mhdsiddikhasibuan@gmail.com

Abstract. Gas is a molecule that is not bound, formless, invisible and can turn into search or solid at certain temperature pressures. LPG gas is one of the needs for industry and household needs, namely for cooking. LPG gas in Indonesia is the cause of many fires, a factor that often causes LPG gas fires is a damaged gas regulator. Therefore prevention and security are needed to minimize fires. Seeing this and given the technological developments, an intelligent Arduino-based device was created that was able to overcome this problem. This tool is equipped with MQ-2, SIM800L, and buzzer gas sensors. In an embedded device the system that can convert input data received from the sensor Mq-2 sensor works to detect propane and butane gas, then the system will send the actual data in the form of short messages (SMS) to the mobile number that has been registered into the system. Besides being able to send SMS the system also emits a sound that is generated from the buzzer

1. Introduction

The use of gas fuel in Indonesia is now a major requirement for household appliances. LPG (Liquefied Petroleum Gas) / LPG is used as fuel for cooking, besides that the use of LPG in the present is very economical, especially for households, not only cheap but also easy to use and easy to use, easy to use, LPG gas for the Indonesian people. The LPG gas usage program is the government's movement to divert subsidies from Kerosene to Gas because of cheaper gas prices. Using LPG gas in their daily lives, especially in terms of cooking, can make cooking products healthier, because the fire produced from LPG gas itself is a clean fire, so it does not need to produce as needed using resistant oil [1]. For this reason, there are things we need to question when using LPG gas in terms of safety. This results in a very volatile gas that is very arising, which is very likely to be the cause of a fire. Normal leakage gas may be easily removed by smell, but if the leaky gas seeps under the carpet or electrical installation it will be very difficult to detect by the sense of smell of humans. In addition, LPG gas is also very sensitive and flammable, so that it is easy to occur which is not indisputable by human olfactory, then the fire source is lit, then this will be the cause of increasing hazards obtained [1] [2]. The number of deaths from the explosion of gas cylinders has increased in recent years. The cause of the explosion was because the old valve, the regulator was damaged and also the biggest reason for causing damage due to gas which was a challenge in the gas cylinder. For this reason, there is a serious problem with this gas leak [3]. The biggest gas accident in the world is the Bhopal gas tragedy in India in 1984 which killed thousands of lives [4]. Due to the very sensitive and flammable nature of the gas, special attention is needed for this type of fuel. Therefore we need a system that can help save the gas when sending information immediately [5]. Therefore an LPG Gas Leak Detection Smart Tool was made with SMS Notification, which is equipped with gas sensors, buzzers and GSM modules that will help prevent gas leaks and provide sirens then send a notification message to the owner so that the dangerous gas leaks can be overcome and anticipated as early as possible. In making this tool is needed, among others.

1. Sistem Embedded

Embedded systems or embedded systems are computer systems designed to carry out certain tasks that are integrated with hardware [6]. One platform that supports embedded systems is Arduino. Arduino was created to simplify the circuit process and microcontroller programming so that it becomes easier to learn. Arduino is compiled on an Arduino IDE software. This software is the one that proclaims machine language to be the language of logic to do the tasks we command. In this study, the Arduino that we use is a type of Arduino Uno R3. With Arduino, researchers can assemble devices that have systems that can detect gas.

2. Gas Sensor MQ-2

he uses of the Mq-2 sensor is for the detection of LPG gas, LPG gas consists of propane gas and butane. The sensitive material used in the MQ-2 gas sensor is a tin oxide (SnO2). If a gas leak is detected, the sensor conductivity rises with the detection range of the MQ-2 sensor is 200 - 5000 ppm [2] [4]. This sensor is used to detect gas, i-butane, propane, methane, alcohol, hydrogen, and smoke. In this system, sensors are used to detect LPG gas.

3. GSM sim800l Modul

The GSM module with type SIM800L is a module for sending actual data from the system to the SIM card with a GSM network. This module can be used in conjunction with Arduino, which makes it easier for users to create programs related to sending and receiving SMS messages.

4. Buzzer

In addition to sending data in the form of short messages or SMS, the tool will also issue a sound generated by the Buzzer. The buzzer is a device that can emit a sound when active. In this series, the buzzer is used as a marker indicating that will sound when the gas leak detector detects a gas leak.

2. Methodology

In this study it can be described how the tool can work to capture the gas received by the sensor, then the sensor is processed in an Arduino program that can display data in the LCD, emit sound and send a short message to the registered mobile number. The scheme can be seen in the block diagram in figure 1.

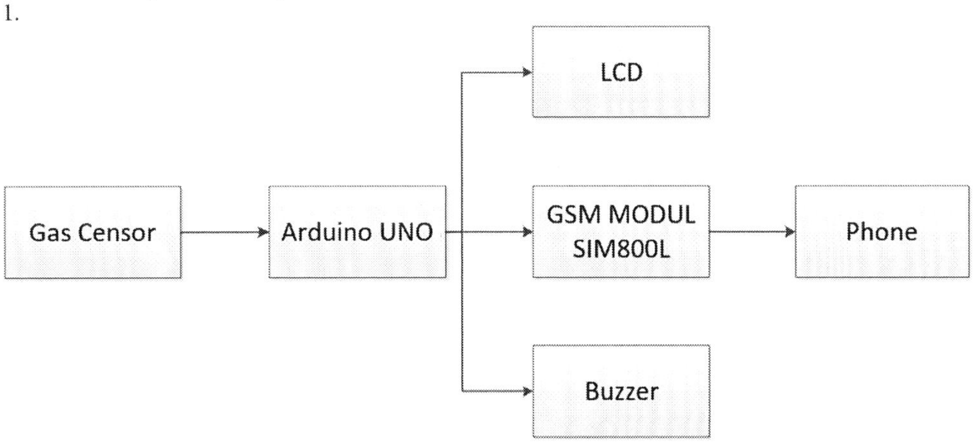

Figure 1. Block Diagram

From the picture above it can be seen that this tool has an input device (input device) consisting of a gas sensor and an output device (output device) which consists of an LCD, GSM Module, and buzzer. The way this tool works is when it is turned on, the sensor will work to transfer LPG gas and send data according to the LPG gas level it detects. The higher the LPG gas is detected, the higher the voltage released. When the sensor output is moved The presence of gas, then Arduino will activate, and activate the buzzer and display the writing on the LCD stating the gas is high (high), which means there has been a gas leak, then the GSM SIM800L module will send a notification message to the handphone number specified in the program. However, if the sensor does not detect a leak, the sensor will not remove the output, and the sensor will continue to work until it is proven that there is an LPG gas leak. The system design in this study is described in the form of a flowchart to facilitate the reading and understanding of the system that will be made in this study. When the program is run the system will immediately detect LPG gas detected by the sensor. Then the Arduino microcontroller will read LPG gas through an LPG gas sensor. If it detects a gas leak, the red LED will light up, the buzzer will activate, then the system will send a notification message stating that there has been an LPG gas leak. If no LPG gas leak is detected, the system will continue to detect the gas level through the LPG gas sensor until it detects an LPG gas leak. System flow can be seen in Figure 2.

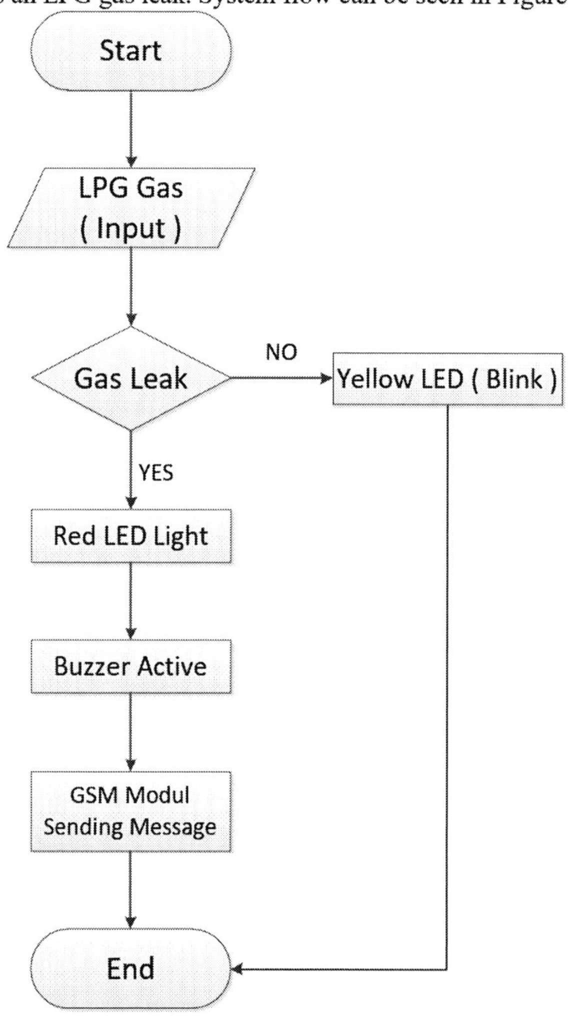

Figure 2. Flowchart Program

3. Result and Discussion

After conducting research and analyzing to produce, make and test the cases submitted in this final project, the results of this case study on the system and this tool can help humans to increase proven LPG gas reserves and provide assistance to their owners that they have LPG gas leak in the form of an alarm is also sent a message in the form of an SMS to the owner. While the circuit and cabling of this tool will open a box that will be displayed as follows. This tool is equipped with a 16x2 LCD as a display to display the work process of the tool itself. Then the buzzer is active as an alarm marker in the event of a gas leak. And also equipped with the gas sensor itself, in this case, the MQ-2 gas sensor. This tool is also equipped with 3 complete LED lights - also a green power light that indicates the device is on, a yellow LED that indicates the active device is detecting the existing LPG gas levels, also a red LED that indicates the danger that gas leak LPG. Then this tool is also equipped with a GSM module, SIM800L, which functions to send SMS messages in the event of an LPG gas leak. The form and format of sending SMS messages from this tool can be seen in Figure 3.

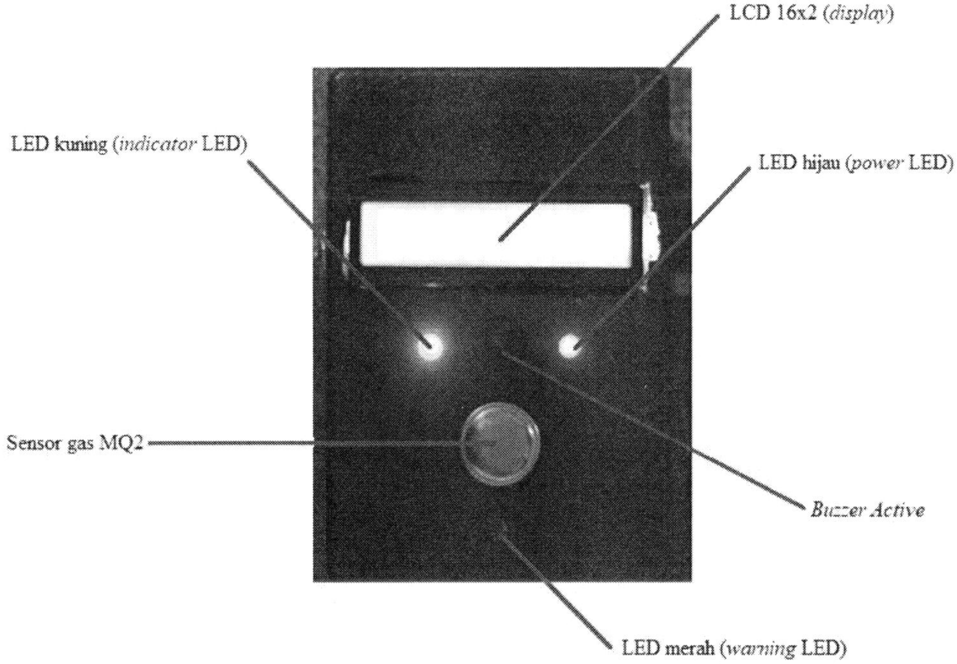

Figure 3. visual tool

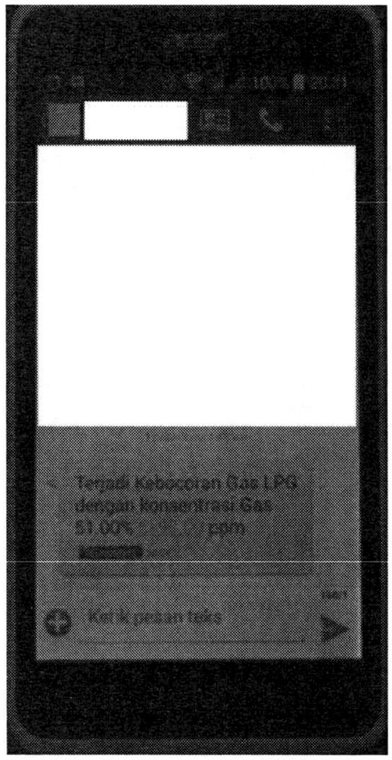

Figure 4. SMS Send Phone

The way the system works and this tool is, first when the system is turned on and this tool will immediately detect the gas content, using a sensor that is designed to be able to detect LPG gas, namely the MQ-2 gas sensor. Then each LPG gas level detected by the sensor is directly processed or converted into an analog signal. Then the analog signal will be sent directly by the MQ-2 sensor to Arduino. Because the analog signal to be sent is a number of LPG gas levels detected by the MQ-2 sensor. Then this analog signal will later become the Arduino working parameter. Does the gas level exceed the limit or not.

If the level of LPG received by Arduino exceeds the predetermined limit of 5000ppm (part per million), then the Arduino will directly control the other connected ones, namely relay, buzzer, and SIM800L, by sending commands to the relay to turn on the LED (Light Emitting Diode) red which indicates danger. Then Arduino sends commands to control the buzzer to be active, to give an alarm signal as a marker in the form of sirens that there has been a leak of LPG gas that has exceeded a predetermined limit. Finally, Arduino will instruct the SIM800L module to send an SMS message to the owner, to provide information about leaked LPG gas.

However, if the gas content in the form of an analog signal received by the Arduino from the MQ-2 gas sensor does not exceed the limit of 5000ppm, then Arduino will not control other components. Or in other words, the system and this tool will work normally as when the initial system was turned on, which is detecting the existing level of LPG gas.

Why in this case did the author make the gas parameter limit leak at the level of 5000 ppm? Because based on the MQ-2 gas sensor technical data, the range that can be measured by the MQ-2 gas sensor against LPG type gas ranges from 200ppm to 5000ppm. Therefore, in this case, the authors set a maximum limit of the level of LPG gas leakage at 5000ppm.

Table 1. Sensitivity sensor gas MQ-2

<table>
<tr><th>Symbol</th><th>Parameter name</th><th>Technical parameter</th><th>Remarks</th></tr>
<tr><td>Rs</td><td>Sensing Resistance</td><td>3KΩ-30KΩ
(1000ppm iso-butane)</td><td rowspan="4">Detecting concentration scope:
200ppm-5000ppm LPG and propane
300ppm-5000ppm butane
5000ppm-20000ppm methane
300ppm-5000ppm H_2
100ppm-2000ppm Alcohol</td></tr>
<tr><td>α
(3000/1000)
isobutane</td><td>Concentration Slope rate</td><td>≤0.6</td></tr>
<tr><td>Standard Detecting Condition</td><td colspan="2">Temp: 20℃ ±2℃ Vc:5V±0.1
Humidity: 65%±5% Vh: 5V±0.1</td></tr>
<tr><td>Preheat time</td><td colspan="2">Over 24 hour</td></tr>
</table>

In this system and tool to enter the cellphone number for sending SMS messages, the author makes it by entering the destination number directly on the syntax code created in the Arduino IDE. In other words, the destination number cannot be replaced automatically except by directly replacing the code syntax that has been created (loading code).

In addition, the cellphone number that is the destination for sending SMS messages in the event of an LPG gas leak can only be entered by one cellphone number. In other words, the system and this tool will send an SMS message that the LPG gas leak notification is limited to only one destination mobile number

In testing the sensor is done in two stages, namely, the testing phase in the room (closed) and the outdoor testing stage (open), to see the performance of the gas sensor MQ-2 itself. Given that the MQ-2 gas sensor is the main indicator of the work of this system and tool. The results of the testing of the MQ-2 gas sensor are as follows.

1. Indoor testing

Table 2. Indoor testing

No	Distance	Gas Level Detected	Percentage of Gas Detected
1	1 – 1,5 Meters	200 – 457 ppm	2 – 4 %
2	0,5 – 1 Meters	237 – 1027 ppm	2 – 10 %
3	0 – 0,5 Meters	459 – 7102 ppm	4 – 70 %

2. Testing in an open room)

The results of this test are that the sensor is unable to detect the presence of leaky gas levels unless the gas leak point is attached to the MQ-2 gas sensor, the sensor will detect the presence of gas with a level of 200-4000 ppm, or in percentage 2 - 40%.

This literature review aims to identify and analyze the trends, datasets, methods and frameworks used in the topic of attribute independence assumption assumptions on NB between 2010 and 2018. Based on the inclusion and exclusion criteria designed, it shows 71 study studies of attribute independence assumptions on the published NB between January 2010 and December 2018 are investigated in this literature review have been conducted as a review of systematic literature. A systematic literature review is defined as the process of identifying, assessing, and interpreting all available research evidence in order to provide answers to specific research questions.

The results of this study identified three of the most commonly used and influential framework methods in the topic of attribute independence on the NB. They are Menzies et al. Framework, Lessmann et al. Framework, and Song et al. Framework. They are Langley et al [20], Friedman et al [2], and Wu et al [28].

4. Conclusion

Based on the results of research, design, and testing of systems and tools, the authors obtain conclusions as follows:

1. This system and tool works as expected, and is able to detect LPG gas leaks that occur within the detection radius of the MQ-2 gas sensor which is attached to the system and the device, which is between 50 cm - 80 cm and placed in a relatively closed
2. With this system and tool, the people or people who use it feel safe and comfortable. In other words, do not worry about the danger if there is an LPG gas leak that is unknown to the owner. Because the system and tool are able to work in accordance with the second objective of the research that the author wrote in chapter 1, this system and tool can send information to the owner if an unknown LPG gas leak occurs through an SMS message sent to the owner's handphone.

References

1] J. Christian and N. Komar, "Prototipe Sistem Pendeteksi Kebocoran Gas LPG Menggunakan Sensor Gas MQ2, Board Arduino Duemilanove, Buzzer, dan Arduino GSM Shield pada PT. Alfa Retailindo (Carrefour Pasar Minggu)," *Jurnal TICOM,* pp. 1-7, 2013.

2] T. H. Mujawar, V. D. Bachuwar, M. S. Kasbe, A. D. Shaligram and L. P. Deshmukh, "Development of wireless sensor network system for LPG gas leakage detection system," *International Journal of Scientific & Engineering Research,* vol. 6, pp. 1-6, 2015.

3] M. F. Putra, A. H. Kridalaksana and Z. Arifin, "RANCANG BANGUN ALAT PENDETEKSI KEBOCORAN GAS LPG DENGAN SENSOR MQ-6 BERBASIS MIKROKONTROLER MELALUI SMARTPHONE ANDROID SEBAGAI MEDIA INFORMASI," *Jurnal Informatika Mulawarman,* vol. 12, pp. 1-6, 2017.

4] H. Santoso, Arduino untuk Pemula, Trenggalek: elangsakti, 2015.

5] Widyanto and D. Erlansyah, "Rancang Bangun Alat Deteksi Kebocoran Tabung Gas Elpiji Berbasis Arduino," *Seminar Nasional Teknologi Informasi & Komunikasi Terapan (SEMANTIK 2014),* vol. 4, 2014.

6] L. I. Ramadhan, D. Syauqy and B. H. Prasetio, "Sistem Pendeteksi Kebocoran Gas LPG Menggunakan Metode Fuzzy yang Diimplementasikan dengan Real Time Operating System (RTOS)," *J-ptiik.ub.ac.id,* vol. 1, 2015.

7] W. Jatmiko, P. Mursanto, G. Jati and D. M. Purnomo, "RTOS Teori dan Aplikasi," *Fakultas Ilmu komputer Universitas Indonesia,* vol. 12, 2015.

Journal of Physics: Conference Series **1424** (2019) 012021 doi:10.1088/1742-6596/1424/1/012021

IoT Based Smart Parking System

Mohd Mustari Syafiq Ismail[1], Muzammil Jusoh[1], Thennarasan Sabapathy[1], Mohamed Nasrun Osman[1], Hasliza Abdul Rahim[1], Mohd Najib Mohd Yasin[2], Ainur Fasihah Mohd Fazilah[1]

[1]Bioelectromagnetics Research Group (BioEM), School of Computer and Communication Engineering, Universiti Malaysia Perlis (UniMAP), Pauh Putra, 02600 Arau, Perlis, Malaysia
[2]Bioelectromagnetics Research Group (BioEM), School of Microelectronic Engineering, Universiti Malaysia Perlis (UniMAP), Pauh Putra, 02600 Arau, Perlis, Malaysia

muzammil@unimap.edu.my

Abstract. Difficulty to find parking space is major problem for users that want to park car especially at crowded spot. The old technique to find parking space is manual by finds free space area by self. This makes the users take more time and energy that end up lead to traffic jam. Due to these problems, based smart parking system is developed to ease user to find parking space. This method is developed to approach the user to using application to review available parking space and pay parking through mobile phone application. The requirement for this project is a Raspberry Pi 3, Python IDE for software, an ultrasonic sensor and a Pi Camera. There are two objectives for this project. Firstly, to develop monitoring system that detect presence vehicle using ultrasonic sensor and camera artificial intelligence. The second objective is to propose the parking fee payment can be made through mobile applications via PayPal. The project of the based smart parking system is work when the users reserve the parking and made a payment via PayPal through mobile application. While ultrasonic sensors and camera detect the present of the vehicle.

1. Introduction

Ongoing development in the economy and because of the accessibility of low value cars in the market, even poor or average class individual can buy a car. This may be a good thing because now people can easily travel more comfortably to any destination without need to aware of weather. Besides a good thing, yet the result of this maybe create concern. Some of the concerns are overwhelming congested driving conditions, less accessibility of street and place to drive the cars.

One of the critical concern is that issue of parking those vehicles. The increasing personal vehicles than public transport make parking slot may not enough to support all those vehicles. Even when there have enough parking slots, there still take a long time to find available slot because the people do not know the exact location of the available parking slot without round it. Consequent of these will increase fuel consumption and not environmentally friendly. It is very difficult and frustrating as well to find parking space in most metropolitan areas, especially during the rush hour such as mall, hospital and so on.

IoT based smart parking system in paper [1] ultrasonic sensor had been used to detect the presence of the vehicles. The sensor is attached to the microcontroller. The microcontroller will process the data received and send it to cloud via Ethernet shield. Cloud can be access using desktop or application to reserve the parking space.

In paper [2], the authors propose car parking and reservation system. The system would find free parking slot to customer that enter the parking lot. Image detection camera is used to detect car number plate to verify its cars owner and find the nearest parking slot. Image processing is being used by camera and the system also provided reservation service for parking lot and the payment can be done through payment wallet.

Parking is easier by booking online through mobile application or web portal as state in paper [3]. This paper used RFID reader to detect car plate number, owner name and it also can check theft list from number plate given by police. If the plate number is matched with the list given, a SMS will be sent to the police. Thus system also provide offline and online payment depend on customer needs.

Intelligence parking IoT application using IR sensor and crystal oscillator in paper [4] as an input while Wi-Fi module and LED indicator as an output. IR obstacles detector sensor fed as input to the microcontroller and microcontroller processed the data and transmit it over the Wi-FI using mobile phone. The availability of parking slot will be displayed on webpages through mobile phone.

Android application for smart parking booking system in paper [5] make customer book or reserved parking easily. The booked can be done by select space and time. Customer also can cancel their booking space anytime. While in paper [6] Raspberry Pi was used as microcontroller and connect the IR sensor as input. The IR sensor will detect the present of vehicles and send data to Raspberry Pi. Green LED indicator will light up when the parking available and red LED will light up when the sensor detects the presence of vehicles. Android application is used to update the status of the parking space in real time.

2. Methodology

A. Proposed System Block Diagram

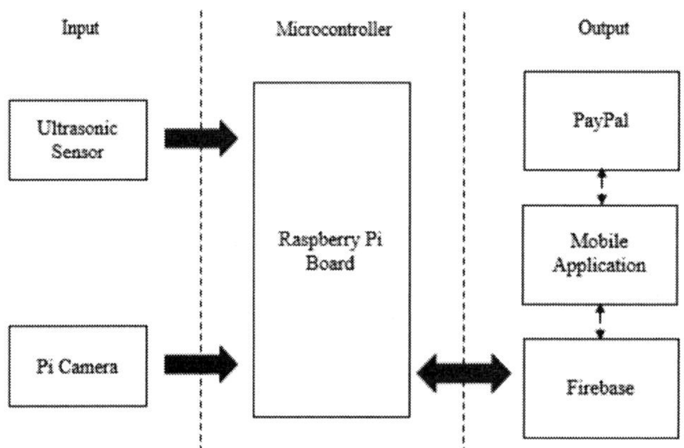

Figure 1. System Block Diagram

Figure 1 shows the system block diagram of proposed work it major constituent of it are:
Ultrasonic sensor and Pi Camera: Act as an input to collect the data from it reading and viewing sensor.
Raspberry Pi: The input from parking lot sensor is given to microcontroller system. Upon any changes in input, microcontroller will update the information collected from sensor to Firebase server.
Firebase: Information regarding registered users and status of the parking slot is maintained in firebase. The information regarding availability of parking lot is shared among all users who use mobile application. Firebase is updated from two terminals. One is from sensors and microcontroller. Another one is from user.

Mobile Application: An android application running on user and administrator mobile device for parking monitoring. User can check an empty parking slot and book it before making a payment.
PayPal: Medium for user to make a payment in the mobile application.

Advantages considered while selection of Ultrasonic sensor for vehicle detection are

- Low voltage and power consumption.
- High accuracy that can detect an object in the range of 25cm until 4.5m.
- Ultrasonic sensor can work in critical condition such as rain.

B. System Flow Chart

Figure 2 depicts developed mobile application flow chart. It starts when the user enters the system by looking at the screen whether the parking is available or not. If yes, the user needs to arrive at available parking space. When the user already arrives, the user then can make payment of what parking space the user stays now. Then, the user can add extra time when park time about finish.

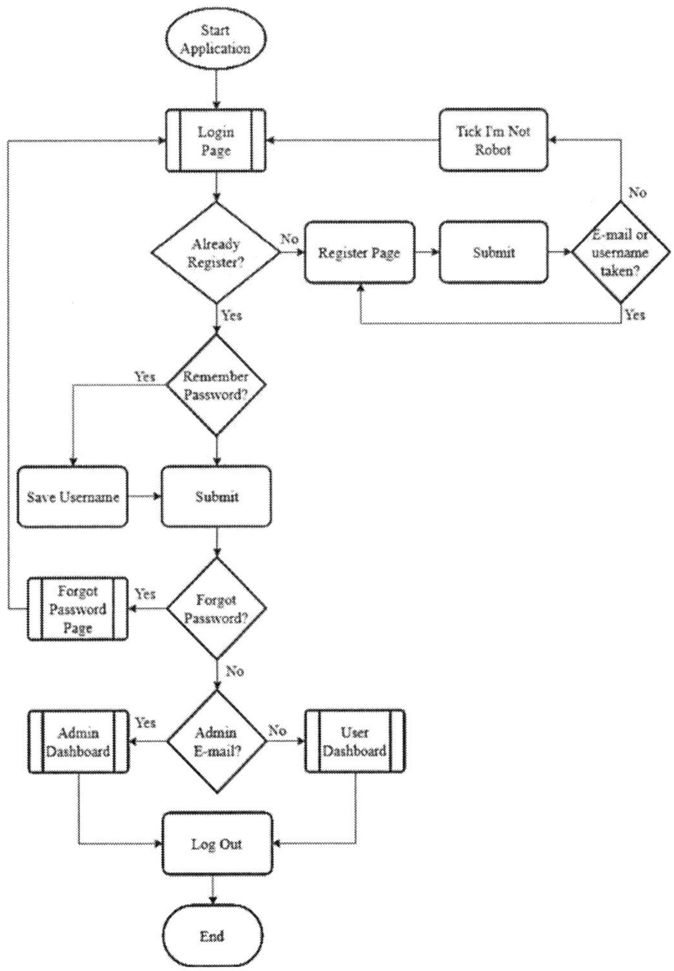

Figure 2. Mobile Application Flowchart

Firstly, user start the application and will directly to login page. For the first time user, they can proceed to register page and make sure e-mail and username is not taken. After successfully register the application, the user can now login to the application. When user login the user e-mail will be check by

Journal of Physics: Conference Series **1424** (2019) 012021 doi:10.1088/1742-6596/1424/1/012021

the system. If the e-mail entered is same with admin e-mail that being set in android studio, the user will be directly go to admin dashboard and if the e-mail is not same to admin e-mail, the user will enter to normal user dashboard. Lastly, admin or user can log out from the mobile application if they had finished their activities.

There are 10 menus that can be view in User Dashboard as listed below:

- Emergency: If there have an emergency occur in parking slot
- Message: User can check message from admin
- Map/Report: To view the parking and report any illegal parking.
- Profile: View user own profile
- Reserve: To reserve the parking with payment.
- Help: For more information regarding the parking.
- History: To see user past booking.
- Status: To view the status of booking.
- Parking View: To view the number of spot and occupied parking
- Log out: For user to log out from the application.

While for Admin Dashboard will have 7 menus that can be used which are:

- Emergency: To view any emergency being reported by user.
- Message: Compose a message to all user of this mobile application.
- Map/Report: To view illegal parking or reserve a spot.
- Price Setting: To change price per hour for the parking.
- Review: To view a user review.
- Parking View: To view number of spot and occupied parking
- Log Out: To end the session.

3. Mobile Application

In this section the working of "iParking" mobile application with the help of screen shots is explained in detail.

User Login: Figure 3 shows layout facilitates for existing user to login. During login user or admin need to write their preregistered username and password. For new user, they can register by tap on "Get registered now!" and the apps will bring the new user to layout as shown in Figure 4. New user need to provide their first name, last name, email, plate number, username and password and need to tick the agree check box before press submit.

Menu Dashboard: After successful login, user or admin will directly enter to dashboard menu as shown in Figure 5 and Figure 6. User or admin can choose which menu depends on they need. The administrator dashboard is different with user because admin can chance the parking price per hour on Price Menu as mention in Section 3 (B).

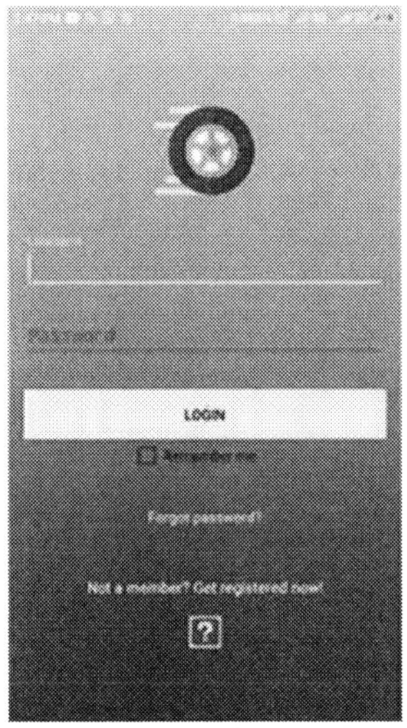

Figure 3. Login window

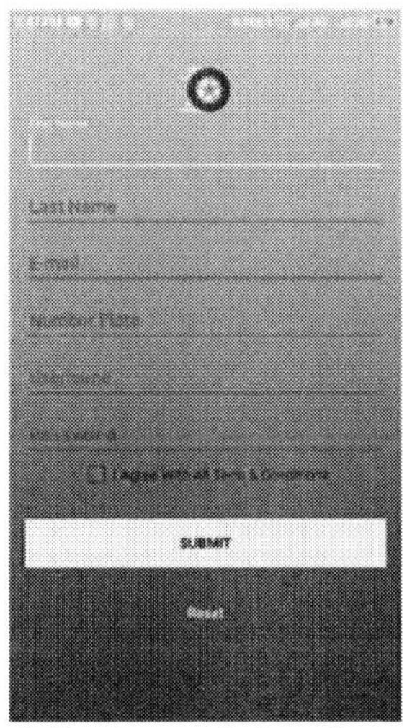

Figure 4. Register window for new user

Figure 5. Menu dashboard for user

Figure 6. Menu dashboard for administrator

4. Result and Discussion

1. Hardware Prototype

Figure 7 represent the prototype of the propose system. The ultrasonic sensor was used to detect the vehicles and gives the information about the vehicle in mobile application. The circuit had been design to give the response of the sensors for several of object within range of 30cm. The output from ultrasonic sensor will be sent to firebase Real-Time databased to specify the illegal parking whether it true or false. Led in prototype indicates the present of vehicle. Red LED will turn up when ultrasonic sensor detects the vehicle while green LED turns up when the sensor not detecting the vehicle.

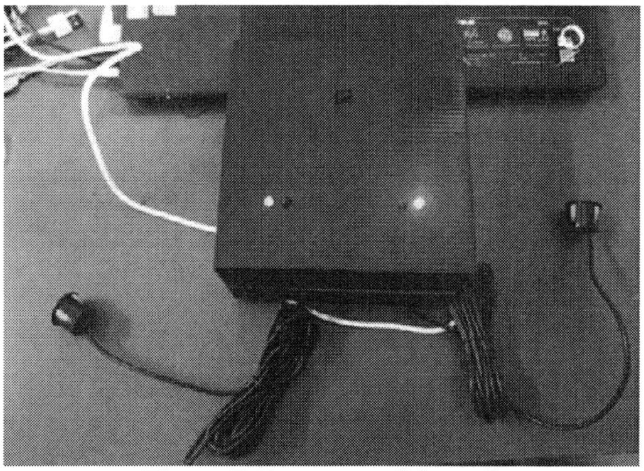
Figure 7. Prototype of the project

Picamera with OpenCV implementation as shown in Figure 8 will detect the vehicle in drawing line and will send the data to Firebase Real-Time database. Drawing line will be in green colour if there is no vehicle detect and will be count as free spot while red colour will be count as occupied when there is vehicle detect on drawing line.

Figure 8. Picamera detect the vehicle

2. Mobile Application Output Result

When the user uses the application, every chance will be sent to Firebase Real-Time databased. Figure 9 show the application retrieve the data from database. When the user reserve parking slot, the parking slot will change colour from green to yellow on mobile application. While the user report the parking slot, that parking slot will change the colour from green to red. This situation can be seen as in Figure 10.

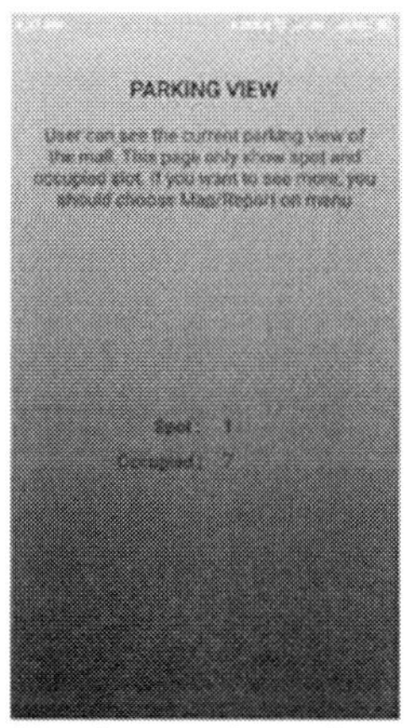

Figure 9. Application retrieve data from Firebase

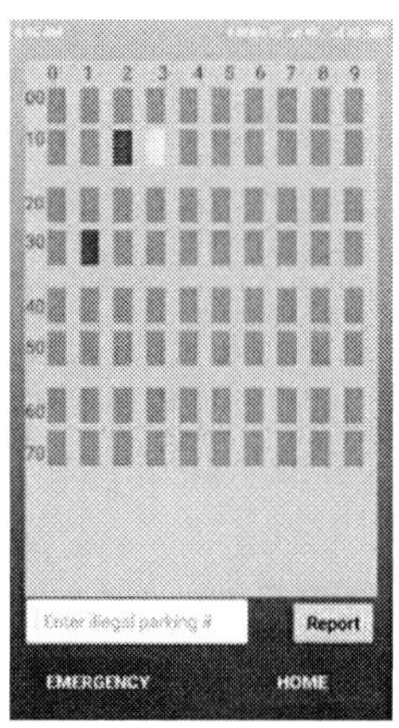

Figure 10. View of parking slot in mobile application

3. Firebase Output Result

User and administrator activity on mobile application will be record in Real-Time database. The database that been record are user registration info, parking lot reservation, administrator messages, emergency history and user comments as shown in Figure 13. While in Figure 11 and Figure 12 shows the Real-Time database data change by ultrasonic sensor and data recorded send by Picamera.

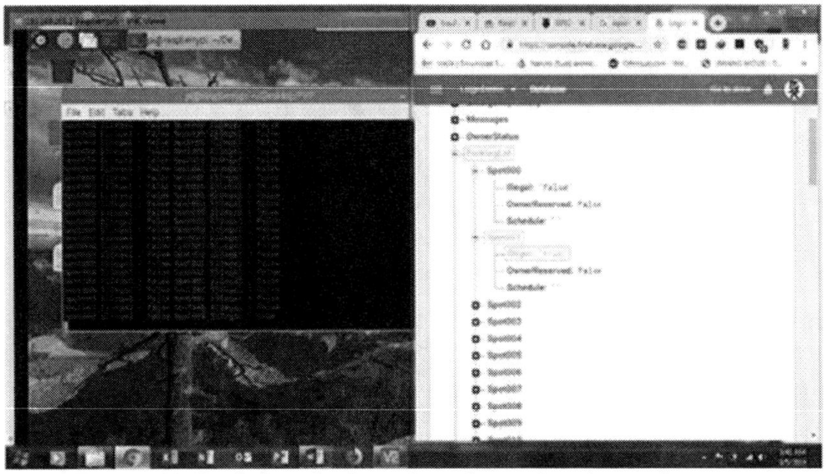

Figure 11. Real-time database data change by ultrasonic sensor

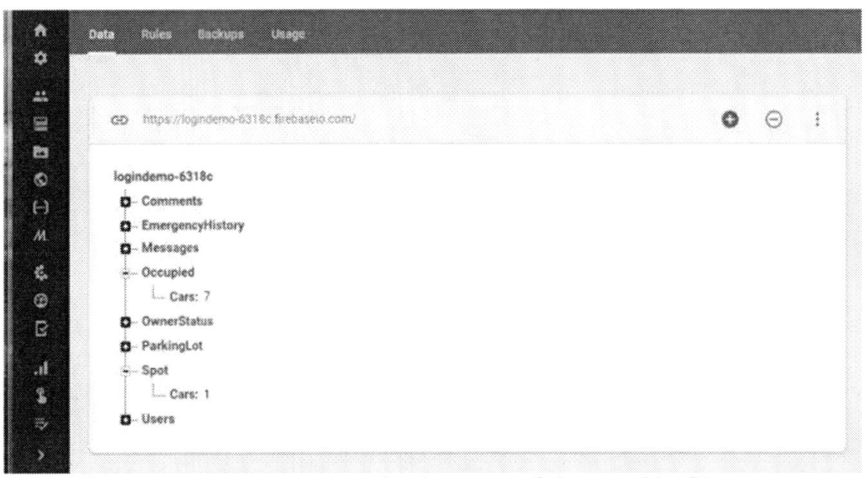

Figure 12. **Firebase real-time database record data send by Picamera**

Figure 13. Firebase real-time database output results

4. PayPal Payment

PayPal has been used as medium for the user to make a payment in the mobile application. The user can make a payment by entering e-mail and password or by entering details card bank number and the payment fee will be send to the admin. Figure 14 shows the PayPal interface on mobile application.

To test the PayPal payment, we need to enter Sandbox mode and create sandbox account. Figure 15 shows the history of the payment made by the user while Figure 16 shows the receipt of the payment.

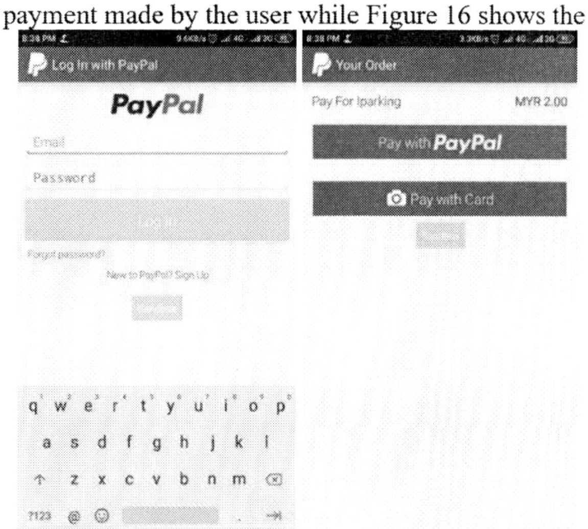

Figure 14. PayPal interface on mobile application

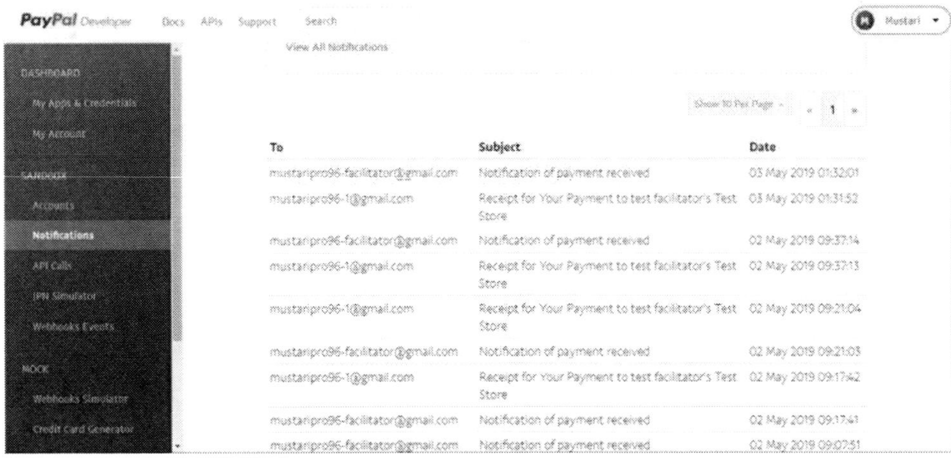

Figure 15. History of payment made by user

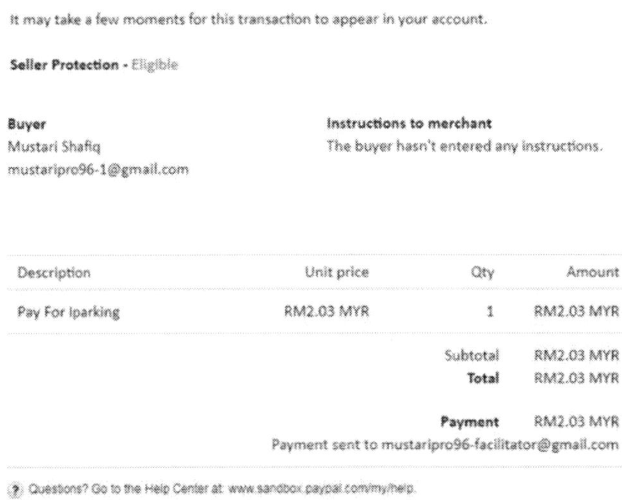

Figure 16. Receipt of the payment

5. Conclusion

Issue of parking in smart city is the main proposed of this system. This system is implemented using low cost ultrasonic sensor JSN-SRO4, Picamera, Raspberry pi microcontroller, "iParking" mobile application and Firebase server. This system provided real time information of availability of parking slots and allows user to reserve the parking slot through mobile application. User also can make a payment for their parking online by using PayPal without need to line up in ticket counter or machine. Proposed work not only to avoid from traffic congestion, it is also providing smart users management, no additional devices and fast payment. In the future, user can view the parking slot by using camera on mobile application and user can use popular online banking such as Maybank2u and CIMB Click for payment instead of using PayPal.

References

[1] Gupta, A., Kulkarni, S., Jathar, V., Sharma, V., & Jain, N. (2017). Smart Car Parking Management System Using IoT, (4), 112–119.

[2] Hans, V., Sethi, P. S., & Kinra, J. (2016). An approach to IoT based car parking and reservation system on Cloud. Proceedings of the 2015 International Conference on Green Computing and

Internet of Things, ICGCIoT 2015, 352–354.

[3] Gandhi, B. M. K., & Rao, M. K. (2016). A Prototype for IoT based Car Parking Management System for Smart Cities, 9(May).

[4] Rao, Y. R. (2017). Automatic Smart Parking System using Internet of Things (IOT). International Journal of Engineering Technology Science and Research IJETSR Www.Ijetsr.Com ISSN, 4(5), 2394–3386.

[5] Kharde, P., Pal, S., & Kawle, S. (2016). Smart Parking System. International Journal of Scientific Research in Computer Science, Engineering and Information Technology, 3(1), 9–13.

[6] Gavali, P. A., Kunnure, P., Jadhav, S., Tate, T., & Patil, V. (2017). Smart Parking System Using the Raspberry Pi and Android, 5(2), 48–52.

Smart Electronic Pigeon Hole System

Seah Kah Hean[1], Mohd Nazmi Mohd Nasir[1], Muhammad Faiz Muhammad Asri Nandan[1], Mohamad Syafiq Abd Rahman[1], Nurul Hanim Shafiqah Sidik[1], Muzammil Jusoh[1], Mohamed Nasrun Osman[1], Hasliza Abdul Rahim[1], Thennarasan Sabapathy[1], Mohd Najib Mohd Yasin[2], Ainur Fasihah Mohd Fazilah[1]

Bioelectromagnatics Research Groups (BioEM), [1]School of Computer and Communication Engineering, [2]School of Microelectronic Engineering, Universiti Malaysia Perlis (UniMAP), Pauh Putra, 02600 Arau, Perlis, Malaysia

muzammil@unimap.edu.my, nasrun@unimap.edu.my

Abstract. Throughout this modernization world, the demands on Internet of Thing keep increasing. Most mailbox or pigeon hole are designed with poor efficiency system and commonly suffer from some possible flaws such as manually checking the letter by the user. The main problem of this project is to improve the existing pigeonhole system. In the case of staff at School of Computer and Communication System, Universiti Malaysia Perlis, each academic staff has been allocated a pigeon hole for any letters or memos from within or inter faculty. Unfortunately, the current conventional pigeon hole system is unable to inform and notify the staff on any urgent letter and this lead to significant delay in responding the letter. The main weakness of the current system is that staffs need to check their respective pigeon hole every day. However, due to the routine commitment or unforeseen circumstances the pigeon hole cannot be possibly checked every day. Hence, to fulfil this requirement, this project named as Smart Electronic Pigeon Hole System is introduced in order to make the pigeon hole more efficient. The purpose of this project is to design a smart electronic pigeon hole system with some detecting sensors such as infrared and ultrasonic sensors. Wi-Fi module acts as a tool to communicate via email between pigeonhole's system and its owner. Arduino board is used to be programmed in such a way to process the sensors when detecting the upcoming document or letters and to continuously check the status of capacity of the pigeon hole. This project work with the administrative staff, who collected the documents from the post center, will entered and select the designated number of the pigeon hole users. Infrared sensors, located at the front, will detect the upcoming documents or letters. Ultrasonic sensor will detect the fullness of the letter in the pigeon hole. When there is arriving of document or letters, the user will get the notification email. The user will also get the warning notification email when their pigeon hole is full with documents. Red LED will light on when the capacity of the pigeon hole is full. The sensor will continually sense the condition of the pigeon hole as long as the system is activated. This project is suitable for the staff or the apartment residents as it has the capability to alert the user when there is a letter to prevent the urgent letter's being ignored.

1. Introduction

Pigeon hole is a small compartment for filling letters or mail. Pigeon hole also known as a message box or internal mail system and commonly used for communication in organizations, workplace and education institutes. Documents and messages are placed in a person's pigeon hole for them to collect. Each staff in most big organization has his/her own pigeon hole to receive any important letter or memo related to the official duty. Commonly pigeon hole used by the lecturers or staff in a university but however residents in some place or regular people who work in an office are also regular user of pigeon hole or others might called it as mail box. In order to receive an important document without worrying it will be lost or stolen, it must be able to alert the owner. The pigeon holes with notification system acts like a pigeon hole but it is different from normal pigeon holes that are sold in the market.

Although there is lots of expensive pigeon hole in the market still, it is not guaranteed that it will satisfy most of the customer's needs. With the aid of the technology that will be incorporated into the pigeon hole, this will improve and expand lot of functions and also reduced problems that most users will encounter such as forget to collect the important documents.

In the case of staff at School of Computer and Communication System, Universiti Malaysia Perlis, each academic staff has been allocated a pigeon hole for any letters or memos from within or inter faculty. Unfortunately, the current conventional pigeon hole system is unable to inform and notify the staff on any urgent letter and this lead to significant delay in responding the letter. The main weakness of the current system is that staffs need to check their respective pigeon hole every day. However, due to the routine commitment or unforeseen circumstances the pigeon hole cannot be possibly checked every day. In order to overcome this problem, the Pigeon Hole with Notification System is introduced as an innovation from conventional pigeon hole to notify user through email. The notification email is included the information such as status of the capacity of pigeon hole.

When any documents are put into the pigeon hole, it will send a notification email to inform the user that it has received some documents. If the pigeon hole is filled up with tons of documents, the pigeon hole will send a notification email to the user that it cannot receives mails and documents anymore unless the user clear up the documents inside it. At the same time, a red led at the top of pigeon hole will also light on, to show that the pigeon hole is full.

The internet of things, or IoT, is a system of interrelated computing devices mechanical and digital machines, objects, animals or people that are provided with unique identifiers (UIDs) and the ability to transfer data over a network without requiring human to human or human-to-computer interaction. A thing in the internet of things can be a person with a heart monitor implant, a farm animal with a biochip transponder, an automobile that has built-in sensors to alert the driver when tire pressure is low or any other natural or manmade object that can be assigned an IP address and is able to transfer data over a network. Increasingly, organizations in a variety of industries are using IoT to operate more efficiently, better understand customers to deliver enhanced customer service, improve decision-making and increase the value of the business.

An IoT ecosystem consists of web-enabled smart devices that use embedded processors, sensors and communication hardware to collect, send and act on data they acquire from their environments. IoT devices share the sensor data they collect by connecting to an IoT gateway or other edge device where data is either sent to the cloud to be analyzed or analyzed locally. Sometimes, these devices communicate with other related devices and act on the information they get from one another. The devices do most of the work without human intervention, although people can interact with the devices -- for instance, to set them up, give them instructions or access the data.

Blynk is a platform with IOS and Android apps to control Arduino, Raspberry Pi and the likes over the internet. It is a digital dashboard where you can build a graphic interface for your project by simply dragging widgets. Blynk also works over USB. This means you can linker with the app by connecting it to your laptop or desktop while waiting for some internet shield to arrive.

2. Methodology

A. Proposed System Block Diagram

Figure 1 shows the system block diagram of proposed work it major constituent of it are:

LCD: Display "Enter no. mailbox".

Keypad: Act as input, used to input the pigeon hole number of the user

Infrared Sensor: Act as input. IR sensor will detect and consequently send a notification emails to inform the user that it has received some documents.

Ultrasonic Sensor: Act as input. Its detect the capacity of the pigeon hole.

Wi-Fi Module: Wi-Fi module will send a notification via email to the owner of pigeon hole when their pigeon hole's full.

White and Red LED: Indicate the status of pigeon hole either its full or not.

Blynk Apps: Notify the user whenever there is arriving of documents or when the pigeon hole is full

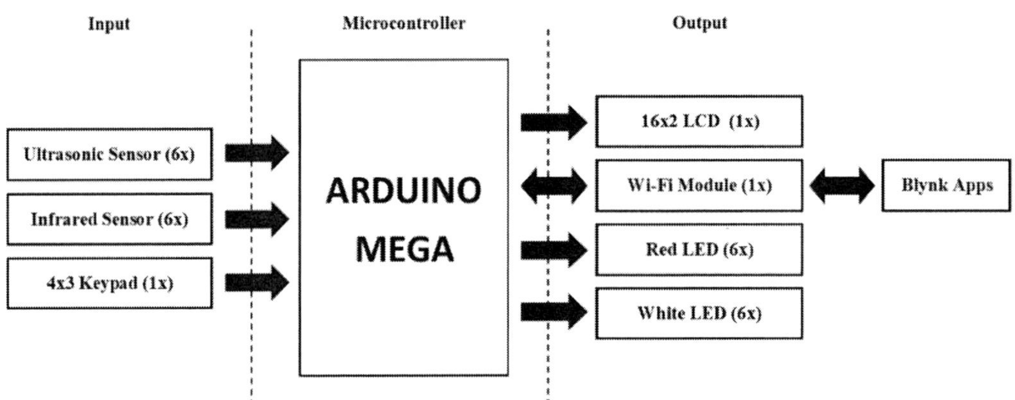

Figure 1. System block diagram

B. System Flow Chart

Figure 2 depicts overall system flow chart. Firstly, the LCD will display "Enter no. mailbox". The user need to enter the number of mailbox by using keypad. After the number of mailbox has been inputted, the chosen number will show up on the LCD. Then, there was a delay of 10 seconds for the mail to be inserted into the pigeon hole. If no mail is detected by the IR sensor, "No letter input" will show up and it will loop back to entering the number of mailbox. If the letter reaches the capacity of ultrasonic that set by the admin, red LED will turn on and an email will be sent to the owner of the pigeon. While white LED will turn on if unreached the capacity set by ultrasonic sensor.

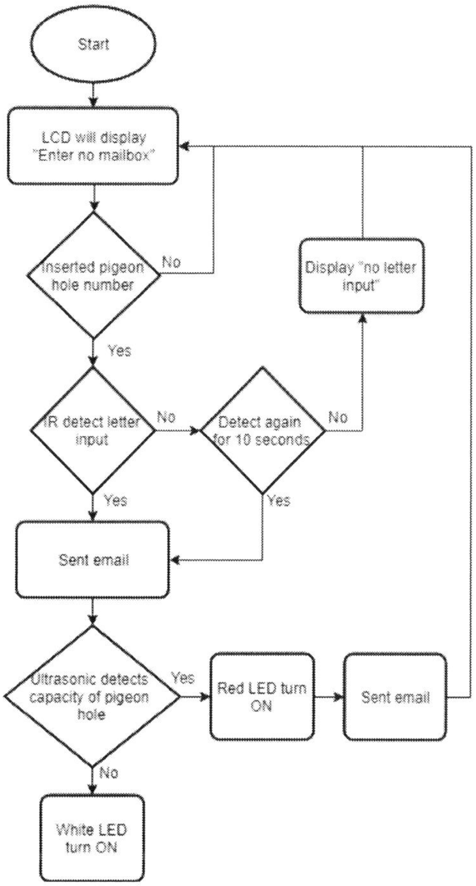

Figure 2. Overall system flow chart

3. Result and Discussion

Pigeon Hole with Notification System is a project that will ease the user of a pigeon hole by sending notification whenever a mail is inserted into their respective pigeon hole and also to notify the maximum capacity of the pigeon hole by the mean of an email. The system is built with an Arduino Mega as its main processing unit and a Node MCU and both of these devices communicate through serial communication.

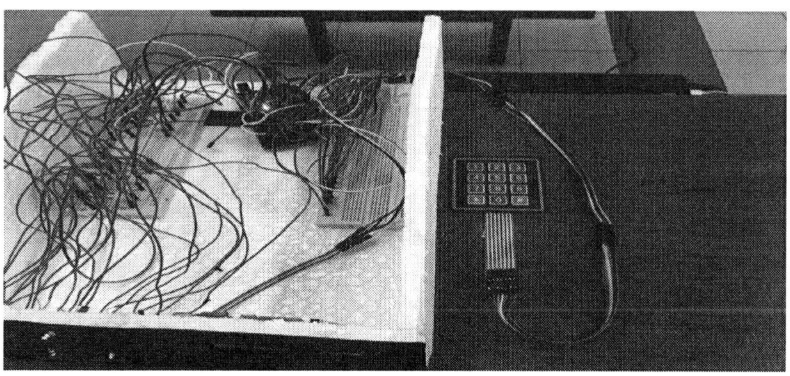

Figure 3. Connection of Arduino with input and output

Figure 3 shows the connection of Arduino Mega with the input and output such as LED, Keypad, IR sensors, Ultrasonic sensors, Node MCU and also LCD.

Figure 4. LCD displays "Enter no. mailbox"

For Figure 4, the LCD will display "Enter no. mailbox" to indicate for the admin to enter the number of mailbox which the mail will be inserted. After the number of mailbox has been inputted through the keypad, the chosen number will show up on the LCD. Then, there was a delay of 10 seconds for the mail to be inserted into the pigeon hole. If no mail is detected by the IR sensor, "No letter input" will show up and it will loop back to entering the number of mailbox.

Figure 5. IR sensor and ultrasonic sensor

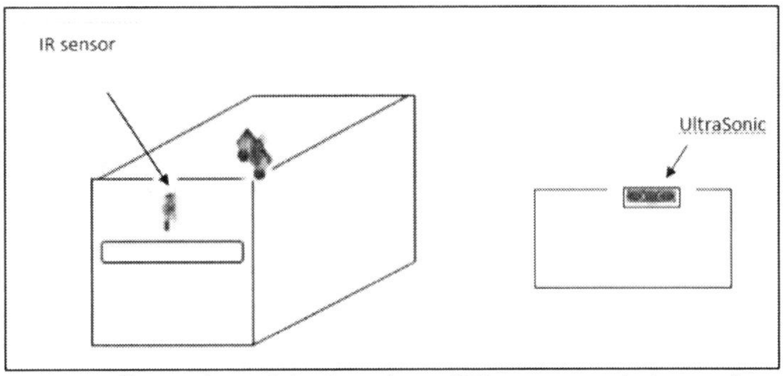

Figure 6. Illustration of the location of IR sensor and ultrasonic sensor

Figure 5 and Figure 6 shows the connection of IR Sensor and Ultrasonic Sensor. The IR sensor is located at the door of the pigeon hole. This is to detect the insertion of mail into the pigeon hole. After every single new insertion, an email notification will be sent to the respective pigeon hole owner to notify that there is new letter for the owner. Next, the ultrasonic sensor is located on the middle top of the pigeon hole. This is to detect the capacity of the pigeon hole. When it reaches the maximum capacity, it will notify the owner through the mean of an email that the pigeon hole is full and the owner need to clear it as soon as possible.

Figure 7. White LED turns ON

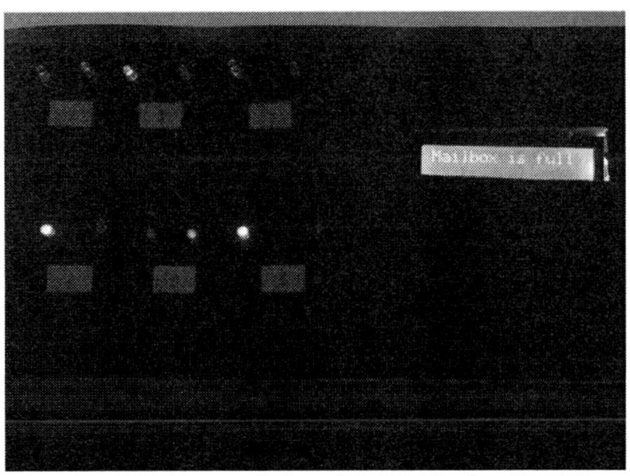

Figure 8. Red LED turn ON and LCD display "Mailbox is full"

For the LED condition, whenever there is space for another mail, and the ultrasonic sensor is not triggered, the white LED will turn on as shown in Figure 7. If the ultrasonic sensor is triggered, the red LED will then turn on to indicate the pigeon hole is full as in Figure 8. Whenever the maximum capacity has been reached, the admin cannot select that particular mailbox anymore. If the admin chooses to do so, the LCD will display "The mailbox is full" and "Enter no. mailbox" again.

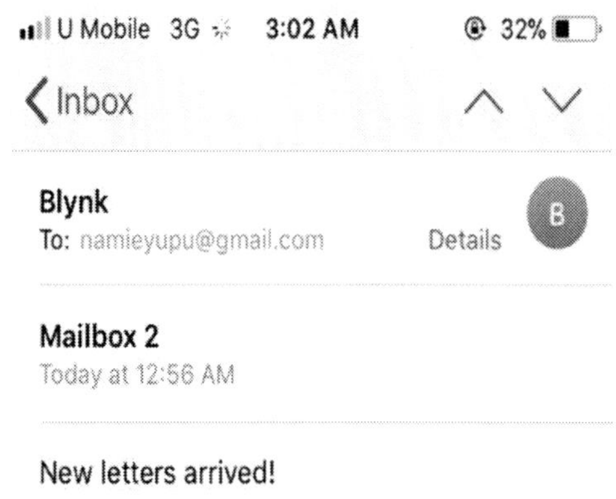

Figure 9. Notification for infrared sensor

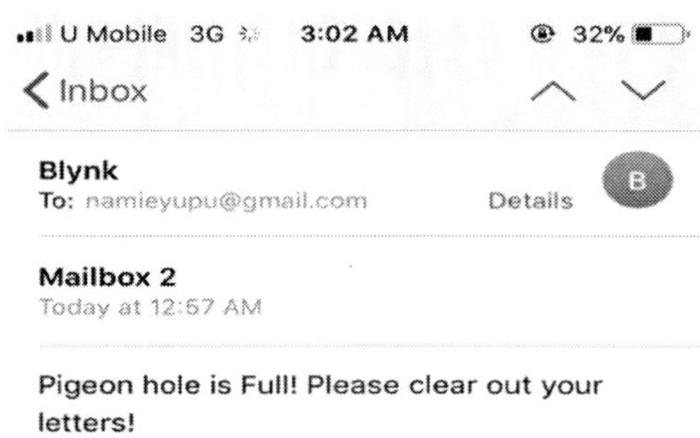

Figure 10. Notification for ultrasonic sensor

Figure 9 shows the email notification that has been sent to the pigeon hole owner whenever the IR sensor detects a new mail. Meanwhile, Figure 10 depicts the email notification when the ultrasonic sensor detects the capacity of the pigeon hole that was filled with tons of documents.

4. Conclusion

In a conclusion, the "Smart Electronic Pigeon Hole System" has been designed and tested successfully. It has been developed by integrated features of all the hardware and software components used. Presence of every module has been reasoned out and placed carefully, thus contributing to the best working of the unit. Thus, the Smart Pigeon Hole System has been designed and tested successfully. The system has been tested to function automatically. Thus, the functionality of the entire system has been tested thoroughly and itis said to function successfully. The users of the pigeon hole (lecturers/workers) at SCCE office does not need to carry out manual checking anymore. The users also will not forget to retrieve their mails as they have been notifying via email about mail availability. The users also notified whenever their pigeon holes were fully occupied and does not have any space for mails anymore.

5. Future Work

Even though the overall project is well functioning and work successfully, there is a room of improvement to make the project better. In the future work, the project will be improved by considering the following:

- Adding another sensor that is able to accurately display the number mail received, as well as the ability to distinguish unwanted mails from important letters.
- Find other software that can run the email function automatically without the need of additional work force of supervision.
- By using a touch sensor, the system can be turn of when need to be used and turn off after the user finish using it.
- By using weight sensor, the system will have better range of detection for incoming objects, including small parcel
- Link smartphones to our sensors so that a simple glance can be advanced to a simple text message or an application notification.
- Provide efficient indicator for LED such as red, amber and green displayed at the top of the pigeon hole. Each lighted LED indicates the amount of mails inside the pigeon hole, with green representing the presence of a mail, amber meaning that the pigeon hole is half full and red meaning that it is fully filled.
- To provide security system such as implementing simple lock system with motor, so only open up when the user key in the correct password and for avoiding missing documents or letters.
- By using more advanced microcontroller or processor, the system can be scale up to provide service for more users.
- Monitoring via apps.
- Database system can be implement and the admin able to change the owner of pigeon hole without needing to change the whole coding.

References

[1] SUN Zhixin, LUO Bingqing, LUO Shengmei, ZHU Hongbo. Security Model of Internet of Things Based on Hierarchy [J].Computer engineering, 20111-7(20).
[2] Curran, K., Craig, R. (2001) "A Short Message Service Online Application for Delivering Urgent Information to Students", In: 1st Joint IEI/IEE Symposium on Telecommunications Systems Research (2001)
[3] Agrawal, P., Famolari, D.: Mobile Computing in Next Generation Wireless Network. In: Proceeding of the 3rd International Workshop on Discrete Algorithms and Methods for Mobile Computing and Communications, pp. 32–39 (1999).
[4] Nazariah bt Abdullah. Smart Pigeonhole System by Sending Notification through Short Messaging System, Universiti Malaysia Pahang: Tesis (2015)
[5] Fadzilah bt Kusin. SMS Based Child Monitoring System, Kolej Universiti Teknologi Tun Hussein Onn: Thesis (2005)
[6] Mohamad Eliyass bin Jamaruppin. Pigeonhole Notifiation System by Utilizing Telegram Messenger, Universiti Malaysia Pahang: Thesis (2014)
[7] Nurliyana bt Azizan. Intelligent Receiving Box Integrated with GSM Network, Universiti Teknikal Malaysia Melaka: Thesis (2015)
[8] Simmons, B. (2015, January 9). Advantage of plywood. Retrieved from https://fennerschool-associated.anu.edu.au/fpt/plywood/advply.html.
[9] Ahmed Hamdy. (2016, March 10) Article about Blynk. Retrieved from countryhttps://www.hackster.io/shakram02/enter-the-iot-world-now-getting-started-with-blynk-/

.

Implementation of Green Building Concept in Office Building Jakarta

Firma Purbantoro*, and Marsul Siregar

Universitas Katolik Indonesia Atmajaya Jakarta

*Purbantoro.firma@gmail.com

Abstract. Green Building Concept has become popular in Indonesia. The purpose of this study was to determine the extent to which the implementation of green building principles in Office Buildings. Research Method using a GREENSHIP Rating Tools for existing building which consists of six categories; Appropriate Site Development (ASD), Energy Efficiency & Conservation (EEC), Water Conservation (WAC), Material Resources & Cycle (MRC), Indoor Air Health & Comfort (IHC) and Building & Environments Management (BEM). The results show that implementation the Green Building Concept could also made energy performance more efficient, after Implementing through Retrofitting of the Chiller System, Recycle Waste Water, Replacement of Conventional lamp to Energy Saving LED and also Training and Education to all employees and tenants. From comparing Data research before implementation of Green Building concept in 2014 and after Implementation and retrofitting in 2017 & 2018, found that Savings Cost from electricity bill payments is IDR. 466,803,325.67/month (18%) and roughly will Break Event Point (BEP) for 3.86 Years and Energy Efficiency Index (EEI) from 238.8 kwh/m^2/Years to 123.65 kwh/m^2/Year

1. Introduction

The concept of Green Building is become popular in the world; there are several institutions that issue certifications including USGBC LEED [1] [2], Singapore BCA-Green Mark [3] and GBCI [4], this research using tool rating from green building council Indonesia (GBCI) for existing buildings. Green Building Council Indonesia (GBC Indonesia) is an independent (non-government) institution that is fully committed to educate of community in applying environmental best practices and facilitating the transformation of a sustainable global building industry [3]. GBCIndonesia has now issued 5 types of *GREENSHIP*, namely: *GREENSHIP* New Building, GREENSHIP existing building, interior green space, GREENSHIP homes and GREENSHIP neighbourhoods. GREENSHIP for existing Buildings is applied to a building that has been operating at least one year after construction. The implementation of green buildings in existing buildings is mostly related to the management and maintenance of buildings. This study took a case study of an office building in Jakarta Indonesia that has two towers; each Tower has 32 floors and 3 basements. The majority of electricity consumption is from the Ventilating Air Conditioning system 51%, Lighting System 20%, Sewage Treatment Plant 9%, Elevator 8%, Pumps 7%, Computer and others 5%. GREENSHIP Rating Tools for existing Building is grouped into six categories: Appropriate Site Development (ASD), Energy Efficiency & Conservation (EEC), Water Conservation (WAC), Resources & Cycle Materials (MRC), Indoor Air Health & Comfort (IHC), and Building & Environment Management (BEM). The award for the existing building can be seen in Table 1.

Table 1. award for existing buildings VERSION 1.1 [4]

Rating	Percentage	Minimum Point
Platinum	73%	86
Gold	57%	67
Silver	46%	54
Bronze	35%	41
Total	100%	117

This research took a case study of an office tower in Jakarta, which in 2012 received a Gold award and planned to get a Platinum award, the achievement point in 2012 was 72 points. Detail points achieved can be seen in table 2.

Table 2. Ratings obtained by buildings in 2012 [4]

Category	Point	Available Point
	2012	
Appropriate Site Development	8	16
Energy Efficiency & Conservation	17	36
Water Conservation	19	20
Material Resources & Cycle	6	12
Indoor Air Health & Comfort	14	20
Building & Environment Management	8	13
Point Achieved	72	117
Rating	GOLD	

Detail of Rating Tools and point available developed by GBCI for Existing Building can be seen in table 2, table 3, table 4. Table 5, table 6, table 7 and table 8.

Table 3. Point Available for Appropriate Site Development (ASD)

No's	Code	Criteria	Point Available
1	ASD P1	Site Management Policy	P
2	ASD P2	Motor Vehicle Reduction Policy	P
3	ASD 1	Community Accessibility	2
4	ASD 2	Motor Vehicle Reduction	1
5	ASD 3	Bicycle	2
6	ASD 4	Site Landscaping	3
7	ASD 5	Heat Island Effect	2
8	ASD 6	Storm Water Management	2
9	ASD 7	Site Management	2
10	ASD 8	Building Neighbourhood Recognize	2
		Sub Total	16

Table 4. Point Available for Energy Efficiency & Conservation (EEC)

No's	**Code**	**Criteria**	**Point Available**
1	EEC P1	Policy and Energy Management Plan	P
2	EEC P2	Minimum Building Energy Performance	P
3	EEC 1	Optimized Efficiency Building Energy Performance	16
4	EEC 2	Testing, Re-commissioning or retro-commissioning	2
5	EEC 3	System Energy Performance	12

Journal of Physics: Conference Series **1424** (2019) 012023 doi:10.1088/1742-6596/1424/1/012023

6	EEC 4	Energy Monitoring and Control	3
7	EEC 5	Operation and Maintenance	3
8	EEC 6	On Site Renewable Energy	5 (Bonus)
9	EEC 7	Less Energy Efficiency	5 (Bonus)
10		**Subtotal**	**36**

Table 5. Point Available for Water Conservation (WAC)

No's	Code	Criteria	Point Available
1	WAC P1	Water Management Policy	P
2	WAC 1	Water Sub-Metering	1
3	WAC 2	Water Monitoring Control	2
4	WAC 3	Fresh Water Efficiency	8
5	WAC 4	Water Quality	1
6	WAC 5	Recycled Water	5
7	WAC 6	Potable Water	1
8	WAC 7	Deep Well Reduction	2
9	WAC 8	Water Tap Efficiency	2 (bonus)
		Sub Total	**20**

Table 6. Point Available for Resources & Cycle Materials (MRC)

No's	Code	CRITERIA	Point Available
1	MRC P1	Fundamental Refrigerant	P
2	MRC P2	Material Purchasing Policy	P
3	MRC P3	Waste Management Policy	P
4	MRC 1	Non ODS Usage	2
5	MRC 2	Material Purchasing Practice	3
6	MRC 3	Waste Management Practice	4
7	MRC 4	Hazardous Waste Management	2
8	MRC 5	Management of Used Good	1
		Sub Total	**12**

Table 7. Point Available for Indoor Health Comfort (IHC)

No's	Code	CRITERIA	MAX POIN
1	IHC P	No Smoking Campaign	P
2	IHC 1	Outdoor Air Introduction	2
3	IHC 2	Environmental Tobacco Smoke	2
4	IHC 3	CO2 and CO Monitoring	2
5	IHC 4	Physical and Chemical Pollutants	6
6	IHC 5	Biological Pollutant	3
7	IHC 6	Visual Comfort	1
8	IHC 7	Acoustic Level	1
9	IHC 8	Building User Survey	3
		Sub Total	**20**

Table 8. Point Available for Building Environment Management (BEM)

No's	Code	Criteria	Point Available
1	BEM P	Operation & Maintenance Policy	P
2	BEM 1	Innovations	5

3	BEM 2	Design Intent & Owner's Project Requirement	2
4	BEM 3	Green Operational & Maintenance	2
5	BEM 4	Green Occupancy/Lease	2
6	BEM 5	Operation and Maintenance Training	2
		Sub Total	**13**

Electricity Bill per month in 2014 before retrofitting can be seen in Table 9 below.

Table 9. Electricity Bills 2014

Months	kWH		TOTAL kWH	Bills (IDR)
	LWBP	WBP		
January	1.892.760	307.800	2.200.560	2.796.413.194
February	1.912.320	294.480	2.206.800	2.459.192.381
March	2.059.440	325.920	2.385.360	2.458.753.699
April	2.046.600	289.320	2.335.920	2.590.922.198
May	2.040.120	310.440	2.350.560	2.661.675.274
June	2.087.400	306.240	2.393.640	2.960.947.907
July	1.902.480	306.240	2.208.720	3.007.494.036
August	2.113.680	321.480	2.435.160	2.732.377.851
September	2.117.760	315.000	2.432.760	3.150.719.192
October	2.190.000	327.600	2.517.600	3.065.388.481
November	2.056.080	320.520	2.376.600	3.140.022.359
December	2.059.440	300.360	2.359.800	2.966.604.401
TOTAL	24.478.080	3.725.400	28.203.480	33.990.510.973
Average	2.039.840	310.450	2.350.290	2.832.542.581

Researcher use Table.9. as a Base line Energy Consumption is 28,203,480 kWh / Year and the Average Electricity Payment is IDR 2,832,542,581 / month

Table 10. Energy Efficiency Index (EEI) standard's

Sources	**EEI (kWH/m²/Year)**	**Years**
ASEAN – USAID	240	1987
ESDM & JICA Electric Power Development Co Ltd	198,2	2008
GBCIndonesia	250	2010
Governor's regulation / *Pergub DKI Jakarta No.38 tahun 2012 Tentang Bangunan Gedung Hijau*	210-285	2012

Building's EEI before retrofitting was 238.8 kWH/m^2/year

2. Methodology

The Methods of this research can be seen on figure 1, the researcher was using *Greenship* rating tool developed by Green Building Council Indonesia [4] and by collection of data from January 2016 to December 2018 from an office's building when they awarded by a *Greenship*'s silver rating and then made an efforts to repair (retrofit) the Ventilation Air Conditioning (VAC), Energy Management System, Water Recycle and Improving indoor air quality, utility report, electricity bill, operational expenses and capital expenditure. The research method used is qualitative and quantitative. Qualitative methods were carried out through interviews with the Building Management, literature studies, observation, and documentation. While the quantitative method aims to find relationships or explain the causes of change based on measurable facts based on numerical data, can be observed and measured. The data used in this study are primary data and secondary data. Primary data is building planning data from the consultant building planner, GREENSHIP assessment data from GBC Indonesia, and data from direct research obtained by researchers through surveys, interviews, and field measurements. Secondary data is data collected by other parties that have been documented so that it can be used by other parties (researchers)

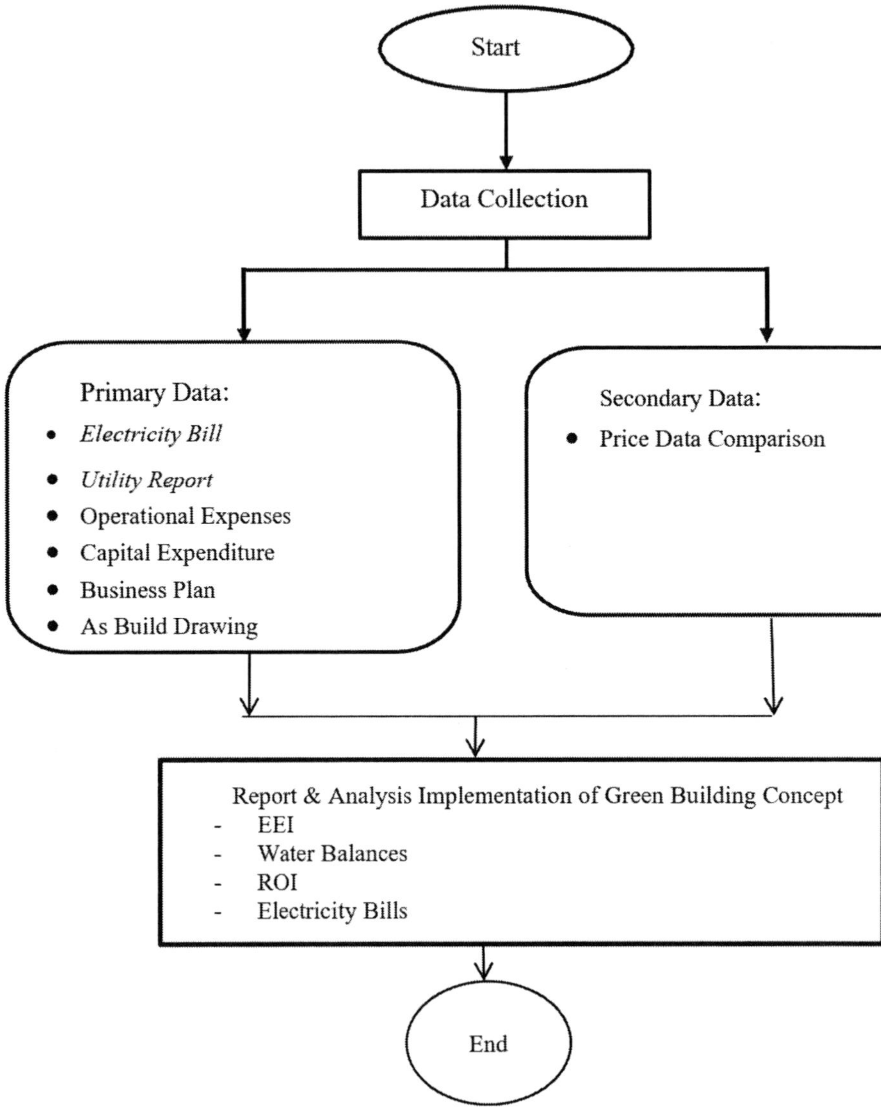

Figure 1. Methodology

3. Result and Discussion

3.1 Repair / Retrofit

3.1.1 Physical / infrastructure improvements by retrofitting

Improvements are made to improve the Energy Efficiency Index (EEI) [5], water consumption index [5] and air circulation system

Item Repair & Retrofit and its costs can be seen in table 11 below

Table 11. Item Repair & Retrofit and its costs

Item Retrofit	Category	Cost (IDR)
Replacement of 4 Chiller Units (3 1000 TR Units and 1 Unit 500 TR) Type Screw Compressor with more energy-efficient centrifugal type	Energy Efficiency & Conservation	16.577.000.000
Improvements to the wastewater system so that it can be	Water Conservation	1.800.000.000

used for cooling tower's make up water		
Replacement of conventional lights with energy-efficient LEDs [6]	Energy Efficiency & Conservation	3.250.000.000
TOTAL COST		21.627.000.000

3.1.2 Energy Efficiency Index (EEI)

Result of EEI can be seen in table 11 below

Table 12. Energy Efficiency Index

Area (m^2)	kWH/Years		EEI ($kWH/m^2/Year$)		Remark
	2014	2018	2014	2018	
203.195	28,203,480	25,125,360	238.8	123.65	

From table 12 EEI can be reduce from 238.8 kWH/m^2/year to 123 kWH/m^2/year, it means building has more efficient in energy consumption

3.1.3 Chiller Performances

Retrofit results can be seen from the performance of the chiller efficiency which can be seen in Figure 1 below, from the picture it is found that the old COP chiller is 3.8 and the new COP chiller is 5.0 or there was an improvement of about 25%

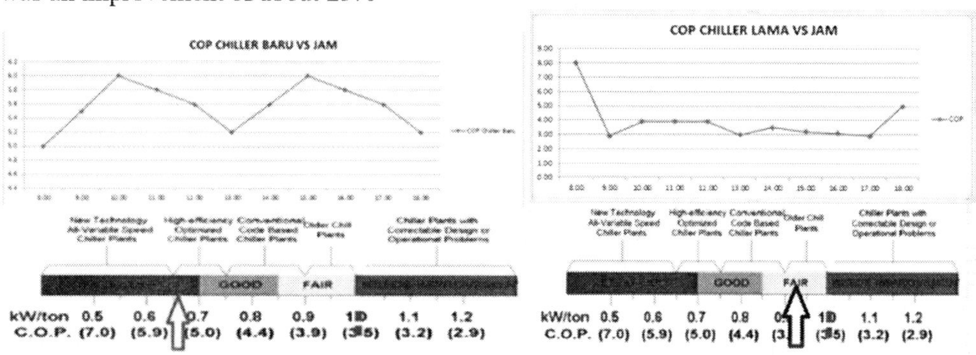

Figure 2. Comparison of Coefficient Of Performance (COP) Old Chiller & New Chiller

Chiller Efficiency Comparison can be seen in table 13 below, from the table it is known that the old Chiller to get a cooling of 1 Ton of Refrigeration (TR) requires energy of 1.63 kW, while for the new Chiller to get a cooling of 1 Ton of Refrigeration (TR) energy is needed at 0.89 kW, meaning that with the new chiller we get savings of around 50% compared to the old chiller.

Table 13. Chiller Efficiency Comparison

Chiller Plant	Efficiency	
	Old Chiller (kW/TR)	New Chiller (kW/TR)
Chiller (A)	0,95	0,53
Pump CHWP (B)	0,3	0,15
Pump CWP (C)	0,3	0,15
Cooling Tower (D)	0,08	0,04
Sistem Effisiensi (A+B+C+D)	1,63	0,89

3.1.4 Waste Water Recycle and Water Balances

Recycling results of wastewater can be seen in the building water balance, where an average of 71% of waste water is reused for Make-up water cooling Tower and gardening.

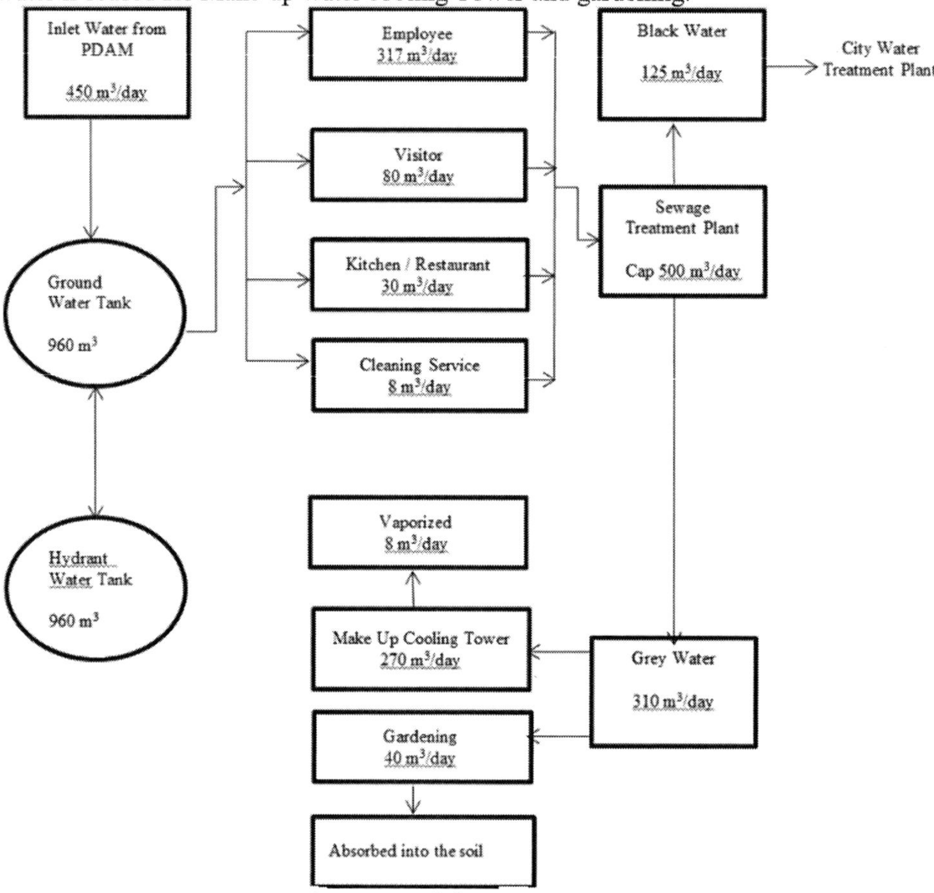

Figure 3. Building Water Balance

3.2 Electricity Bill after Retrofit

Electricity bills can be compared before and after retrofit, electricity bills after retrofit in 2017 can be seen in Figure 5 and figure 5, electricity bills continue to decline due to continued retrofit of conventional lamp replacements with energy-saving lamps (LED). Researcher using electricity bill to calculate Return on Investment (ROI).

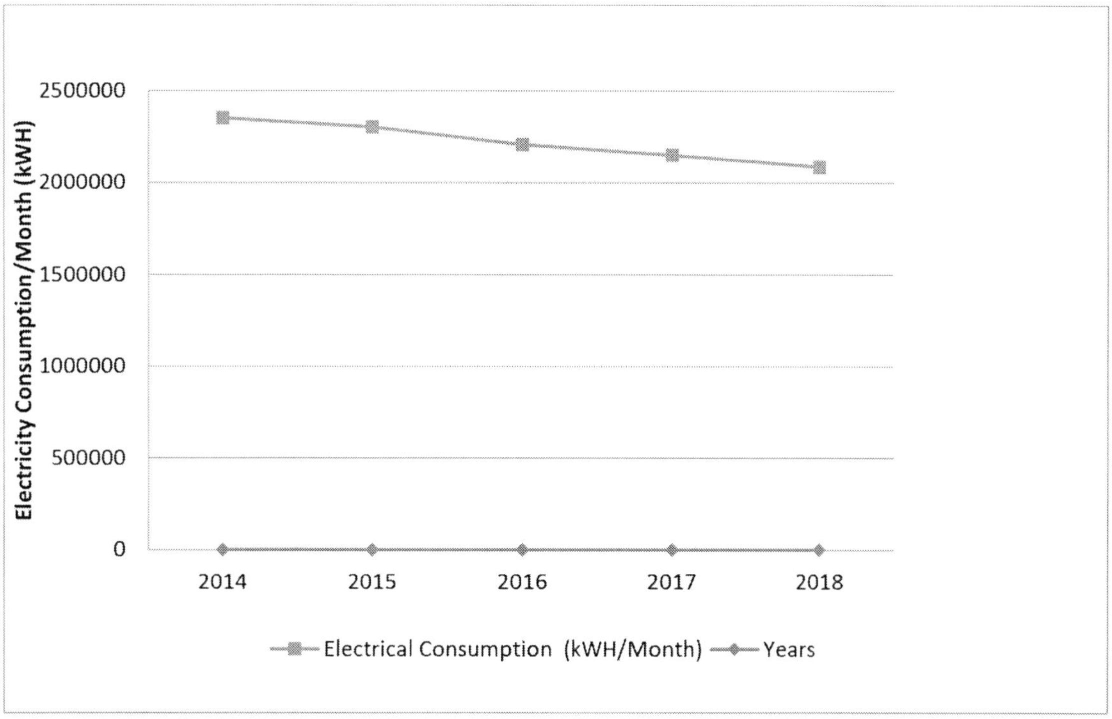

Figure 4. Average Electricity Consumption kilowatt Hour (kWh)/month

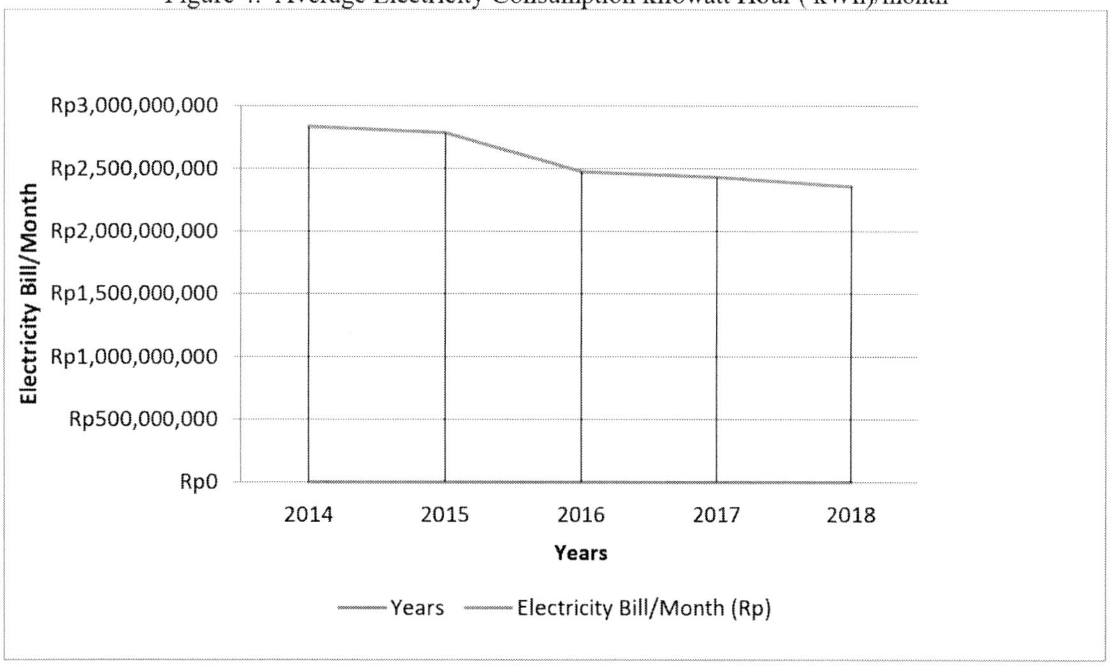

Figure 5. Average Electricity Bill (Rp)/month

Table 14. Electricity bills in 2018

Months	kWH		TOTAL kWH	Bills (IDR)
	LWBP	**WBP**		
January	1.932.360,00	283.200,00	2.215.560,00	2.500.101.165,00
February	1.688.400,00	252.840,00	1.941.240,00	2.193.040.256,00

March	1.878.120,00	287.520,00	2.165.640,00	2.449.439.060,00
April	1.887.960,00	286.200,00	2.174.160,00	2.457.775.681,00
May	1.839.600,00	285.480,00	2.125.080,00	2.405.337.702,00
June	1.572.960,00	236.880,00	1.809.840,00	2.045.208.432,00
July	1.878.720,00	275.760,00	2.154.480,00	2.431.365.776,00
August	1.840.080,00	281.160,00	2.121.240,00	2.398.973.869,00
September	1.785.840,00	277.320,00	2.063.160,00	2.335.335.546,00
October	1.948.680,00	278.520,00	2.227.200,00	2.509.959.105,00
November	1.818.480,00	270.960,00	2.089.440,00	2.359.836.300,00
December	1.773.240,00	265.080,00	2.038.320,00	2.302.498.172,00
TOTAL	21.844.440,00	3.280.920,00	25.125.360,00	28.388.871.064,00
Average	1.820.370,00	273.410,00	2.093.780,00	2.365.739.255,33

By comparing Payment of Electricity bills before and after Retrofit, we can compare it from table 9 and table 14, it can be seen that the difference in electricity payments = IDR. 2,832,542,581 - IDR. 2.365.739.255, 33 = IDR. 466.803.325,67 / month, From Table 11 it is found that the retrofit Cost is IDR 21,627,000,000.00 so Estimated Return of Investment = (IDR 21,627,000,000 / IDR 466.803.325,67) = 46,33 months or 3.86 Years

3.3 *GREENSHIP* Rating

Table 15. Greenship Award

No's	**Category**	**Years**		
		Point 2012	Point Plus	Point 2017
1	Appropriate Site Development	8	-1	7
2	Energy Efficiency & Conservation	17	13	30
3	Water Conservation	19	0	19
4	Material Resources & Cycle	6	-1	5
5	Indoor Air Health & Comfort	14	2	16
6	Building & Environment Management	8	2	10
	Point Achieved	72	15	87
	Award	GOLD		PLATINUM

After retrofitting, GREENSHIP Rating increased and received the Platinum Awards with 87 Points.

4. Conclusions

Implementing of the green building concept not only strengthen the brand image of the building by achieving platinum award but also saving operational costs for building management, in the case study of an office building in Jakarta, after retrofitting the VAC system by replacing the *chiller*, Recycle waste water, Replacing less efficient light with Energy Saving lights and also training to all employees and tenants can reduce the electricity bills average Rp. 466,803,325.67/month (or around 18% from building consumption) and roughly Return on Investment (ROI) is 3.86 Years.

Future Research

High-rise buildings still have a lot of potential for further research by installing renewable energy plants such as the installation of solar power plants (solar panels) [7] by utilizing the roofs or facades of buildings and also utilizing the gravitational potential of clean water systems and VAC systems (micro hydro).

References

[1] https://new.usgbc.org/ accessed on May 06th 2019
[2] Shawn Roy, Prcscott C. Ensign dan Tom Brzustowski, 2016, An assessment of LEED certification's impact on net rental rates for commercial office space in Toronto, Ontario, Portland International Conference on Management of Engineering and Technology (PICMET), Honolulu, HI USA, 4-8 Sept. 2016, Date Added to IEEE Xplore: 05 January 2017
[3] https://www.bca.gov.sg/greenmark/green_mark_buildings.html, accessed on May 06th 2019
[4] http://gbcindonesia.org/ , diakses 06 Mei 2019, accessed on May 06th
[5] Peraturan Gubernur DKI Jakarta No.38 tahun 2012 Tentang Bangunan Gedung Hijau
[6] SNI No. 6197-2011, Konservasi energi pada Sistem Pencahayaan, 2011
[7] Pamungkas dany, Manisa Pipattanasomporn, Saifur Rahman, Nanang Hariyanto, Suwarno, 2018, ICUE 2018 on Green Energy for Sustainable Development Thavorn Palm Beach Resort Karon, Phuket, Thailand. 24 – 26 October 2018

Strategic Planning Information System Using Ward and Peppard Method with Anita Cassidy Method

Ninik Sri Lestari*, Andrew Ghea Mahardika, Ahmad Sujana, Nurcaweda Riztria Adinda, Ie Day Lie

Sekolah Tinggi Teknologi Mandala, Bandung, Indonesia

*ninik4lestari@gmail.com

Abstract. The company strategy must be integrated with the preparation of business strategies and IS / IT strategies. In today's global competition innovation is very much needed. The results of the analysis of the overall identification of the business processes that are very important to be able to map completely applications that will be applied for that support and following the mission and vision of the company that will be able to help produce the most appropriate architecture adapted by the company. Appropriate strategic planning can also minimize costs. To improve efficiency, effectiveness, improve the image and corporate social responsibility, standard principles of Good Corporate Governance (GCG) are required, namely transparency, accountability, responsibility and fairness, and to recognize and protect rights and the obligations of other stakeholders. Strategic information system planning that is in line with the company's business strategy so that it can be highly competitive. By analyzing Ward and Anita Cassidy's method, it will produce a portfolio of information systems development applications. Recommendations are given by analyzing the problem using the SWOT method and information needs both internal and external using the ward and Peppard method. The results of the study are the SI business strategy, IT strategy and information system application portfolio and recommendations for the number of costs needed as well as applications that can be implemented.

1. Introduction

Organizations always change along with strategic changes both external and internal. To face challenges in the era of globalization The ability to exist and be able to compete is needed to face challenges and make the most of available opportunities. Information Technology as a very decisive factor in supporting operations and decision making for management. Need a structured planning, communication, and management as well as good infrastructure technology. The principles that direct and control the company or Good Corporate Governance (GCG) as a form of accountability to stakeholders. To o improve efficiency, effectiveness, improvement of corporate image and social responsibility, transparency, accountability, responsibility and fairness principles are needed as well as recognizing and protecting the rights and obligations of other stakeholders to achieve the effectiveness and efficiency of information systems. Management of strategic information systems planning that is appropriate and in line with the vision and mission of the institution is achieved. so that they are able to compete with other tertiary institutions. Data and information needs that are fast, precise and accurate are needed. Facing increasingly fierce competition needs to be supported by strategic information systems planning. To be able to increase value and create competitive advantage, a framework that can provide a strategic role for information systems or information technology is good for the Mandala STT Bandung.
The performance of tertiary institutions in the organization is aimed at stakeholders, especially student competencies. Higher education performance indicators in addition to academic achievement are also to achieve education Tri Dharma namely education, research and community service.[1]

2. Material and method

The ward and Peppard model in strategic planning is described as follows

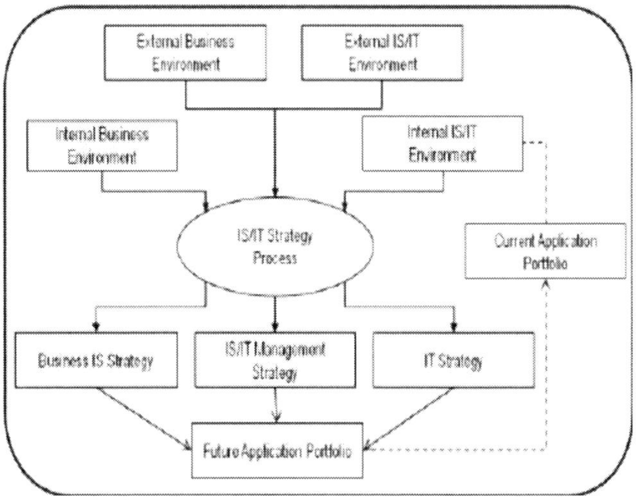

Figure 1. Ward & Peppard Analysis.

The deep framework develops a system strategic planning information based on this methodology, requires an analysis of four input (input), [2] as follows:

1. Internal Business Environment
2. External Business Environment
3. Internal IS / IT environment
4. External IS / IT environment

Output produced from design This strategic information system produces three outputs, as follows:

1. Information systems business strategy (Business IS / IT Strategy)
2. Management strategy (IS / IT Management Strategy)
3. Information technology strategy (IS / IT Strategy)

The final result of planning strategic information system is a portfolio application in the future come (future application portfolio). Once implemented, this portfolio will be the application that will be the basis for improvement or development of the next system.

A strategic plan is more than a statement of strategy. Although it might seem obvious that merely stating strategy is not enough, many strategies stop with a statement of intent.[3]

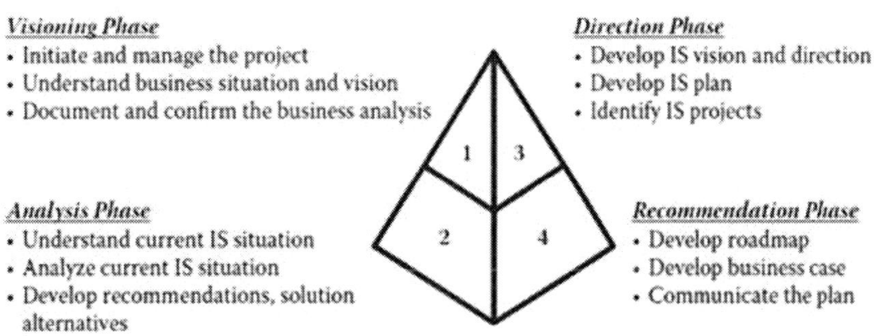

Figure 2. Anita Cassidy Method.

The purpose of the Analysis phase is to identify information systems that already exist and are already running in the organization, its role, and its relation to existing information systems and future development. which influences the organization of the present and the future. The process of comparing against competitors and what information system they use also form part of the analysis phase.
Stages in the Direction Phase by forming strategic goals and objectives of information systems that refer to the vision and mission of information systems based on the analysis of organizational conditions. Direction Phase consists of various processes such as:
1) determine how to measure the value or progress of the SI on an ongoing basis
2) determine the direction of business applications and special projects needed,
3) determine the technical computer architecture and projects needed to achieve the objectives, including changes in the PC, server, network, and telecommunications fields.
4) determine the desired SI service architecture, which includes the people and processes needed by the system,
5) determine how to allocate resources and the role of outsourcing or other alternatives.

2.1. Data Collection and Analysis

The stages were carried out by interviewing the relevant sections at STT Mandala. In the initial stages of identification and data collection on profiles, visions mission and goals and problems and also a work plan for 5 Next year. Perform data collection by observation directly, as well as gathering topics Related to research.
Related to the STT Mandala system in the form of information or oral documents; observation or observation of research objects to understand the needs of the system being built as well as the study of documents. Data analysis was carried out to produce descriptions and interpretations.

2.2. Analysis with Ward and Peppard

The analysis phase, at this stage, is carried out analysis based on framework Ward Peppard which includes environmental analysis business and environmental analysis SI / TI STT Mandala. The explanation is as follows:
a. External Environmental Analysis by using PEST Analysis and Porter. Strategy planning is done to increase competitiveness. Analysis of external environmental factors which are covering the political, economic, social and technology is done by PEST analysis.[4]
b. Internal Environmental Analysis by using Critical Success Analysis Factor and Value Chain Analysis.
c. SI / IT External Environmental Analysis conducted to determine the use of the latest technology by paying attention to the outside environment. The influential towards a strategic plan that will be made.
d. Internal / IT Environmental Analysis is to map business processes and the existing portfolio for portfolio planning going forward. Analysis of IS / IT assets and infrastructure computer network.

A strategic plan is a key component to the success of any IS function and an important factor in assisting a company in meeting its business objectives.[5]

3. Results and discussion

3.1. Internal Business Environment Analysis

Business environment analysis table Internal SWOT analysis results are as follows:

Table 1. Internal SWOT analysis results

Strenght	Weakness
Have legality from the government. Most units already have a system governance information the good one. Have a partner company inside good cooperation internship as well as alumni placement. Most alumni absorbed into the world work Already have a planning system,	Evaluate input performance, process and output not maximal Number of Research and Lecturer devotion Still not optimal Not optimal human resources Performance is different Large units are still slow Information technology has not been used fully deep support performance.

implementation and reporting that library Adequate.
Have a lecturer and employees still young – young so that his ability can be raised
Having IT personnel experienced
Has a structure institutional ones in accordance with the rules
Always done curriculum review in accordance with the needs
Have funds good operational to develop information Systems
Having a commitment to always developing
in following development information Technology and communication (ICT)

3.2. Internal SI Environmental Analysis

Some trends technology that can be utilized by STT Mandalas include:

1. Web technology. Help performance as marketing media, student services and improve relationships with stakeholder.
2. Mobile applications. Utilization of application needs mobile is very necessary, Mobile technology applications can be used to get students new, expanding marketing target in a run and enhance the brand
3. QR (Quick Responses) Barcode Police System QR code is used as an application identity and becomes a means effective, simple and modern promotion.
4. Integrated CCTV. Very needs important for the company, by level good funding, controlled employees, and also student services can be monitored
5. Virtual Private Network (VPN).Data sent encrypted so it is protected from outsiders even though data was sent from the public network. Network analogy in the network and VPN technology can secure data.
6. Datawarehouse and Business Intelligence (DBWI).In order to determine the policy in the form Executive Information System portal to the leadership
7. Datamining.For find out the latest trends that are currently is being favored by students.
8. Document Management. Storage technology and this update can be very beneficial for the quality of accreditation and quality improvement.
9. BlockChain. Centralized approach become decentralized, at the principle of blockchain technology condition each server run this software form network consensus automatically to mutually replicate transaction data and verify each other's existing data, therefore when one server hacked, the server can ignore because it was considered to have data which is different from the majority of networks another server.[6]

3.3. IT Strategy

Potential IT strategy needs to be obtained later mapped in gap analysis to get an overview of IT. [7]

Table 2. IT Needs Gaps Analysis

Business Needs	Information
Server purchase	New systems
Network quality improvement	Upgrade
Addition of employees to the department	Upgrade

4. Conclusions

Company leaders must develop aspects of communication technology to be able to compete. With the right strategic planning, it can minimize costs for developing the technology.

Proper analysis to determine internal and external factors that must be done to find out the right SI / IT portfolio and infrastructure. Upcoming application portfolio and the right SI / IT infrastructure for deep business processes face challenges in the future and sustainability.

System strategic planning information has referred to vision, mission STT Mandala. Information system strategic planning has been designed using Ward and Peppard analysis, analysis SWOT and Anita Cassidy's method. STT Mandala Strategic Plan has been a target system development information.

References

[1] Faslah R, Haris A. 2017. Perencanaan Strategis Sistem Informasi. Jurnal ELTIKOM, 1(1): 31-8
[2] Ward, John; Joe Peppard. 2002. “Strategic Planning for Information System” Third Edition, John Wiley & Sons Ltd, West Sussex
[3] Cassidy, Anita, 2006; “A Practical Guide to Information Systems Strategic Planning" Second Edition, uerbach Publications
[4] Respati RB. 2013. Persepsi Pengguna Terhadap Kinerja Online Public Accsess Catalog (OPAC) Badan Perpustakaan dan Kearsipan Propinsi Jawa Timur (Studi Deskriptif Analisa Sistem Informasi OPAC Badan Perpustakaan dan Kearsipan Propinsi Jawa Timur dengan Menggunakan Analisis Pieces). Journal Universitas Airlangga, 2(2)
[5] Setiawan A, Ilman B. 2011. Perencanaan Strategik Sistem Informasi pada Perusahaan Penerbitan dengan Metode Ward & Peppard: Studi Kasus pada Penerbit Rekayasa Sains Bandung. Journal of Technology Management,11(3)
[6] Sutarman. 2012. Buku Pengantar Teknologi Informasi Jakarta: Bumi Aksara
[7] Sutabri, Tata. 2012. Analisis Sistem Informasi. Yogyakarta: Andi.

.

Analysis of Time Acceleration Costs in Level Building Using Critical Path Method

Andrew Ghea Mahardika[1]*, Hetty Fadriani[1], Muntiyono[1], Salamatul Afiyah[2], Givy Devira Ramady[1]

[1]Sekolah Tinggi Teknologi Mandala, Bandung, Indonesia
[2]UIN Sunan Gunung Djati, Bandung, Indonesia

*andrewhinata@gmail.com

Abstract The multi-story building construction project was carried out according to the schedule at the time of the initial session's work. There was a delay in the agreement for the receipt of land which caused work to be delayed. The researcher's aim is to find out the work time using the Critical Path Method (CPM) method, in this analysis the time from the acceleration of the duration obtained from the CPM, so that in analyzing the estimated normal and up-normal costs of labor wages and the estimated amount of the budget due to the acceleration of implementation time. By making direct observations, and interviewing project team members, the data obtained are used to base analysis. Based on the results of the data obtained and analyzed using CPM. The results of the normal duration of work with the CPM method produce 119 working days, the results of the acceleration of the time done by the CPM method obtained 104 working days. Therefore, using labor analysis, the total cost of normal labor costs is IDR 30,901,500.00 and the total up-normal costs of labor costs are IDR 26,037,000.00. So the results of the analysis, save labor costs of Rp.4,864,500.00. Thus, the total normal cost budget is Rp. 1,337,999,697.00 and the total up-normal cost budget is Rp. 1,333,135,197.00.

1. Introduction

The multi-story building construction project was done on time as scheduled. In reality, the problems faced by the multi-story building construction project experienced delays in the submission of land documents that caused work to start late, within 2 weeks. Initial development project planning for 119 working days. But at the time of implementation, there was a delay of 14 working days so, it was carried out 105 working days. To pursue the target time that is by the 006 / GD-KRW / PPK-DISTARCIP / 201 contract, the researcher applies the CPM (Critical Path Method) so that planning according to schedule is efficient to produce project costs according to the contract and minimize the time to avoid occurring project delay. The analysis states What is the duration of the work analyzed using CPM and based on the Critical Path, what is the duration of the possible time acceleration of a multilevel development project? If the acceleration of the project duration uses labor analysis what is normal and up to the normal cost of labor costs. [1]

Critical Path Method (CPM) is a technique of analyzing network activities/activities when running a project to predict the total duration. A project's critical path is a sequence of activities that determines the fastest time possible for a project to be completed. A critical path is the longest path in a network diagram and has the fewest errors. Some things to note about this critical path: Delayed work on the critical track will delay the completion of the overall project track. Completion of the project as a whole can be accelerated by accelerating completion of work - work on the critical path. The critical path slack work is equal to 0 (zero). This allows the relocation of resources from non-critical work to critical work. [2]

2. Material and method

Objects in research analyze time consistently in the critical path cpm to make it more efficient, in the method of data collection. [3]

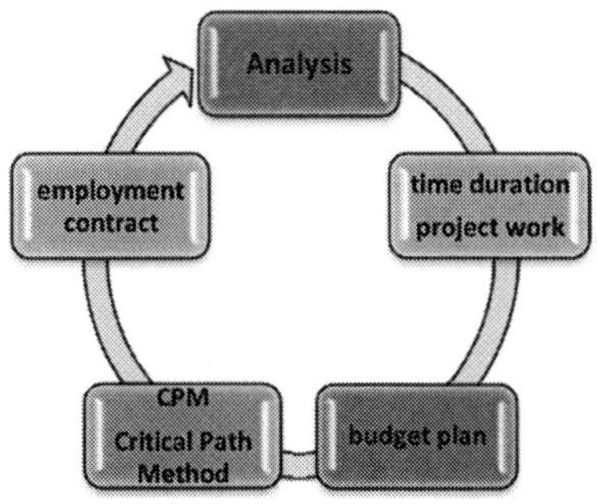

Figure 1. Model of network development life cycle

2.1 Analysis Budget Plan Data

Data is analyzed per item and per project from beginning to end according to the work contract agreement and using time schedule data and cost budget plans in accordance with the unit price analysis, in analyzing the data collected is done by drawing a network with the CPM method the data used in the Budget Plan Data (RAB) and Work Unit Price Analysis.[4] In the data analysis is done by drawing a network with CPM (Critical Path Method) which consists of a total time count of completion of multi-storey building construction projects and determine the critical path in accelerating some activities to achieve the scheduled targets and adding workforce to be known whether the overall cost will be cheaper or more expensive. [5]

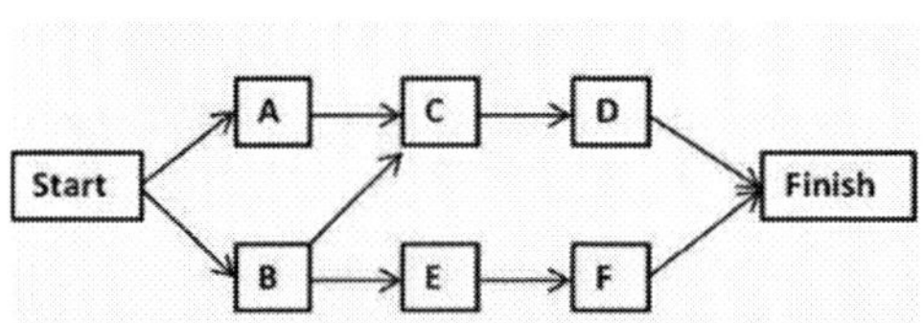

Figure 2. Critical Path Method

2.2 CPM Critical Path Method

Three Basic Assumptions in calculating the critical path method:

1. The project has only one initial event (start) and one terminal event (finish).
2. The fastest time for the initial event to occur is day zero.
3. The slowest time for a terminal event to occur is LS = ES

The technique to calculate the critical path method Forward Pass

Starting from the Start (initial event) to Finish (terminal event) to calculate the fastest completion time of activity (EF), the fastest time for the occurrence of an activity (ES) and the fastest start of an event (E).[6]

2.3 Rules for forwarding Pass

Except for the initial activity, then the activity can only begin if the predecessor activity has been completed and the earliest time to finish an activity is the same as the earliest start time, plus the time period of the activity that preceded it.EF (i-j) = ES (i-j) + t (i-j) If an activity has two or more previous activities that combine, then the earliest start time (ES) of the activity is the same as the earliest finish time (EF) which is the largest of the previous activities.

1) Backward Pass
   Starting from Finish to Start to identify the time the slowest occurrence of an activity (LF), the time the slowest occurrence of an activity (LS) and the slowest time an event occurs (L).

2) Rules of Backward Pass
   The last start time of an activity is the same as the last finish time minus the time period for the activity concerned. LS (i-j) = LF (i-j) t If an activity is divided into 2 or more activities, then the last time (LF) of the activity is the same as the last starting time (LS) of the smallest next activity.

If both calculations have been completed, the Slack or Float value can be obtained which are a number of time and elasticity allowances in a network. [7]

The earliest time to finish an activity is the same as the earliest start time, plus the period of the activity in question. Forward count formula EF = ES + D or EF (i-j) = ES (i-j) + D (i-j). The result of the work duration with forwarding Calculation is 119 working days. Calculation formula backward The most recent start time of an activity is the same as the most recent completion time minus the time period for the activity concerned or LS = LF-D Network with forwarding and backward calculations for multi-story building projects can be seen and Figure 1. [8]

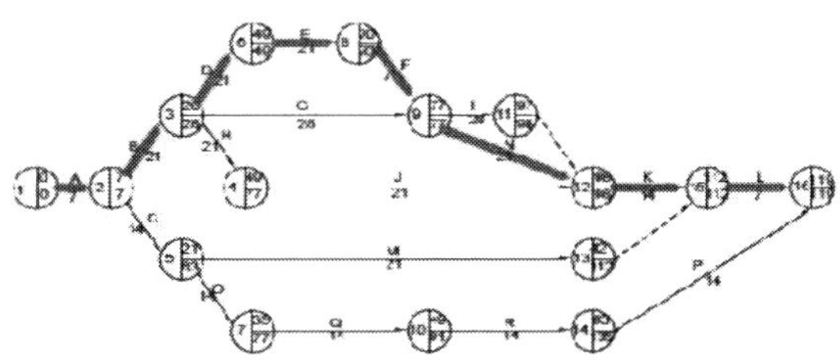

Figure 3. Forward and Backward Calculation Network

3. Results and discussion

3.1. BOQ Time Period

Estimated time for each activity is the time needed to complete each of these jobs using the CPM (Critical Path Method) network. For this multi-story building project uses several analyzes, namely determining the amount of time required as can be seen in Table 1 For building an area of 250 m2 determine the cost with the Bill of Quantity (BOQ) such as the Details of Activities Implementation Budget can be seen in Table 2

Table 1. Time needed for Activities

No.	Job description	the activities	Duration(Day)
1	Preparatory work	A	7
2	Maturation Land Work	B	21
3	soil work	C	14
4	Basic Floor Structure Work	D	21
5	Upper Floor Structure Work	E	21
6	Roof work	F	7
7	Wall / Plastering Work	G	28
8	Sills, Doors & Windows Work	H	21
9	Ceiling Work	I	20
10	Floor work	J	21
11	Painting wall Work	K	14
12	Sanitation Work	L	7
13	Plumbing work	M	21
14	Electrical work	N	21
15	Building Environment Road Work	O	14

16	Landscaping Road Work	P	14
17	Side & Rear Fence Work	Q	14
18	Front Fence Work	R	14

Table 2. Budget Details of Activity Implementation

No.	Job Description	Budgeting	Duration (Day)
1	Preparatory work	29,135,500.00	7
2	Maturation Land Work	156,303,981.00	21
3	Soil Work	3,467,620.00	14
4	Basic Floor Structure Work	107,814,154.00	21
5	Upper Floor Structure Work	293,555,688.00	21
6	roof work	58,415,768.00	7
7	Wall / Plastering Work	144,233,733.00	28
8	Sills, Doors & Windows Work	80,695,692.00	21
9	Ceiling Work	51,331,228.00	20
10	Floor work	56,907,053.00	21
11	Painting wall Work	31,252,390.00	14
12	Sanitation Work	13,027,905.00	7
13	Plumbing work	32,946,992.00	21
14	Electrical work	96,868,596.00	21
15	Building Environment Road Work	58,073,605.00	14
16	Landscaping Road Work	6,931,250.00	14
17	Side & Rear Fence Work	57,776,704.00	14
18	Front Fence Work	59,261,831.00	14
		1,337,999,697.00	

3.2 Speed up the duration

From the planning of activities to the estimated time discussed, it is known that the results of scheduling exceed the targets set by management, so it is necessary to accelerate in several activities. Based on the critical path of the results of the calculation, it is determined to accelerate several activities including activities (AB-DE-FNKL) which include preparatory work, land maturation work, ground floor structure work, top floor structure work, roof work, electrical work, work painting, and sanitation work. Due to the accelerated time duration of 14 days, the acceleration method was chosen using the addition of manpower so that some of the accelerated activities included activities (A-B-K) which included preparatory work, land maturation work and painting work.

3.3 Activity A (Preparatory Work)

In preparatory work there are several sub-jobs to be carried out including the following:

a. Measurement of the situation again.
b. Demolition of existing barrier walls.
c. Field cleansing & leveling.
d. Building bow plank measurements & boards.
e. DPT bow plank measurement & board.
f. CBR soil test.
g. Water supply and work electricity.
h. Keet Direction
i. Project nameplate
j. Administration, Documentation & As-built drawing.

Not all preparatory sub-works are analyzed, because in the preparatory work items there is a critical path that can be accelerated, namely the keet direction work items. Direction keet is a temporary office/warehouse for storing building materials and tools. there is a direction keet unit price analysis, such as the coefficient of wages obtained from productivity results. Productivity is the amount of energy to complete a job. For example 2.00 OH workers (person/day) in this case shows labor productivity to complete an area of 1 m2 temporary / warehouse office, in this case, the unit price value follows the classification owned by the company for building construction. B ZX>,.Coeficient x Work volume / duration Remarks coefficient = 2.00 OH Work volume = 12 m2 Duration (days) = 7 days So the result of the calculation is Coefficient x Work volume / duration (days) = (2.00 x 12 m2 / 7 days) = 3.43 OH. In this case, all unit price values follow the classification owned by the company for building construction. The results of the analysis of direction keet energy duration of 7 days From the results of the analysis of labor painting the wall for a duration of 7 days When the results of the total cost of completing the work of painting the wall area of 966.24 m2 for 7 days amounted to Rp.7,063,000.00.
So it is needed:
a. 3 workers.
b. 9 craftsmen.
c. 1 head mason.
d. 1 foreman.

So there is an additional workforce that is 1 worker and 4 workers. Activities (A-B-K) include preparatory work, land clearing work, and accelerated painting work so that, the results of the analysis of the CPM (Critical Path Method) method have met the outlined target of 105 days. So there is no need to accelerate again. To accelerate the project network For preparatory work, land clearing work and painting work required the number of workers in normal duration is 12 people. The number of crash duration workers is 16 people. The number of carpenters of normal duration is 9 people. The number of carpenter crash duration is 15 people. The number of masons with a normal duration is 5 people. The number of masons crash duration is 7 people. The number of head masons of normal duration is 3 people.

3.4 Activity B (Land Maturing Work)
In the land maturation work, there are several sub-jobs that will be carried out, among them are the DPT (Soil Retaining Wall) earth excavation work, 10 cm thick sand fill work. DPT pair camp split 1: 5 work, landfilling work from outside and soil compaction work every 20 cm. Not all sub-ripening sub-works are analyzed because in the land-ripening work items there is a critical path that can be accelerated, namely DPT work items. From the analysis of the Batu Belah DPT workforce for the duration of 16 days listed in Table 4.14 for completion of the foundation stonework. When the results of the total amount of cost for completion of the foundation stonework as much as 72.77 m3 for 16 days amounted to Rp.14,200,000.00. So it is needed: 7 workers, 4 masons, 1 head mason, 1 foreman.

3.5 Activity K (Painting Work)
In the painting work, there are several sub-jobs that will be carried out, including the following: Ground floor, interior wall / concrete painting work, wall painting / concrete weather shield, ceiling painting work. The upper floor includes interior / concrete wall painting work, wall painting work, ceiling painting work. Coefficient x Work volume / duration Remarks: coefficient = 0.02 OH Work volume = 966.24 m2 Duration (days) = 14 days So the result of the calculation is Coefficient x Work volume / duration (days) = (0.02 x 966, 24m3 / 14 days) = 1.38 OH. In this case, all unit price values follow the classification owned by the company for building construction. From the results of the analysis of wall painting workers for 7 days are listed in Table 4.17. When the results of the total cost for completing the work of painting the wall area of 966.24 m2 for 7 days amounted to Rp.7,063,000.00. So it is needed: 3 workers, 9 masons, 1 head mason, 1 foreman, So there is an increase in labor, namely 1 worker and 4 masons. Activities (A-B-K) include preparatory work, land clearing work, and p

4. Conclusions

Based on the results of the study it can be concluded that the results of the normal duration of work with the Critical Path Method are 119 working days, the same as the planned duration. And with the acceleration of the duration carried out by the Critical Path Method using the addition of labor in activities (A-B-K) including preparatory work, land maturation work, and painting work so as to obtain 104 working days. Therefore, by using labor analysis, the total cost of normal labor costs is Rp.30,901,500.00 and the total cost of labor wage crashes is Rp.26,037,000.00. So that, save labor costs in the amount of Rp.4,864,500.00. So that the normal budget is Rp. 1,337,999,697.00 and the total cost of the crash are Rp. 1,333,135,197.00. Suggestions that can be given based on research that network planning must pay attention to the systematic preparation of the network and the preparation of dependency logic and conduct further analysis in the Evaluation and Review Technique) and PDM (Precedence Diagram Method).

References

[1] Antuli, N. (2014) Optimization of the Revitalization Building BPS City Schedule in the Gorantalo Building Using CPM and PDM Methods, *Civil Engineering Scientific Journal*, vol.1, no.1
[2] National Standardization Agency (2013) Analysis of Work Unit Price (AHSP) in Cipta Karya
[3] Spatial Planning and Cipta Karya (2014) Main Tasks and Functions, Organizational Structure.
[4] Ervianto, W.I. (2004) Construction Project Management Application Theory, ANDI, Yogyakarta
[5] Frederika, A. (2010) Analysis of the Acceleration of Implementation by Adding Optimum Working Hours to Optimum in Construction Projects (Case Study: Super Villa Construction Project, Casket Tenget - Badung), *Scientific Journal of Civil Engineering*, vol.14, no. 2
[6] Iwawo, E.R.M. (2016) Application of the CPM Method in Construction Projects (Case Study: New Building Construction of the Eben Haezar Complex in Manado), *Static Civil Journal*, vol. 4, no. 9
[7] Ministry of Public Works (2013) Regulation of the Minister of Public Works Number 11 / PRT / M / 2013 Regarding Guidelines for Analysis of Unit Work Price in Public Works
[8] Suputra, I.G.N.O. (2011) Project Scheduling with Precedence Diagram Method (PDM) and Ranked Position Weight Method (RPWM), *Scientific Journal of Civil Engineering*, vol.15, no.1

Journal of Physics: Conference Series **1424** (2019) 012026 doi:10.1088/1742-6596/1424/1/012026

Optimizing Wireless Distribution System Network Infrastructure in Hybrid Topology using PCQ Method

Givy Devira Ramady*, Rahmad Hidayat, Anung, Andrew Ghea Mahardika, Wisnu Hidayat

Sekolah Tinggi Teknologi Mandala, Bandung, Indonesia

*givy.d.ramady@gmail.com

Abstract. A good and efficient network infrastructure design must be considered when first building a computer network system. The position of buildings that are far from each other sometimes becomes a separate problem that results in the quality and performance of computer network signal services not being optimal. The implementation of the Wireless Distribution System (WDS) on the STTM campus network infrastructure will be combined with the use of the Peer Connection Queue (PCQ) method, which is one of the features of the Mikrotik device. The use of the PCQ method aims to optimize internet bandwidth, through a mechanism for distributing bandwidth evenly to each user that has been set on the management bandwidth, also we can observe the throughput, delay, and jitter of the built network infrastructure. The study was carried out following the stages of designing schemes and network topology, designing and configuring network infrastructure, then in the last stage testing connectivity and performance was done through several clients connected to the network. The data obtained from the test results will be analyzed and then evaluated whether it is by the expected target of 0% packet loss, delay below 150ms, and jitter under 75ms.

1. Introduction

Planning for computer network infrastructure in an institution or agency that has a high dependence on information technology needs to be done, it aims to make the performance and quality of internet connection services become more stable and optimal. In some conditions, especially in the environment of institutions or institutions that have several separate buildings, it will be very inefficient if you have to connect it using cable media.

One effort to overcome this wiring obstacle is to implement a Wireless Distribution System (WDS) technology. WDS is a technique of connecting multiple access points through wireless media in a Wireless Local Area Network (WLAN) and only uses one network cable as a backbone path on a central Access Point device. Whereas at other Access Points (clients), the backbone path originates from the central Access Point that is distributed through the wireless network.

With the implementation of WDS, the work area (coverage) of a WLAN can be expanded without having to connect the Access Point with a cable backbone system. Another advantage of the WDS is that the MAC Address header of the packet traffic does not change between the access point links because the connection between clients is connected when in a WDS state, using the MAC Address as the target destination. not based on the IP address. However, the WDS technique also has weaknesses, namely, the throughput that can be passed between networks is usually only half of the total throughput because the access point works double (sending and receiving data packets) so that it also impacts on bandwidth reduction.

Earlier in [1], a Wireless Distribution System service quality (QoS) service was tested using the parameters of output, delay and packet loss. Where in the test carried out a comparison of service quality based on the network topology used including Bus, Star, Mesh, Ring, and Tree. Then in [2][3][4], a Peer Connection Queue (PCQ) method is added to further optimize internet bandwidth. From the test results obtained an average value of 85% better throughput, and 200% delay and jitter 150% shorter than without using PCQ. This improved performance allows other service features that require stable network connectivity such as Voice Over Internet Protocol (VOIP) as in [5]to be more optimal.

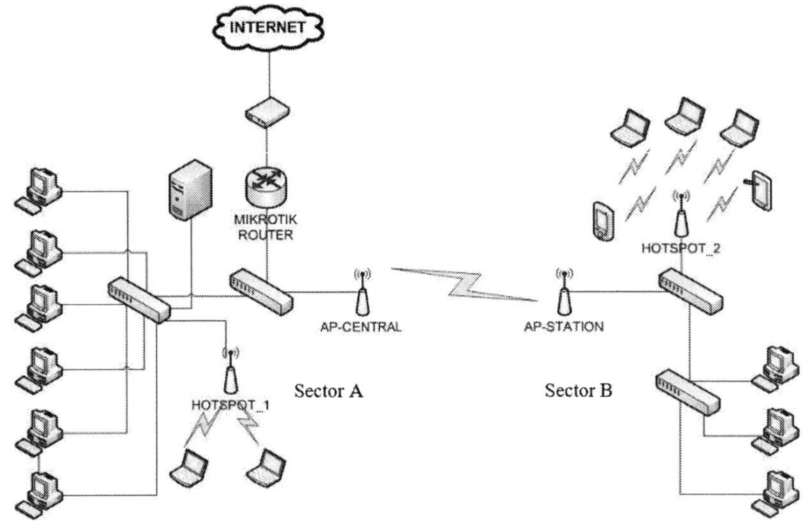

Figure 1. WDS Network Infrastructure Design

2. Material and method

Network Development Life Cycle (NDLC) is a method that depends on the previous development process so that at each stage it will determine the success of the next stage and this is a key in building a computer network infrastructure.[6]

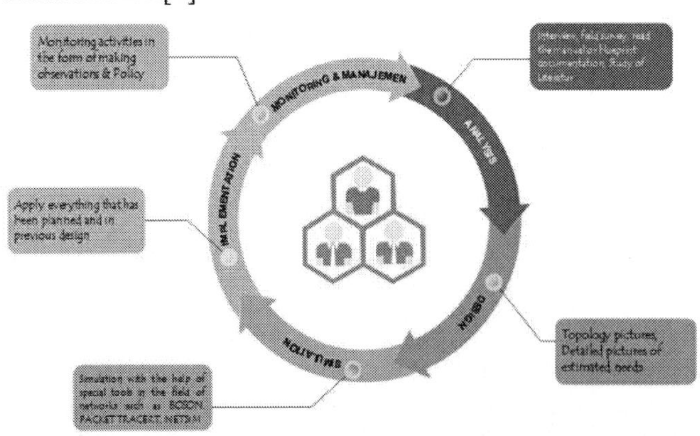

Figure 2. Model of Network Development Life Cycle

In the initial stage, an analysis of needs, problems that arise, user desires, as well as topology/network analysis that already exists today. The interview involves the management structure top to the bottom level in order to get concrete data and complete. Next step is to conduct a field survey to see the real conditions so that things that might be a factor in the decline in network quality can be traced. From the data obtained in the previous stage, the next step is to create a network topology design drawings that will be built, the results of the topology design will then be simulated using network simulation software to see the extent of its performance and reliability. Then the next step is to implement the topology that was designed in the previous stage to be realized in the field, the last is monitoring and management of the network infrastructure that has been built.

Peer Connection Queue (PCQ) is one of the features found in Mikrotik router OS that is useful for managing traffic rates and packet traffic in a network. The main purpose of this method is to make bandwidth sharing automatic and evenly distributed to multiple clients.

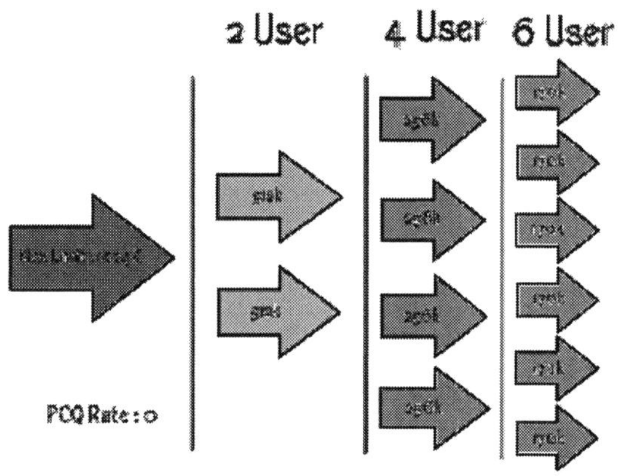

Figure 3. Peer Connection Queue system

PCQ Working Principle by implementing queue trees where there is only one active client that uses bandwidth, while other clients are in an idle position, the active client can use the maximum available bandwidth, but if another client is active, then the maximum bandwidth can be used by both clients (bandwidth or the number of active clients) so that bandwidth can be distributed fairly for all clients.

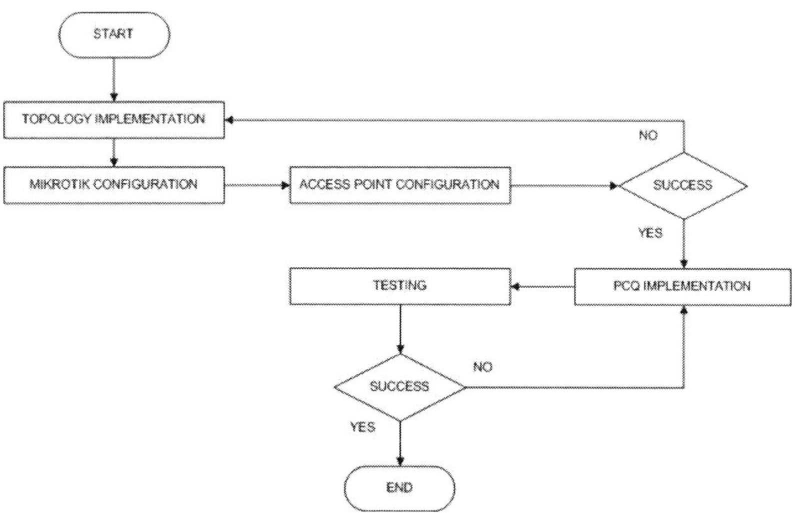

Figure 4. Steps of system design [7]

In this study, the proxy device used was the RB 951-U2nd router board version with router level 4 license OS. Testing was carried out by the client with a laptop operating system Windows 7 and an Android mobile device. Measurements and analyses are carried out in the form of connection quality tests on sector A network (AP_CENTRAL) and sector B network (AP_Station), while for QoS observations are made on packet loss, delay, and jitter.

Table 1. TIPHON QoS parameter standarization

Grade	Packet Loss	Delay	Jitter
Perfect	0 %	< 150 ms	0 ms
Good	< 3 %	< 250 ms	< 75 ms
Medium	< 15 %	< 350 ms	< 125 ms
Poor	< 25 %	< 450 ms	< 225 ms

Source : ETSI TR 101 329 v.2.1.1, 1999-06, page 26. [8]

3. Results and discussion

Network performance testing is done through a monitoring application found on the Winbox remote desktop. From there we can see the results of the configuration settings that have been made in the PCQ in the form of bandwidth management that controls the use of internet bandwidth, besides that it can also be seen the amount of packet loss, delay, and also its jitter as shown in the picture below :

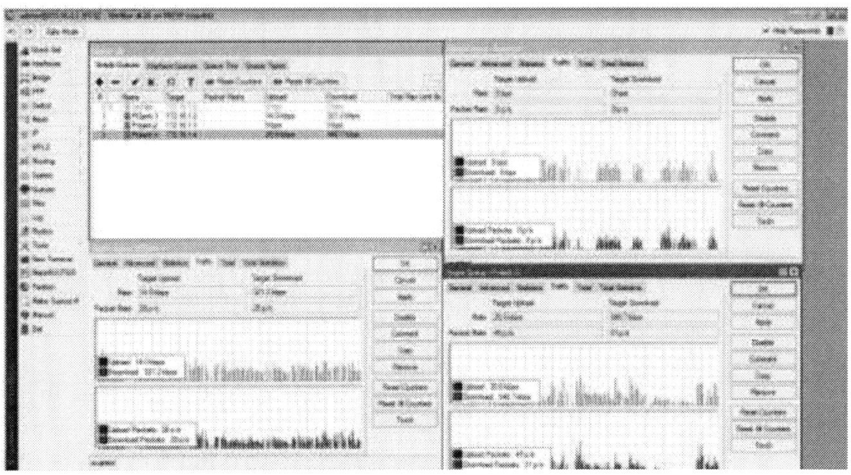

Figure 5. Graph monitoring results using win-box

3.1 Quality of service (QoS) monitoring in sector A

In testing conducted in sector A which is the main backbone network, where it can be seen that the implementation of Peer Connection Queue has made the use of internet bandwidth more controlled and between client 1 and other clients there is no bandwidth tug because the system has divided its bandwidth each client is equally equal. Network performance and stability testing are done by doing a youtube streaming test because this activity most requires a connection with large and stable bandwidth. From the observations obtained packet loss value of 0%, delay 81 ms, and jitter 13 ms.

Table 2. QoS monitoring result Sector A

	Value	Grade
Packet Loss (%)	0	Perfect
Delay (ms)	81	Perfect
Jitter (ms)	13	Good

3.2 Quality of service (QoS) monitoring in sector B

Then in testing conducted in sector B which is a WDS client network, it can be seen that the implementation of Peer Connection Queue has also succeeded in making the use of internet bandwidth more controlled and between client 1 and other clients there is no bandwidth tug because the system has divided the bandwidth to each client is even. Testing the performance and stability of the network is still carried out through the YouTube streaming test, and the results of observations made are 0% packet loss, 142 ms delay, and 25 ms jitter.

Table 3. QoS monitoring result Sector B

	Value	Grade
Packet Loss (%)	0	Perfect
Delay (ms)	142	Perfect
Jitter (ms)	25	Good

4. Conclusions

Based on an analysis of the results of testing the implementation of the wireless distribution system and the use of the PCQ method in the network infrastructure created, it was concluded that the WDS network has a wide range of coverage but is economically cheaper than the use of cable media, the use of the PCQ method is proven to have helped the distribution of bandwidth evenly on each client through bandwidth management. The difference in the value of the test results in sector A and sector B, is indicated to be the influence of the weak point of WDS on the position of AP Station (Client) in sector B, which in sector B AP functions as a sender and receiver of data packages so that the ability to pass throughput is reduced . But from the test results, the value of packet loss, delay, and jitter in sector B is still in the area that is standardized by TIPHON.

References

[1] D. I. Haerudin, L. M. B. Aksara, and M. Yamin, "Implementasi Wireless Distribution System (WDS) pada Hotspot (Studi Kasus: SMK Negeri 1 Kendari)," *SemanTIK*, vol. 3, no. 2, 2018.

[2] S. Supiyandi, "Rancang Bangun Jaringan Menggunakan Mode PPOE Client Mikrotik dan Metode PCQ di Jaringan Internet SMA Negeri 20 Medan," *Comput. Eng. Sci. Syst. J.*, vol. 3, no. 1, pp. 28–35, 2018.

[3] I. Faisal, "Analisis QoS Pada Implementasi Manajemen Bandwidth Menggunakan Metode Queue Tree dan PCQ (Peer Connection Queueing)," *J. Teknol. dan Ilmu Komputer. Prima*, vol. 1, no. 1, pp. 137–142, 2018.

[4] H. P. Situmorang and J. C. Chandra, "Implementasi Manajemen Bandwidth Menggunakan Metode Peer Connection Queue pada SMK Budi Mulia Tangerang," *IDEALIS Indones. J. Inf. Syst.*, vol. 2, no. 3, pp. 202–208, 2019.

[5] R. Hidayat, N. Sri Lestari, A. Sujana, Herlina, and G. Devira Ramady, "Optimizing Branch Telephone Networks for Campus VoIP with Mobile Clients," in *Journal of Physics: Conference Series*, 2019, vol. 1175, no. 1.

[6] J. E. Goldman and P. T. Rawles, *Applied data communications: a business-oriented approach.* Wiley, New York, 2004.

[7] T. U. Di Cafe and A. Yogyakarta, "Perbandingan PCQ pada Simple Queue dan Queue Tree Mikrotik Untuk Pengelolaan Bandwidth."

[8] European Telecommunications Standards Institute (ETSI) 1999 Telecommunications and Internet Protocol Harmonization Over Networks (TIPHON); General aspects of Quality of Service. ETSI TR 101 329 V2.1.1

Jigsaw using animation media for elementary school

Rahmatina[1], **A K Kenedi**[2], **R Eliyasni**[1], **R Fransyaigu** [2]
[1]Universitas Negeri Padang, Jl. Prof Dr. Hamka Air Tawar, Padang, Indonesia
[2]Universitas Samudra, Jl. Meurandeh, Langsa, Indonesia

E-mail: rahmatina61@gmail.com

Abstract. This research is motivated by the need for innovation in the learning process in the 4.0 Industrial Revolution era. The purpose of this study is to determine the effect of jigsaw cooperative learning using animation media for elementary school students. This research is quasi-experimental using posttest only. The results show a significant positive effect of the application of jigsaw cooperative learning by using animation media towards the learning outcomes of elementary school students. The implication of this study is that elementary school teachers can use cooperative learning models using animation media as an effort to improve the learning outcomes of elementary school students in the 4.0 industrial revolution era.

1. Introduction

The 4.0 industrial revolution era is the era of change from analogous life systems into digital life systems [1]. This change is accompanied by the development of technology use in every community life [2]. This can be seen that many people make technology as a foundation in the daily life process.
4.0 industrial revolution era also has an impact on the education system [3]. The education system in the era of 4.0 industrial revolution requires new innovation in the learning process in the classroom [4]. This is due to the challenge in the industrial revolution which demands a change in the learning system from traditional to the modern learning system. This challenge must be carried out by the teacher to produce graduates who have competitiveness in line with the expectations. Included in the learning process in elementary school.

Elementary school is the first formal education level in Indonesia [5]. Therefore learning innovation in the era of 4.0 industrial revolution can be started from learning in elementary school. Thus students get used to and are trained in adjusting themselves in the process of changing the learning system in elementary schools.

One of the innovations that can be developed in class learning systems is learning models and technology-based learning media [6] [7] [8]. The learning model is a series of learning processes that have patterns in each process. The learning model is a planned and systematic procedure that aims to organize the learning process through learning experiences to achieve learning objectives [9]. The learning model is a systematic process in the learning process that aims to achieve learning objectives. One model that can be used by teachers is a cooperative learning model. Cooperative learning model is identical to the learning process by the way students learn in groups [10]. Cooperative learning model is learning using small groups of students who are mutually responsible and interact in the learning

---

1 To whom any correspondence should be addressed.
2 To whom any correspondence should be addressed.

process together [11]. The cooperative learning model provides various advantages such as being able to improve student achievement, student self-esteem, student motivation, love of the learning atmosphere, student retention and student socialization in the learning process [12] [13] [14]. The cooperative learning model has a variety of types, one of which is a jigsaw.

Jigsaw is a cooperative learning model implemented in groups and consists of several heterogeneous students who are responsible for learning certain problems from the material given and conveying the material to other groups [15]. Jigsaw is a type of cooperative learning model that requires students to take responsibility in the learning process [16]. This is because each group has the responsibility to mastering the learning material and in charge of delivering the material to other groups. If one group member is not responsible for understanding the material, the other learning process will not be carried out properly.

Jigsaw aims to train students to develop their abilities and facilitate teacher in delivering material [17]. In addition, a jigsaw can also improve student learning outcomes, positive attitudes, mutual respect for differences and significantly to increase responsibility [18] [19] [20]. This is in line with the challenge given by the 4.0 industrial revolution era that the graduates must have responsibilities and strong competitiveness in the globalization era market.

Besides the learning model, another innovation is learning media. Learning media is a set of tools that help the learning process with the purpose to facilitate the delivery of meaning and the message to be conveyed [21]. Learning media is used by teachers to compose communication and interaction with students in the learning process. Learning media aims to convey learning messages, make learning interesting, make learning interactive, time efficiency, improve the quality of learning, enhance students' positive attitudes and make teacher's role more positive [22] [23] [24] [25] [26].

A learning media following the characteristic of 4.0 industrial revolution era is animation learning media. This is because animation is an audio-visual product of technological development. Animation media is an interesting form of pictorial presentation because it is a simulation of a moving picture depicting the movement of an object [27]. Animation can be interpreted as a process of changing the shape or property of an object displayed within a certain period of time. The animation is considered capable of being an innovation in elementary school learning because the media of animation can display objects that do not exist physically (imagery) [28]. Besides, animation can combine virtual elements such as text, images, video, audio into an integrated product [29]. Animation can also harmonize student learning styles such as visual, audio, kinesthetic or others and animation facilitates teachers in developing the learning activity.

Based on the facts above, it can be concluded that the animation media is suitable to be applied in the elementary school learning process. This is because elementary school students have concrete operational characteristics [30]. Thus the animation media can connect learning with the mindset of elementary school students.

Therefore, the learning model and learning media can be used as a new innovation in the learning process in the era of 4.0 industrial revolution. The use of jigsaw in the learning process has often been carried out in elementary schools and has a positive impact on learning outcomes. But further research into the application of jigsaw using animation media in elementary school learning is needed. Therefore the purpose of this study is to determine the effect of the jigsaw cooperative learning model by using animation media in the learning process in elementary schools.

2. Methods

This is a quasi-experimental research. This is post-test research. The technique used in this research is simple random sampling. The samples are grade V students of Elementary School number 11 Koto Gadang and grade V students of Elementary School number 12 Koto Tinggi, amounting to 40 people who have been tested previously for normality and homogeneity with normal distribution results. The experimental class is a class given a jigsaw using animation media while the control class is a class that only uses jigsaw without using animation media. In this research, the researcher uses tests in collecting data.

3. Result and discussion

This study starts with the implementation of the prerequisite test, namely the normality test and homogeneity test.

Table 1. Results of Post-Test Normality Test for Experimental and Control Classes

No	Class	N	L_{count}	L_{table} α 0.05	Conclusion
1	Experimental	20	0.04143	0.190	Normal Data
2	Control	20	0.05108	0.190	Normal Data

The results of the posttest normality test of the experimental and the control class in the study. It is known, L_{count} of the experimental class is 0.04143 and L_{count} of the control class is 0.05108 with the number of samples each 20. At the significance level $\alpha = 0.05$, the L_{table} value is 0.190. Thus the experimental class obtained $L_{count} < L_{table}$, $(0.04143 < 0.190)$, and the control class also obtained $L_{count} < L_{table}$ $(0.05108 < 0.190)$. Normality test aims to see whether the residual value is normally distributed or not [31]. Based on these data, it appears that the experimental class and control class data are normal.

Next is the homogeneity test. Homogeneity test can be seen in the table below.

Table 2. Results of Post-test Homogeneity Test for Experimental and Control Class

No	Class	N	Variance	F_{count}	F_{table}	Conclusion
1	Experimental	20	197.853	1.2773	2.12	Variant
2	Control	20	154.895			Homogeneous

The results of the Post-test Homogeneity Test for Experimental and Control Class in this study are known that the calculated experimental class and control class are 1.80299 with a sample of 20 each. At a significance level of $\alpha = 0.05$ obtained F_{table} of 2.12. Thus we get $F_{count} < F_{table}$, i.e. $(1.2773 < 2.12)$.

Thus, it can be concluded that the data in the experimental class and the control class have a homogeneous variant. The homogeneity test is done by examining whether the two samples come from the population with the same variance or not. This test is conducted to discover whether the assumption of homogeneity in each data category has been fulfilled or not [32]. If the assumption of homogeneity is proven, the researcher can carry out the advanced data analysis stage. However, if it is not, the researcher must make methodological corrections.

The next step is to use a hypothesis test. A hypothesis test is carried out on the post-test value of the experimental class and the control class. Based on the prerequisite data analysis, post-test data for both groups is normally distributed and has a homogeneous variant. Hence hypothesis test can be done using a variant polled t-test. The T-Test results for the experimental class and the control class are as follows.

Table 3. Results of T-Test for Experimental and Control Classes

No	Class	N	Mean	T count	T table 0,05/2	Conclusion
1	Experimental	20	83,4	13,8024	11,6859	H_a Accepted
2	Control	20	73,6			H_o Rejected

It is known that the calculated t count is 13.8024 and t table is 11.6859 at the level of significance $\alpha = 0.05$. Because 13.8024> 1.6859 hen Ho is rejected, and Ha is accepted. Consequently, it can be concluded that there is a significant influence on the use of Jigsaw Cooperative learning model using animation media in elementary school learning. Nevertheless, the average of the control class is categorized as quite good.

Jigsaw is a learning model requires students to work together in groups and each group is responsible for understanding the material delivered to each other group members [33]. Those who study in groups will improve learning outcomes. In group work, students will be trained to explore both existing and non-acquired knowledge. Students will easily exchange information and ask questions about the subject matter between peers. This makes the cognitive development of students trained in finding information. Furthermore, the jigsaw can improve student activity in the learning process [34]. Activeness is a part of social knowledge (Vygotsky social development theory) [35]. Student activity is a manifestation of social knowledge development of elementary school students. The activeness arising from the collaboration will affect the learning process such as the ability to understand and solve problems [36]. Moreover, the activeness builds students' curiosity thus students eagerly look for the information they want. Student responsibility given to jigsaw learning will make students stronger and more focused on finding the information given to other members. This sense of responsibility makes students continue to practice their ability to understand the material delivered [37] [38] [39] [40]. Therefore the social development produced in the jigsaw learning process has a high impact on the process of student learning outcomes. This what makes the control and experimental classes have high learning outcomes.

However, in the experimental class, the student learning outcomes are higher than the control class. This is supported by the fact that the experimental class gets jigsaw learning using animation media. Animation media plays an important role in the process of increasing learning outcomes. Animation media can motivate students in the learning process and improve the positive attitude of students [41]. This is because animation is a motion illusion product that can present information in an entertaining way [42]. Children in elementary school tend to use the principle of playing while learning in line with the animation media. Therefore, students will be motivated to learn without pressure.

Animation media can also improve student learning outcomes [43]. Animation media knows how to present visualization that cannot be raised in a real way in the classroom, making it easier for students to understand difficult abstract concepts permanently [44]. This is what stimulates the formation of a concrete mindset of elementary school students thus elementary school students can imagine the information. This is because of elementary school students considered as children in the concrete operational phase.

As a result, the application of jigsaw using animation media is very appropriate to be used for elementary school students. This proves that this innovation will have a positive impact on student learning outcomes in elementary school.

4. Conclusion

This study concluded that the application of jigsaw using animation media had a positive effect on the learning outcomes of elementary school students.

Acknowledgment

Thank you to Universitas Negeri Padang for permitting collaborative research. Thank you to the Universitas Samudra that permitted research collaboration.

References

[1] Lasi H, Fettke P, Kemper, H G, Feld T, and Hoffmann M 2014 *Business & information systems engineering* **6** 239
[2] Kinzel H 2017 *Journal of Urban Culture Research* **15** 70
[3] Jones C and Pimdee P 2017 *Asian International Journal of Social Sciences* **17** 4
[4] Puncreobutr V 2016 *St. Theresa Journal of Humanities and Social Sciences* **2** 2
[5] Kenedi A K, Helsa Y, Ariani Y, Zainil M and Hendri S 2019 *Journal on Mathematics Education* **10** 69
[6] Serdyukov P 2017 *Journal of Research in Innovative Teaching & Learning* **10** 4
[7] Subramani P N and Iyappan V 2018 *Journal of Applied and Advanced Research* **3** 2

[8] Ala-Mutka K, Punie Y and Redecker C 2008 ICT for learning, innovation and creativity. *Institute for Prospective Technological Studies (IPTS)* (European Commission, Joint Research CenterTechnical Note: JRC)
[9] Ahmad Zaki 2018 *J. Phys.: Conf. Ser.* **1028** 012165
[10] Singh Y P and Agrawal A 2011 *Indian Streams Research Journal* **1** 2
[11] Yassin A A, Razak N A and Maasum T N R T M 2018 *Advances in Social Sciences Research Journal* **5** 8
[12] Pateşan M, Balagiu A and Zechia D 2016 *International Conference Knowledge-based Organization* **22** 478
[13] Yoyok Soesatyo Y and Dewie Tri Wijayati Wardoyo D 2018 *Interciencia Journal* **43** 124
[14] Cornelius-Ukpepi, Aglazor G N and Odey C O 2016 *Advences In Multidisciplinary & Scientific Research* **2** 67
[15] Tran V D and Lewis R 2012 *International Education Studies* **5** 99
[16] Alabekee E C, Samuel A and Osaat S Dm 2015 *International Journal of Education Learning and Development* **3** 67
[17] Garcia A, Abrego J and Robert R 2017 *Global Journal of Humansocial Science: G Linguistics & Education* **17** 5
[18] Melinamani S, Francis F, George R, Pushpa L M and Vergheese S 2017 *Asian J. Nursing Edu. and Research* **7** 2
[19] Adams F H 2013 *International Journal of Education and Practice* **1** 64
[20] Tran V D and Lewis R 2012 *International Journal of Higher Education* **1** 20
[21] Widodo S A 2018 *Turkish Online Journal of Educational Technology-TOJET* **17** 154 [22] Shinn Y H 1997 *Teaching strategies, their use and effectiveness as perceived by teachers of agriculture: A national study* (Retrospective Theses and Dissertations :12244)
[23] Aini W N 2013 *Journal of English and Education* **1** 196
[24] Naz A A and Akbar R A 2008 *Journal of elementary education* **18** 35
[25] Falahudin and Iwan 2014 *Jurnal Lingkar Widyaiswara* **1** 104
[26] Anjarwati D, Winarno A and Churiyah M 2016 *IOSR Journal of Research & Method in Education* **6** 5
[27] Daly C J, Bulloch J M, Ma M and Aidulis D 2016 *Advances in physiology education* **40** 201
[28] Nurizmawati C, Apriliaswati R and Arifin Z 2013 *Jurnal Pendidikan dan Pembelajaran* **4 3**
[29] Omar M A M and Ishak M S A 2012 Understanding Culture through Animation: From the World to Malaysia. *Jurnal Pengajian Media Malaysia* **13** 2
[30] Kiswanto A 2017 *The Effect Of Learning Methods And The Ability Of Students Think Logically To The Learning Outcomes On Natural Sciences Of Grade IvS Student* (ICSET 2017: Atlantis Press)
[31] Das K R and Imon A H M R 2016 *American Journal of Theoretical and Applied Statistics* **5** 5
[32] Conover W J, Johnson M E and Johnson M M 1981 *Technometrics* **23** 351
[33] Karacop A 2017 *Universal Journal of Educational Research* **5** 420
[34] Suendarti M 2017 *International Journal of Environmental and Science Education* **12** 1617
[35] Shabani K 2016 *Cogent education* **3** 1252177
[36] Sharan Y 2014 *Anales De Psicología/Annals of Psychology* **30** 802
[37] Huang Y M, Liao Y W, Huang S H and Chen H C 2014 *Journal of Educational Technology & Society* **17** 128
[38] Buhr G T, Heflin M T, White H K and Pinheiro S O 2014 *Journal of the American Medical Directors Association* **15** 429
[39] Maden S 2011 *Educational Sciences: Theory and Practice* **11** 911
[40] Berger R and Hänze M 2015 *International Journal of Science Education* **37** 294
[41] Baglama B, Yucesoy Y and Yikmis A 2018 *TEM Journal* **7** 670.
[42] Martinez O O L 2015 *Animation* **10** 42
[43] Siregar U A, Hadi W and Daulay S 2017 *Journal of Education and Practice* **8** 35

[44] Berney S and Bétrancourt M 2016 *Computers & Education* **101** 150

Morphological Segmentation for Identifying Backbone Area

Yuti Dewita Arimbi[1], Lussiana ETP[2], Sarifuddin Madenda[3], Yuhilza Hanum[4]

[1,3,4]Faculty of Computer Science and Information Technology, Gunadarma University, Indonesia
[2]Department of Information System, STMIK Jakarta STI&K, Indonesia

*yuti_dewita@staff.gunadarma.ac.id

Abstract. Segmentation in images of abdominal CT scan is an important step in the processing and interpretation of medical images. This is not easy, because the organs are close together and the color intensity is similar. Various methods have been developed, but the problem of locating the position of the backbone is still difficult to solve. The purpose of this study is to develop a segmentation method to determine the location of the backbone using a morphological method with a binarization approach. Testing of 11 CT Scan images shows that the method developed can determine the area and position of the backbone. Thus, it can be concluded that the method of backbone segmentation using a morphological method with a binary approach is successfully developed so that it can be applied to the advanced analysis process.

1. Introduction

Kidney stones are characterized by the formation of crystals in the urine caused by the concentration of a substance or by genetic disorder. According to [1], kidney disorders caused by kidney stones showed an increase in numbers both in terms of age and sex. In the majority of kidney stone patients, the stones are detected after symptoms such as fever, low back pain, abdominal pain and urinary tract infections occur. The presence of kidney stones has been confirmed by urine testing in the laboratory and abdominal imaging, one of which is by using Computer Tomography (CT), known as CT scan. If kidney stones are found, treatment can use the ESWL (Extracorporeal Shockwave Lithotripsy) method, PNL (Percutaneous Shockwave Litholapaxy), open surgery, and conservative therapy or expulsive medical therapy (TEM). This shows that the accuracy of identifying kidney stones is an important step for making decisions on follow-up actions taken by specialists. Identifying kidney stones based on CT scan images is not easy. This is due to various factors, including the complexity of anatomy and the quality of images produced [2].

Figure 1 is an example of the complexity of abdominal anatomical structures from CT Scan images.

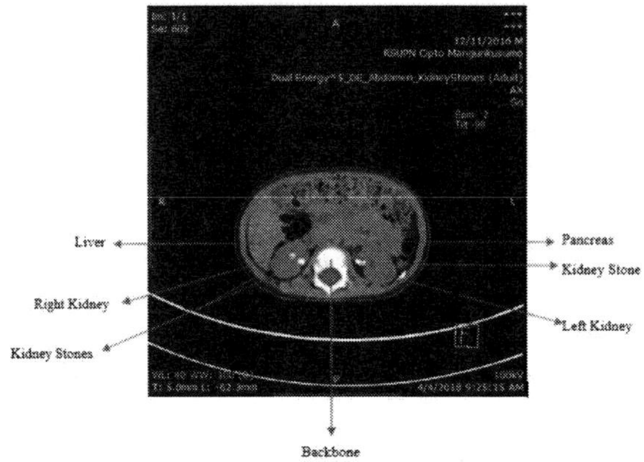

Figure 1. CT Scan image of Abdomen Anatomy

Figure 1 shows the parts of the abdomen including the liver, right kidney, left kidney, pancreas and backbone. Visually, the color of the organs is gray shade, so it can be stated that the separation (segmentation) of the abdomen organs is difficult.

Various methods of segmentation to identify kidneys have been developed. One of them is by Belgherbi [3], who develop semi-automatic kidney segmentation algorithms in abdominal CT images using morphology with a watershed approach. The goal is to remove the backbone so that kidney segmentation can be done. This study resulted in an average sensitivity of 95% but could be below 91% if the CT Scan was not precise and used a different machine. Campadelli [4] segmented the abdominal organs with CT images using morphology to identify the backbone. This study succeeded in segmenting the liver and pancreas, but segmentation of the backbone, kidneys, and heart required further visual examination. Myint [5] compared the effectiveness of the kidney segmentation method manually with a gradient-based approach, and semi-automatic kidney segmentation with the region growing approach. It can be concluded that the region growing technique is not better than the gradient-based approach method. From the 258 CT Scan images, the region growing technique produced an average performance of 70.5% and an average gradient-based technique of 85.5%.

Referring to the research that has been done, identifying the backbone is an important stage in identifying the presence of kidney stones. Based on these conditions, this study aims to develop a segmentation method for identifying the backbone from CT scan images with morphological methods using a binary approach.

2. Literature Review

2.1 CT scan image

CT Scan stands for Computed Tomography or Computerized Tomography Scan. Computed Tomography uses X-Ray technology to detect various conditions and diseases by synthesizing basic images and synthesis units called volume elements [6]. The x-ray tube is rotated and the light is projected in various positions. The amount of electricity received is converted into a digital scale by Analog to Digital Converter (A/DC) which is then recorded by the computer. Next it is processed using the Image Processor, and finally an image is displayed on the TV monitor screen. The resulting image can be made into a film with Multi Imager or Laser Imager [7].

2.2 Morphological Segmentation

One method of image segmentation is the morphological method, which is a mathematical function used to extract and describe the geometric structure of objects in an image [8]. Image processing with a morphological approach is usually done on binary images but does not rule out the possibility of conducting it on gray-scale images. Morphology has two basic operators, namely dilation and erosion, which are used to extract the desired components in an image [9]. Dilation operation is a morphological process that is intended to expand the area or size of an object [10]. Dilation of the image f by the structure of the dilated B matrix element is given by equation (1) below:

$$f \oplus B = \{s \mid (B)_s \cap f \neq 0\} \quad (1)$$

Image dilation, namely *f*. Mathematically, the dilation operator is expressed as the symbol □, with *B* being the structure of the dilated matrix element *B* with a shift of *s*. The implementation of dilated operations in this study uses binary image which can be calculated by equation (2) as follows:

$$f \oplus B \begin{cases} 1 \text{ if } B_s = f = 1 \\ 0 \text{ if } B_s = f = 0 \text{ if only } B_s \neq f \end{cases} \quad (2)$$

Equation (2) shows that the value of the dilation element will be equal to 1 if and only if there is one value of the element $f = 1$ which is equal to the value of element $B = 1$ in the same coordinate position. Accordingly, the value of the dilation element will be equal to 0 if and only if none of the values of the element $f = 1$ which is equal to the value of element $B = 1$ in the same coordinate position.

2.3 Kidney

Kidneys have a very important role as an organ that functions to control fluid balance in the body, especially the urinary system, control salt concentration in the blood and acid-base balance of blood, and secretion of waste materials and excess salt [11]. The kidneys are a pair of nut-shaped organs, which are located on both sides of the vertebral column (backbone) [12]. In CT Scan images, the kidneys are of gray-scale color, similar to other abdominal organs such as the liver, pancreas, stomach, and others.

2.4 Backbone

Bone is a living structure composed of proteins and minerals. The mineral content of bone is calcium phosphate carbonate. Bone becomes hard due to deposition of calcium phosphate crystals [13]. The kidneys are located between the backbone precisely in the retroperitoneal space between the thoracic XII or lumbar I and lumbar vertebrae [14]. When viewed in the CT Scan image, the backbone is white like the color of the bone around the abdomen. The white color is produced from the calcium in the bones.

2.5 Kidney Stones

Kidney stones are solid objects formed by various substances dissolved in urine. The stones can be distinguished based on the composition of the substances that make them up. Based on the composition of the substances that make up the stones, they are divided into calcium stones, struvite stones, gout stones, cystine stones, xanthine stones, triamterene stones, and silicate stones [15]. When viewed on a CT scan, kidney stones produce white color because of the composition of calcium contained in the stone, just like bone color.

3. Methodology

Basically, the research stages are as shown in Figure 2.

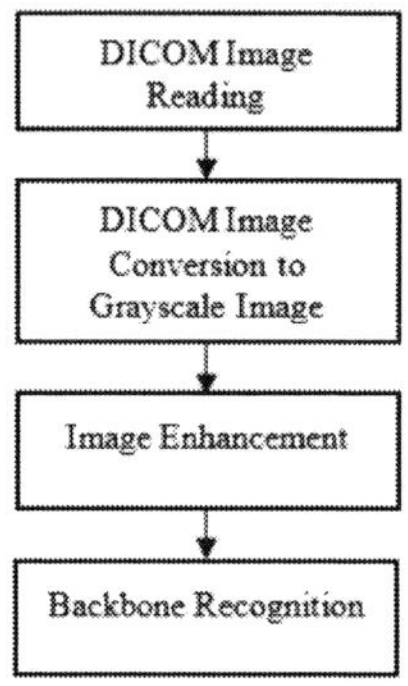

Figure 2. Research Stages

The first step is to read the CT scan image with the DICOM format studied. The next step is to convert the DICOM image into a gray-scale image. This step is done to delete the writing or symbol that is not needed in the subsequent processing. The third stage is image enhancement which consists of (1) reversing the color of the image, this process aims to see objects around the abdomen that are not needed in the area being analyzed, (2) adjusting the intensity of the image color to make sure that unnecessary areas are eliminated. The fourth step is backbone detection by performing the binarization of gray-scale image. This step produces an image with the area sought (ROI), namely the backbone.

4. Result and Discussion

DICOM Image Reading

The input image is in DICOM (Digital Imaging and Communication in Medicine) format. The DICOM format is used for medical imaging that can be exchanged for data and quality needed for clinical use. DICOM has the format extension .dcm with a size of 512x512 pixels. An example of a DICOM format image can be seen in Figure 3.

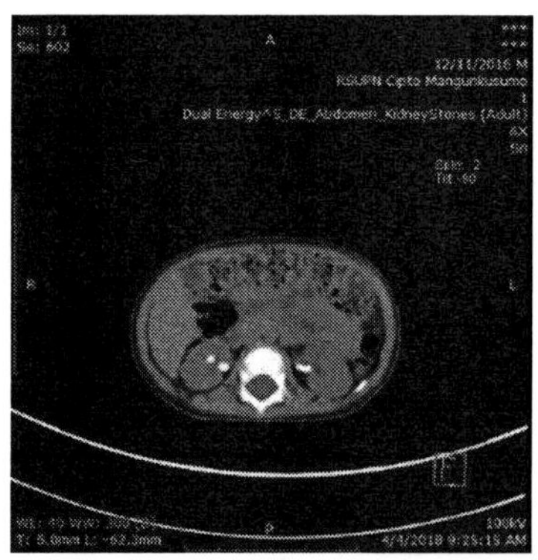

Figure 3. DICOM Image

DICOM Image Conversion

This stage aims to convert DICOM image to gray-scale image so that CT Scan images can be identified and displayed by MATLAB 2018a software. The formulation of RGB image conversion into gray-scale image can be seen in equation (3).

$$\textit{gray-scale} = 0.299 \times R + 0.5876 \times G + 0.114 \times B \quad (3)$$

With R expressing the value of the Red component, G represents the value of the Green component, and B denotes the value of the Blue component.

The results of the gray-scale image conversion can be seen in Figure 4.

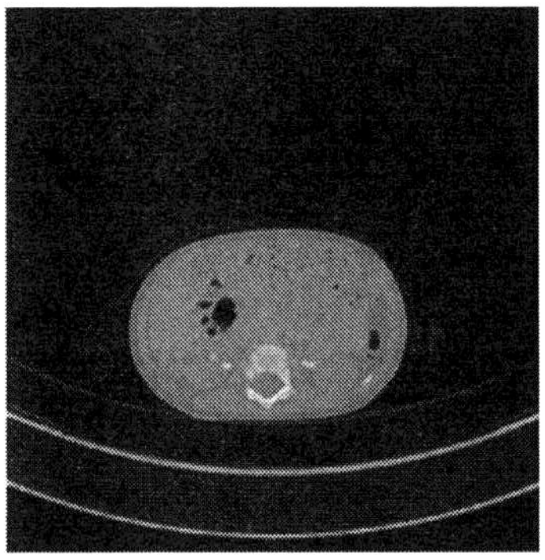

Figure 4. The Result of Converting DICOM Image into Gray-Scale Image

Figure 4 gray-scale image resulting from the conversion of an input image that shows that the patient data information is deleted.

Image Enhancement

Image quality improvement is needed because the image used as the research object has poor color quality, such as too dark or too bright, and to delete unnecessary information (noise). The enhancement stage in this study consisted of color reversal, and setting the color intensity of the image.

Color Reversal (*Imcomplement*)

The operation of reversing the color of the image aims to produce a negative image, so that it is expected to strengthen the color of the object needed and facilitate processing. The process of reversing the color of the image is done using equation (4).

$$g\,(x, y) = 255 - f(x, y) \tag{4}$$

with,

$g\,(x, y)$: image resulted form imcomplementing $f(x, y)$

$f\,(x, y)$: gray-scale value of the input image

The result of reversing the color of the image can be seen in Figure 5.

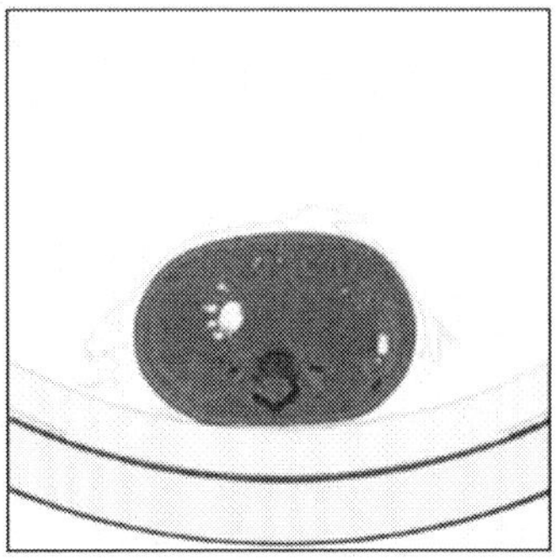

Figure 5. Color Reversal Result

In figure 5, it appears that the background of the image turns white, so that unneeded objects around the abdomen can be spotted. The abdominal area also changes according to the results of the new intensity value. Thus, the object around the abdomen is easy to remove.

Journal of Physics: Conference Series **1424** (2019) 012028 doi:10.1088/1742-6596/1424/1/012028

Image Color Intensity Settings

Increasing image quality can be done by increasing the brightness so as to make the image appear brighter, and the required object becomes clearer. It is known that f (x, y) states the pixel value in gray-scale images at coordinates (x, y), then the new image uses equation 5 as follows:

$$g(x, y) = f(x, y) + \beta \tag{5}$$

with,

g (x, y) : Result of image color intensity settings
f (x, y) : gray-scale image
□ □□variable (If the value is positive then the brightness of the image increases. If the value is negative, then the brightness of the image decreases)

The results of setting the color intensity of the image can be seen in Figure 6.

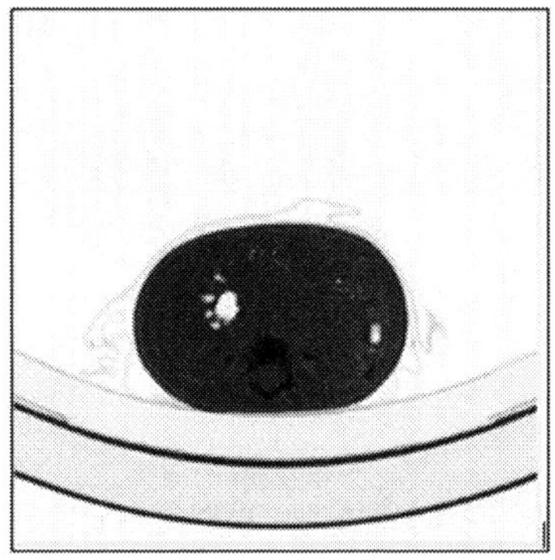

Figure 6. Result of Image Color Intensity Setting

The result of setting the color intensity of the image in Figure 6 shows that the color of the image becomes more black compared to the previous image in Figure 5, and unnecessary objects around the abdomen are clearly visible. This stage aims to clarify the area of the backbone.

Backbone Recognition

Backbone recognition aims to determine the ROI (region of interest) of a CT Scan image by taking the most important part of the image and removing part of the image that is not needed. Prior to making ROI steps, the binarization process is conducted. The binarization process is done by defining the threshold value (T) as the limit to check whether a pixel is included in the desired area or not. The formulation for the binarization process can be seen in the following equation 6:

$$g(x, y) = \begin{cases} 0, & if\ f(x, y) \geq T \\ 1, & if\ f(x, y) < T \end{cases} \tag{6}$$

with,

g (x, y): Binary image
f (x, y) : Gray-scale image
T : *Threshold*

The threshold value (T) is used to determine the intensity to be converted to 0 (zero) or to 1 (one).

The ROI identified as the midpoint area is the location of the backbone. The result of the morphological approach to the image can be seen in Figure 7.

Figure 7. Morphological Results

Figure 7 shows a black area with a white background, indicating that the identification of the backbone has succeeded.

This method was tested on 11 CT Scan images and successfully carried out the identification of the backbone. The results of the study can be seen in Table 1.

Table 1. Image of Backbone Detection Result

Image Code	CT Scan Image	ROI	Morphological Result	Image Code	CT Scan Image	ROI	Morphological Result
01_02	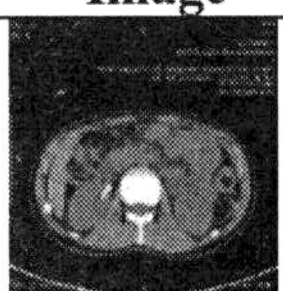	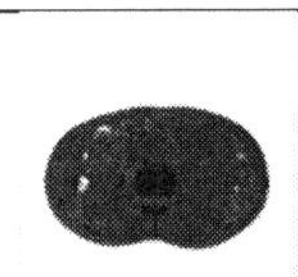	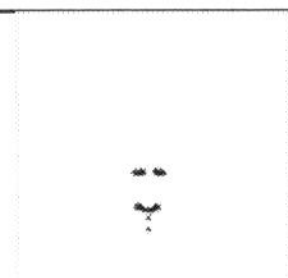	02_14	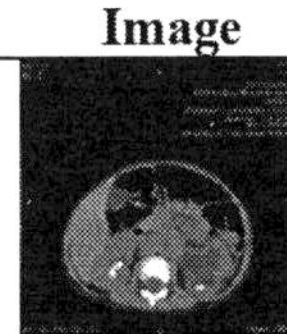		
01_08	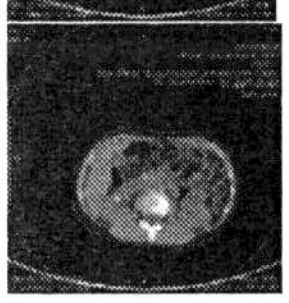			02_18	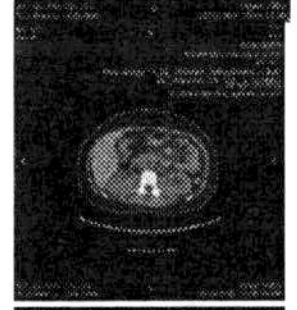		
01_35	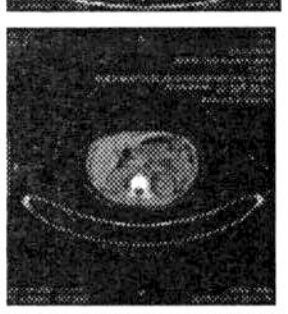	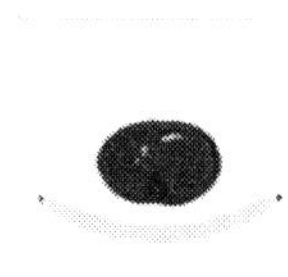		02_22	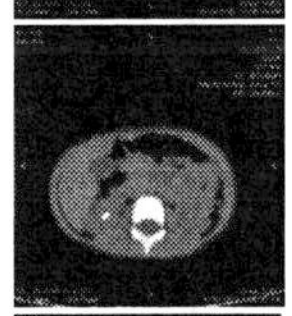	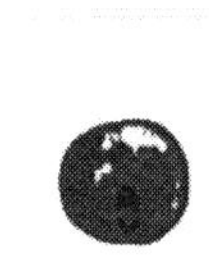	
01_44	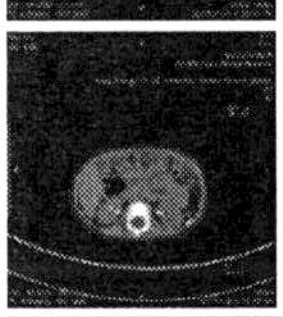			02_23	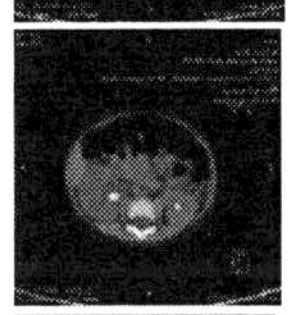		
01_48	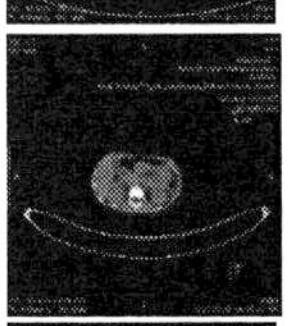	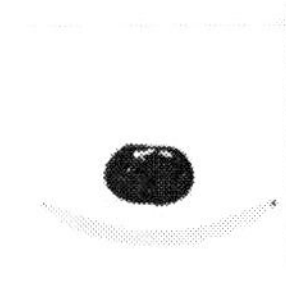		02_33	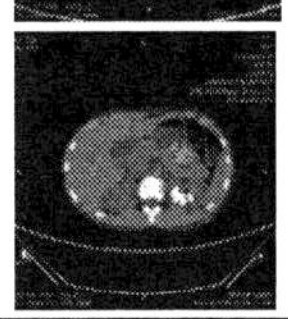		
02_01	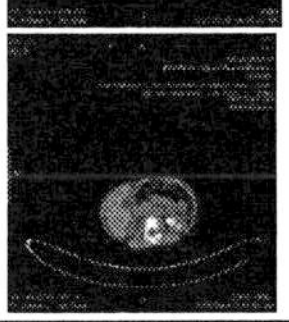						

5. Conclusion

From the results of trials on 11 CT Scan images, it was proven that the method developed successfully identified the backbone. Based on the results of testing in table 1, it can be stated that the development of morphological methods with a binary approach is suitable for conducting backbone recognition.

Future work

The result of the identification of the backbone will be then used as a reference to develop kidney stone identification methods in further research.

References

[1] RISKESDAS, "Riskesdas 2013," 2013.
[2] O. D. Nurhayati, "Analisis Citra Digital CT Scan Dengan Metode Ekualisasi Histogram Dan Statistik Orde Pertama," *J. Sist. Komput.*, vol. 5, no. 1, pp. 1–4, 2015.
[3] A. Belgherbi, I. Hadjidj, and A. Bessaid, "Morphological Segmentation of The Kidneys From Abdominal CT Images," *J. Mech. Med. Biol.*, vol. 14, no. 05, p. 1450073, Oct. 2014.
[4] E. Casiraghi, P. Campadelli, S. Pratissoli, and G. Lombardi, "Automatic Abdominal Organ Segmentation from CT images," *ELCVIA Electron. Lett. Comput. Vis. Image Anal.*, 2009.
[5] M. M. Myint and T. Myint, "Effective Kidney Segmentation Using Gradient Based Approach in Abdominal CT Images," pp. 130–135, 2015.
[6] Suroyo, "Aspek Dasar Penggunaan CT Scan dalam Kedokteran," *J. Med. Sci.*, 1986.
[7] M. Mahesh, "The Essential Physics of Medical Imaging, Third Edition.," *Med. Phys.*, 2013.
[8] R. Noviana, S. Madenda, Rodiah, and S. Widiyanto, "Morphology and Watershed Segmentation for Extracting the Cancer Nodule in Lung CT Scan Image," *Adv. Sci. Eng. Med.*, 2014.
[9] P. Soille, *Morphological Image Analysis : Principles and Applications*. 2004.
[10] S. Madenda, *Pengolahan Citra dan Video Digital*. Erlangga, 2015.
[11] H. Rayner, M. Thomas, and D. Milford, *Understanding Kidney Diseases - Springer*.
[12] F. H. Netter, *Atlas of Human Anatomy, Professional Edition*. 2011.
[13] L. Sherwood, *Fisiologi Manusia dari Sel ke Sistem*. 2013.
[14] F. H. Martini, M. J. Timmons, and R. B. Tallitsch, *Human Anatomy (4th Edition)*. 2002.
[15] Pierce A. Grace and Neil R. Borley, *At a Glance: Ilmu Bedah*. Erlangga, 2006.

A review Cyber of Industry 4.0 (Cyber-Physical Systems (CPS), the Internet of Things (IoT) and the Internet of Services (IoS)): Components, and Security Challenges.

Husham I. AL-Salman *1, Muataz H. Salih 1,2

1School of Computer Engineering, Universiti Malaysia Perlis, 01000 Kangar, Perlis, Malaysia.
2IR4.0 and Intelligent Automation Group, Core Engineering, Plant 5, Flex. 13600 Prai, Penang, Malaysia.

*Correspondence Author: husham.alsalman90@gmail.com

Abstract. The term Industry 4.0 as well as Cyber Physical System has lately gained a lot of consideration among most of the researchers as well as manufacturers while bringing in possible improvement solutions for maintenance of advanced technologies. The term Cyber Physical System or CPS is defined as the addition of computational and physical procedures on the technology of Internet of Things. This prototype helps in ensuring the connectivity of devices and elements with the Internet based protocols which includes the Internet of Services. This pattern of connectivity also enables horizontal services of internal factory through the entire value chain. However, there are several challenges that are faced while maintain the theories and practices of the above protocols. This part of the paper aims at providing a better understanding of the challenges that are faced by these components. The challenges that are faced by CPS are hence described with the help of further researches and are thus summarized from the perspective of energy control, transmission, secure control as well as management control techniques including allocation of system resources and designing of software based models. The paper also reflects the challenges that are faced by IoT from the perspective of communication security, protection of data through sensors and algorithms based on cryptography. Finally the paper also describes the challenges of IoS that are developed through cloud aggregation technology, allowing cloud providers to interact and collaborate in the improvement of cloud security infrastructure including its energy efficiency and reliability.

1. Introduction

This theory comparatively investigates about the smart factories that are controlled with the help of Cyber Physical Systems that executes the simple and iterative activities that were previously carried out by humans. Here the execution process is presumed to be relied on awareness based on machines and their auto agonistics and auto configurations [1]. Here everything is supported by the interconnection of machines and their components via computer based methods. The products and the machines are well equipped with the microchips that help in facilitating the processes while customizing the operations over the Internet and cloud loading while utilizing the data centers, 3D printing services and automated problems related to smart warehouses on a regular basis. The tools and techniques that are used with the Industry 4.0, are considered to save a lot of stretch as well as

costs while improving the flexibility of businesses and manufactures on a basis of large scale including all the major type of risks like hacking of computer systems and misusing the data [2].
The Industry 4.0 could potentially improve the quality of living while increasing the amount of productivity from the workforce and eliminating the monotonous culture of work which includes demanding of jobs physically. The term Industry 4.0 is usually used to denote the current position of digitization that are related to automated process of manufacturing. At ye Hannover Fair held in the year 2013, the concept of Industry 4.0 was first introduced [3]. With this the concept led to the opening of a diverse range of opportunities while putting systems within their specified positions considering all the types of threats that are simultaneously published through data based on industrial system attacks indicating a worst situation with occurrence of more number of attacks directed at the control systems. The Internet Security threat report as published by the Symantec confirmed the fact that "attackers are more likely to choose systems that can be easily compromised and suggests evidence that SCI is particularly vulnerable". The entire section based on this confirmation is published in its report based on Threats Predictions which is issued annually describing the major threats faced by industries. The threats of Cyber-attacks manipulating the workflow system and processes are escalated through the challenges that are faced by the Industrial Internet [3]. This can lead to disruption or outages causing enormous costs. Recently, a survey was conducted by Deloitte shows that the manufacturers believe for the potential rise in the cyber risks as a result of the transformation to Industrial 4.0. For example, as per the report published by the Department of Homeland Security of U.S the cyber-attacks on manufacturing sectors are going to rise at a double rate within one year. This gradual increase in security incidents will cause a great cost leading to financial loss and compromising the security to about 38% than the past years [4].

2. Cyber Challenges in Industry 4.0

The Cyber security in the era of Industry 4.0 has a direct outcome on the Cyber Physical Systems, Internet of Services as well as on the Internet of Things [5].

2.1. Cyber Physical Systems (CPS)

A cyber physical system is a type of scheme that monitors the physical systems while creating a virtual copy and developing decisions in a decentralized way[6]. Within the Cyber physical system, the control elements and the sensors of the system are connected with machines and devices along with installations, networks, vessels and human beings.

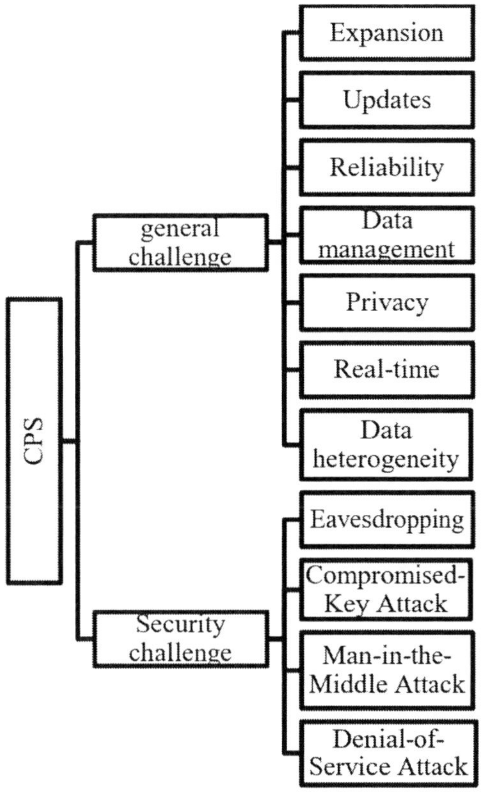

Figure 1. Showing the challenges while monitoring from a group of studies related to Cyber Physical Systems (CPS) general: [7][8][9][10][11] [12] [9][13], securty: [14][15][16][17].

The system refers to the new group of systems which have combined physical and computational capabilities enabling interaction through various new modalities[18].
System and Control researchers have pioneered themselves in various researches conducted over years on the development of science of powerful systems and engineering techniques such as domain methods of frequency and time, analysis of state-space, identification of systems, prediction, controlling robustness, filtering and controlling stochastic [12]. Researchers have also made further studies based on the new type of programming languages, methods of visualizations, techniques for real time computing, compiler designing architectures of embedded systems, software systems as well as various innovative techniques to ensure the reliability of computer systems, their security and fault tolerance capability. Researches have also tried developing varieties of modeling and powerful verification tools. All these challenges mentioned above are now being studied so as to find solution using the modern FPGA technology.

2.2. Internet of Things (IoT).

The Internet of Things is acted upon by the CPS. This protocol helps in ensuring that the strategies and elements are well linked to the Internet based Protocols or the TCP/IP.
The challenges that are being reached under the IoT are represented below:

NO	SECURTY CHALANGE	RF
1	Laws and Regulations regarding security: The laws and regulations of security are still not in place and also there is no such standard technologies related to IoT. IoT is mainly related to the security of national information including business secrets and personal privacy.	[19]

2	Key Management: Key management is one of the basic mechanisms of security and is mostly researched. It is regarded as the most critical aspect of security of cryptography. Still no such solutions have been identified by researchers about its security and the algorithms of cryptography based on high performance sensors are still not applied in many cases. In this case the real networks of large scale sensors are put into practices.	[20]
3	Domain Name System: The main challenge that arises during identification of object is to determine the integrity of the records that are used during naming of architectures. Although the DNS helps in providing translation to the names of services to the Internet users, it is regarded as an unsecured way of naming system as it is vulnerable to various cyber-attacks such as poisoning attack of DNS Cache or man in the middle attack .	[21]
4	Burgeoning Application requirements: With the introduction of RFID, WSNs, technology related to network communication, various computing technologies and various control theory for real time distribution, the CPS is regarded as an emerging form of IoT technology with high risk of security over its system performance.	[22] [23]
5	Privacy and trusts: With the use of monitoring and remote sensors in the Io Technology, the controlling access along with the ownership of data are becoming highly sensitive.	[24]
6	Absence of Global root CA has become one of the challenges for authentication system, causing infeasibility in certifying objects related to IoT. Hence this needs to be taken into consideration as it will also lead the high computational methods to suffer.	[25] [26]
7	High cost security against malware: While shifting from the x86 architecture, the conventional methods of security can be infeasible against malware leading to issues such as divergence of the architecture of hardware among various strategies. Without the proper malware abstraction of IoT, the current solutions related to ad-hoc remains inapplicable.	[27]
8	Crypto-systems: In spite of having advantages of the public key cryptosystem, there remains a lack of root certificate authority over authentication of construction schemes that are actually deployed.	[15] [28]

Since the past few decades, the Internet of Things has been the main area of researches. Security and privacy has been the main issues for the application of IoT including several other challenges facilitating the domains and reviewing the research progresses of IoT paying special attention to the security. While thoroughly analyzing the security features and architectures of IoT the requirements of security are established. Based on theses researches, the technologies including the mechanism of encryption, security of communication, protection of sensor data and algorithms based on cryptography are hence outlined through the challenges.

2.3. Internet of Services (IoS):

The IoS protocol offers a variety of services of internal factory running horizontally throughout the value chain [29]. The challenges that are being reached under the IoS are represented below:

NO	SECURTY CHALANGE	RF
1	Scalability of services: Although there are many services that are static in nature from the view pint of their size but still the services experiences fluctuations due to workload.	[30]
2	Monitoring of Services: Most of the mechanism of billing including payment as offered by providers are based on resources of individual users as per their consumption time based on unit.	
3	Context awareness of applications: One of the key challenges of the future of IoS is to address the increase in the information relevancy within a given context. Of becoming more aware in specific application and services such as social networking, computational environment and various other mobile applications. This context awareness adapts certain behaviors related to the environment such as activity of users, accessing of devices, location or people who are nearby.	[31]

Cloud computing plays a vital part in the upcoming of Internet of Services while allowing provision of on –demand requests, computational infrastructures and platforms. The communal part in cloud needs to address the challenges that are faced technically while turning the visions into reality. Some of the issues that make the cloud service providers to collaborate and improve the security, efficiency and reliability of the cloud infrastructures include elastic service platforms, managing of clouds efficiently and delivering of scalable cloud services.

3. References

[1] Sonntag, D., Zillner, S., van der Smagt, P., and Lörincz, A., 2017 *Industrial Internet of Things*, Springer, pp. 487–504
[2] Zezulka, F., Marcon, P., Vesely, I., and Sajdl, O., 2016 IFAC-PapersOnLine, **49**(25), pp. 8–12
[3] Marcon, P., Zezulka, F., Vesely, I., Szabo, Z., Roubal, Z., Sajdl, O., Gescheidtova, E., and Dohnal, P., 2017 Prog. Electromagn. Res. Symp., **Part F1343**, pp. 1694–1697
[4] Symantec, 2016 *SMARTER SECURITY FOR MANUFACTURING IN THE INDUSTRY 4.0 ERA*
[5] Dulaunoy, A., Iklody, A., and Vinot, R., 2016 APCO interational, **1.0**, p. 24
[6] Brookson, C., 2016 ETSI, (18)
[7] Jirkovský, V., Obitko, M., and Mařík, V., 2017 IEEE Trans. Ind. Informatics, **13**(2), pp. 660–667
[8] Rasmussen, T. B., Yang, G., Nielsen, A. H., and Dong, Z., 2017 2017 IEEE Manchester PowerTech, Powertech 2017
[9] Benias, N., and Markopoulos, A. P., 2017 South-East Eur. Des. Autom. Comput. Eng. Comput. Networks Soc. Media Conf. SEEDA-CECNSM 2017
[10] Flatt, H., Schriegel, S., Jasperneite, J., Trsek, H., and Adamczyk, H., 2016 IEEE Int. Conf. Emerg. Technol. Fact. Autom. ETFA, **2016–Novem**
[11] Humayed, A., Lin, J., Li, F., and Luo, B., 2017 IEEE Internet Things J., **4662**(c), p. 30
[12] Lee, E. A., 2015 Sensors (Switzerland), **15**(3), pp. 4837–4869
[13] Taylor, J. M., and Sharif, H. R., 2017 *Selected Topics in Mobile and Wireless Networking (MoWNeT), 2017 International Conference On*, IEEE, pp. 1–6
[14] Rawat, D. B., White, T., Parwez, M. S., Bajracharya, C., and Song, M., 2017 IEEE Internet Things J., **4**(6), pp. 1987–1993
[15] Ali, S., Al Balushi, T., Nadir, Z., and Hussain, O. K., 2018 *Cyber Security for Cyber Physical Systems*, Springer, pp. 1–10
[16] Jia, Y., Chen, Y., Dong, X., Saxena, P., Mao, J., and Liang, Z., 2015 Comput. Secur., **55**, pp. 62–80
[17] Liu, S., Liu, X. P., and El Saddik, A., 2013 *Innovative Smart Grid Technologies (ISGT), 2013 IEEE PES*, IEEE, pp. 1–6
[18] Habib, H. F., Lashway, C. R., and Mohammed, O. A., 2017 2017 IEEE Ind. Appl. Soc. Annu. Meet. IAS 2017, **2017–Janua**, pp. 1–8
[19] Banafa, A., 2015 OpenMind, (March), pp. 16–27
[20] Wurm, J., Jin, Y., Liu, Y., Hu, S., Heffner, K., Rahman, F., and Tehranipoor, M., 2017 IEEE Trans. Multi-Scale Comput. Syst., **3**(3), pp. 215–227
[21] Murphy, M., 2017 Netw. Secur., **2017**(7), pp. 17–19
[22] Turmel, V., Anderson, E., and Security, I., 2016 *Industrial Security : Applying IoT Security Controls on the Industrial Plant Floor*
[23] Ni, J., Zhang, K., Lin, X., and Shen, X., 2017 IEEE Commun. Surv. Tutorials
[24] Yogita Pundir, M., Sharma, M. N., and Singh, Y., 2016 Int. J. Adv. Res. Comput. Commun. Eng., **5**(3), pp. 960–964
[25] Guan, C., Ren, K., Zhang, F., Kerschbaum, F., and Yu, J., 2015 *European Symposium on Research in Computer Security*, Springer, pp. 203–223
[26] Herder, C., Fuller, B., van Dijk, M., and Devadas, S., 2017 IACR Cryptol. ePrint Arch., **2017**, p. 210
[27] Singh, S., 2015 Int. Conf. Green Comput. Internet Things, pp. 1577–1581

[28] El-Hadedy, M., Skadron, K., Mihajloska, H., and Gligoroski, D., 2015 Proc. Hear.
[29] Moreno-Vozmediano, R., Montero, R. S., and Llorente, I. M., 2013 IEEE Internet Comput., **17**(4), pp. 18–25
[30] Liu, L., 2016 IEEE Trans. Serv. Comput., **9**(5), pp. 661–663
[31] Yurur, O., Liu, C., and Moreno, W., 2014 IEEE Commun. Mag., **52**(6), pp. 24–31

Journal of Physics: Conference Series **1424** (2019) 012030 doi:10.1088/1742-6596/1424/1/012030

An Integrated Multi Criteria Decision Making Method For Fashion Selection

Asyahri Hadi Nasyuha[1*], Zulham[2], Idi Jang Cik[3], Muhammad Amin[4], Sigit Candra Setia[5], Dodi Siregar[6]

[1]STMIK Triguna Dharma, Medan, Indonesia
[2]Universitas Dharmawangsa, Medan, Indonesia
[3]Sekolah Tinggi Teknologi, Pagaralam, Indonesia
[4]STMIK Royal, Kisaran, Indonesia
[5]Sekolah Tinggi Teknologi, Pagaralam, Indonesia
[6]Universitas Harapan Medan, Medan, Indonesia

*ayi.nasyuha@gmail.com

Abstract. ZOOM Fashion is a business engaged in the sale of clothing, which is currently undergoing development and is in great demand in the community. In these companie, performance improvements are needed, so companies must be more careful in buying clothes to be resold in order to find out which clothes are of high quality. TOPSIS uses the principle that the chosen alternative must have the closest distance from the positive ideal solution and the longest distance from the negative ideal solution from a geometric point of view by using the distance between two points to determine the relative proximity of an alternative. this method is very well used for the completion of fashion selection. this research is expected to help in determining quality clothing in accordance with the fashion needed at this time, so that the existence of this system makes it easy to find out the needs of the community in the field of clothing.

1. Introduction

Fashion is a blend of styles with designs that tend to be chosen, favored and used by most people who be able to provide comfort at a certain time. Fashion is formed from a blend of styles with designs that meet the tastes of the wearer. The style of dress is an appearance of the material used that has a difference from the clothes that other people wear, while the design is something more special than style. most people don't know the quality of the clothes they are going to buy. therefore when choosing or buying clothes, the public must know and be careful about the quality of the clothes they want to buy.

2. Methodology

Many have used Multi-Criteria Decision Analysis over the past few decades. Its role has increased significantly in different fields of science, especially as new methods evolve and old methods increase [1]. Two basic approaches to MCDM are decisions taken from multiple attributes and decision making from multiple objectives [2]. Computer-based information systems and can support business or organizational decision making activities are said to be decision support systems. This system provides organizational planning by assisting decision making. [3].

The use of TOPSIS to make the selection of alternatives provided, provided that the alternative chosen must have the shortest distance from the positive ideal solution and the farthest distance from the negative ideal solution [4]. The principle of using the TOPSIS method is that the chosen alternative must have a short distance from the positive ideal solution and the farthest from the negative ideal solution from a geometric point of view using Euclidean distance to determine the relative proximity of an alternative to the optimal solution [5]. The formula used in this method is as follows:

1. Normalize each alternative value (normalized matrix) and weighted normalized matrix [6].

   $$r_{ij} = \frac{x_{ij}}{\sqrt{\sum_{i=1}^{m} x_{ij}^2}}$$

   With i=1.2,..,m; and j=1,2,..,n,
   Where :
   r_{ij} = normalized matrix [i][j]
   x_{ij} = matriks ternormalisasi [i][j]
2. Calculate the weighted matrix value.
   $y_{ij} = w_{i.}r_{ij}$: with i=1,2,..,m; and j=1,2,..,n
   $A^+ = (Y_1^+, Y_2^+, \ldots., Y_n^+)$
   $A^- = (Y_1^-, Y_2^-, \ldots., Y_n^-)$
   With condition:
   y_{ij} = weighted normalized matrix [i][j]
   Y_j^+ = max y_{ij}, if j is a profit attribute
   min y_{ij}, if j is a cost attribute
   Y_j^- = min y_{ij}, if j is a profit attribute
   min y_{ij}, if j is a cost attribute.
3. Calculate weighted distance values for each alternative to positive and negative ideal solutions. For a positive ideal solution [7]

   $$D_i^+ = \sqrt{\sum_{j=1}^{n} (Y_i^+ - Y_{ij})^2} ;$$

   i=1,2,..,m
   D_i^+= the alternative distance of Ai with a positive ideal solution.
   Y_i^+= Positive ideal solution [i]
   y_{ij} = normal matrix weighted [i][j]
   For negative ideal solutions [8].

   $$D_i^- = \sqrt{\sum_{j=1}^{n} (Y_{ij} - Y_i^-)^2} ;$$

   i=1,2,..,m
   D_i^-= jarak alternative Ai dengan solusi ideal negative.
   Y_i^-= Solusi ideal Negative [i]
   y_{ij} = weighted normalized matrix [i][j]
4. Calculating Preference Values from each alternative.

   $$V_i = \frac{D_i^-}{D_i^- + D_i^+};$$

   i=1,2,..,m
   where:
   V_i= The closeness of each alternative to the ideal solution

D_i^+= the alternative distance of Ai with a positive ideal solution.
D_i^-= the alternative distance of Ai with a negative ideal solution.
A larger V_i value indicates that the Ai alternative is preferred.

Fashion is a blend of styles with designs that tend to be chosen, favored and used by most people who be able to provide comfort at a certain time. Fashion is formed from a blend of styles with designs that meet the tastes of the wearer. The style of dress is an appearance of the material used that has a difference from the clothes that other people wear, while the design is something more special than style.

3. Results and Discussion

Analysis of the problems discussed is the implementation of a decision support system in the selection of fashion based on quality. As described in the previous chapter, there are many factors that are taken into consideration in determining the quality of a garment. The activity of determining the quality of a garment is an activity carried out by the community and shop owners. but there are some difficulties in determining the quality of clothing. The criteria used for this study are presented in the following table 1.

Table 1. Weight of interest Each Criteria

Number	Code	Criteria	Weight
1	C1	material	3
2	C2	Stitches	2
3	C3	Pattern	2
4	C4	Price	2
5	C5	Origin of the product	1

The research using several alternatives as in the table 2 below:

Table 2. Alternative

Number	Name	Material	Stitches	Pattern	Price	Origin of the Product
1	Drexel Shirt	Street	Enough	Not Good	140	Local
2	Voresti Shirt	Katun	Neat	Good	130	Import
3	Luccaman Shirt	Beldu	Enough	Not Good	140	Import
4	Eighty Eight Shirt	Flannel	Enough	Good	120	Local

Each criterion is given a range of values, as in the material criteria in table 3 below:

Table 3. Criteria Of Material

Criteria	Name Of Material	Value
	Beldu Fabric	1

Material	Flannel Fabric	3
	Street Fabric	3
	Katun Fabric	4
	Denim/jeans Fabric	4
	Chinos/Drill Fabric	3
	Spandex Fabric	2

Range of values from Criteria of Stitches, presented in table 4:

Table 4. Criteria of Stitches

Criteria	Description	Value
Stitches	Neat	4
	Pretty Neat	3
	Not Neat	2
	Not Neat	1

Range of values from Criteria of Pattern, presented in table 5:

Table 5. Criteria Of Pattern

Criteria	Range	Value
Pattern	Good	4
	Quite good	3
	Not Good	2
	Bad	1

Range of values from Criteria of Price, presented in table 6:

Table 6. Criteria of Price

Criteria	Range (Thousand)	Value
Price	< 200	4
	140 – 110	3
	120 – 90	2
	>90	1

Range of values from Criteria form Origin Of Product, presented in table 7:

Table 7. Criteria of Origin Of Product

Criteria	Range	Value
Origin Of Product	Locally Made	3
	Import	4

Range of values from Rules for Assessing All Criteria, presented in table 8:

Table 8. Rules for Assessing All Criteria

Range	Feasibility Status
0 - 0.25	Bad

0.26 – 0.50	Enough
0.51 – 0.75	Good
0.76 – 1	Verry Good

a. Create a decision matrix. the matrix column states the existing criteria, while the matrix row represents an alternative. The reference to alternative decision matrix which will be evaluated based on criteria.

$$r_{ij} = \frac{x_{ij}}{\sqrt{\sum_{i=1}^{m} x_{ij}^2}}$$

X_{ij} is alternative performance to I for criteria to j.

A_i(i=1,2,3,….,m) are possible alternatives.

X_j(j=1,2,3,…..,m) is a criterion where alternative performance is measured.

In this study, the value of j is as follows:

J = 1 for criteria of material

J = 2 for criteria of Stitches

J = 3 for criteria of Pattern

J = 4 for criteria of Price

J = 5 for criteria of Origin Of Product

The results of the decision matrix formed from the initial data table of each alternatinve can be presented in the following example:

Table 9. Data Normalization Results of Customized Clothing From Table 2

lternative	Material	Stitches	Pattern	Price	Origin Of Product
A_1	3	3	2	4	3
A_2	4	4	4	3	4
A_3	2	3	2	4	4
A_4	4	3	4	2	3

Normalized Decision Matrix can be seen in the following agreement:

$$r_{ij} = \frac{x_{11}}{\sqrt{x_{11} + x_{21} + x_{31} + x_{41}}}$$

$$x_1 = \sqrt{3^2 + 4^2 + 2^2 + 4^2} = 6.708$$

$$r_{11} = \frac{x_{11}}{x_1} = \frac{3}{6.708} = 0.447 \qquad r_{31} = \frac{x_{31}}{x_1} = \frac{2}{6.708} = 0.298$$

$$r_{21} = \frac{x_{21}}{x_1} = \frac{4}{6.708} = 0.596 \qquad r_{41} = \frac{x_{41}}{x_1} = \frac{4}{6.708} = 0.596$$

$$x_2 = \sqrt{3^2 + 4^2 + 3^2 + 3^2} = 6.557$$

$$r_{12} = \frac{x_{12}}{x_2} = \frac{3}{6.557} = 0.457 \qquad r_{32} = \frac{x_{32}}{x_2} = \frac{3}{6.557} = 0.457$$

$$r_{22} = \frac{x_{22}}{x_2} = \frac{4}{6.557} = 0.610 \qquad r_{42} = \frac{x_{42}}{x_2} = \frac{3}{6.557} = 0.457$$

$$x_3 = \sqrt{2^2 + 4^2 + 2^2 + 4^2} = 6.324$$

$$r_{13} = \frac{x_{13}}{x_3} = \frac{2}{6.324} = 0.316 \qquad r_{33} = \frac{x_{33}}{x_3} = \frac{2}{6.324} = 0.316$$

$$r_{23} = \frac{x_{23}}{x_3} = \frac{4}{6.324} = 0.632 \qquad r_{43} = \frac{x_{43}}{x_3} = \frac{4}{6.324} = 0.632$$

$$x_4 = \sqrt{4^2 + 3^2 + 4^2 + 2^2} = 6.708$$

$$r_{14} = \frac{x_{14}}{x_4} = \frac{4}{6.708} = 0.596 \qquad r_{34} = \frac{x_{34}}{x_4} = \frac{4}{6.708} = 0.596$$

$$r_{24} = \frac{x_{24}}{x_4} = \frac{3}{6.708} = 0.447 \qquad r_{44} = \frac{x_{44}}{x_4} = \frac{2}{6.708} = 0.298$$

$$x_5 = \sqrt{3^2 + 4^2 + 4^2 + 3^2} = 7.071$$

$$r_{15} = \frac{x_{15}}{x_5} = \frac{3}{7.071} = 0.424 \qquad r_{35} = \frac{x_{35}}{x_5} = \frac{4}{7.071} = 0.565$$

$$r_{25} = \frac{x_{25}}{x_5} = \frac{4}{7.071} = 0.565 \qquad r_{45} = \frac{x_{45}}{x_5} = \frac{3}{7.071} = 0.424$$

Calculation Result of Normalized Decision Matrix, presented in table 10:

Table 10. Calculation Result of Normalized Decision Matrix

Number	Name Of Alternative	Material	Stitches	Pattern	Price	Origin Of Product
1	Drexel	0.447	0.457	0.316	0.596	0.424
2	Voresti	0.596	0.610	0.632	0.447	0.565
3	Luccaman	0.298	0.457	0.316	0.596	0.565
4	Eighty Eight	0.596	0.457	0.632	0.298	0.424

b. After the normalized matrix is made, then it is to make a weighted Y normalized decision matrix whose elements are determined by the following formula:
Where :

$y_{ij} = w_j . r_{ij}$

y_{ij} are elements of normalized decision matrices R.

Weight $w_j(w_1, w_2, w_3, \ldots \ldots \ldots w_n)$ is the weight of the criteria of the criteria to -j.

r_{ij} are elements of normalized decision matrices R

The Weighted Normalization Decision Matrix presented in table 11:

Table 11. Weighted Normalized Decision Matrix

Number	Name Of Atlternative	Material	Stitches	Pattern	Price	Origin Of Product

1	Drexel	$w_1.r_{11}$	$w_2.r_{12}$	$w_3.r_{13}$	$w_4.r_{14}$	$w_5.r_{15}$
2	Voresti	$w_1.r_{21}$	$w_2.r_{22}$	$w_3.r_{23}$	$w_4.r_{24}$	$w_5.r_{25}$
3	Luccaman	$w_1.r_{31}$	$w_2.r_{32}$	$w_3.r_{33}$	$w_4.r_{34}$	$w_5.r_{35}$
4	Eighty Eight	$w_1.r_{41}$	$w_2.r_{42}$	$w_3.r_{43}$	$w_4.r_{44}$	$w_1.r_{45}$

$y_{11} = w_1.r_{11} = 3\,x\,0.447 = 1.341$ $\quad y_{12} = w_2.r_{12} = 2x\,0.457 = 0.914$
$y_{21} = w_1.r_{21} = 3\,x\,0.596 = 1.788$ $\quad y_{22} = w_2.r_{22} = 2\,x0\,.610 = 1.22$
$y_{31} = w_1.r_{31} = 3x\,0.298 = 0.894$ $\quad y_{32} = w_2.r_{32} = 2\,x\,0.457 = 0.914$
$y_{41} = w_1.r_{41} = 3\,x\,0.596 = 1.788$ $\quad y_{42} = w_2.r_{41} = 2\,x0\,.457 = 0.914$

$y_{13} = w_3.r_{13} = 2\,x\,0.316 = 0.632$ $\quad y_{14} = w_4.r_{14} = 2\,x\,0.596 = 1.192$
$y_{23} = w_3.r_{23} = 2\,x\,0.632 = 1.264$ $\quad y_{24} = w_4.r_{24} = 2\,x\,0.447 = 0.894$
$y_{33} = w_3.r_{33} = 2\,x\,0.316 = 0.632$ $\quad y_{34} = w_4.r_{34} = 2\,x\,0.596 = 1.192$
$y_{43} = w_3.r_{43} = 2\,x\,0.632 = 1.264$ $\quad y_{44} = w_4.r_{44} = 2\,x\,0.298 = 0.596$

$y_{15} = w_5.r_{15} = 1\,x\,0.424 = 0.424$
$y_{25} = w_5.r_{25} = 1\,x\,0.565 = 0.565$
$y_{35} = w_5.r_{35} = 1\,x\,0.565 = 0.565$
$y_{45} = w_5.r_{45} = 1\,x\,0.424 = 0.424$

Calculation of Weighted Normalized Decision Matrix Results, presented in table 12:

Table 12. Calculation of Weighted Normalized Decision Matrix Results

Name Of Alternative	Material	Stitches	Pattern	Price	Origin Of Product
Drexel	1.341	0.914	0.632	1.192	0.424
Voresti	1.788	1.22	1.264	0.894	0.565
Luccaman	0.894	0.914	0.632	1.192	0.565
Eighty Eight	1.788	0.914	1.264	0.596	0.424

c. Calculating Positive or Negative Values.

Formula:

$y_{ij} = w_i r_{ij}$
$A^+ = (Y_1^+, Y_2^+, \ldots, Y_n^+)$
$A^- = (Y_1^-, Y_2^-, \ldots, Y_n^-)$

Under the condition

$$Y_j^+ = \begin{cases} \max\limits_{i} Y_{ij} \; ; If\ j\ isprofit\ attribute \\ \min\limits_{y} Y_{ij} \; ; \ if\ j\ is\ cost\ attribute \end{cases}$$

$$Y_j^- = \begin{cases} \max\limits_i Y_{ij} \; ; If\ j\ isprofit\ attribute \\ \min\limits_y Y_{ij} \; ; \ if\ j\ is\ cost\ attribute \end{cases}$$

1) Ideal Positive Solution
$y1^+ =$ max {1.341; 1.788; 0.894; 1.788}= 1.788
$y2^+ =$ max {0.914; 1.22; 0.914; 0.914}= 1.22
$y3^+ =$ max {0.632 ; 1.264 ; 0.632 ; 1.264}= 1.264
$y4^+ =$ max {1.192; 0.894 ; 1.192 ; 0.596}= 1.192
$y5^+ =$ max {0.424; 0.565 ; 0.565 ; 0.424}= 0.565

2) Ideal Negative Solution
$y1^+ =$ max {1.341; 1.788; 0.894; 1.788}= 0.894
$y2^+ =$ max {0.914; 1.22; 0.914; 0.914}= 0.914
$y3^+ =$ max {0.632 ; 1.264 ; 0.632 ; 1.264}= 0,632
$y4^+ =$ max {1.192; 0.894 ; 1.192 ; 0.596}= 0.596
$y5^+ =$ max {0.424; 0.565 ; 0.565 ; 0.424}= 0.424

And, the results of determining the Ideal Ideal Solution Matrix (A ^ +) and the Ideal Negative Matrix (A ^ -).

A^+	1.788	1.22	1.264	1.192	0.565

A^-	0.894	0.914	0.632	0.596	0.424

Calculate the distance of weighted values for each alternative to positive and negative ideal solutions[9]. Next, determine the alternative distance from the ideal positive solution (D ^ +) and the alternative distance from the negative ideal solution (D^-).

Formula: $D_i^+ = \sqrt{\sum_{j=1}^{n}(y_i^+ - y_{ij})^2}$; i=1,2,...m.

$$D_1^+ = \sqrt{\begin{matrix}(1.788 - 1.341)^2 + (1.22 - 0.914)^2 + (1.264 - 0.632)^2 \\ +(1.192 - 1.192)^2 + (0.565 - 0.424)^2\end{matrix}}$$
$= 0.842$

$$D_2^+ = \sqrt{\begin{matrix}(1.788 - 1.788)^2 + (1.22 - 1.22)^2 + (1.264 - 1.264)^2 \\ +(1.192 - 0.894)^2 + (0.565 - 0.565)^2\end{matrix}}$$
$= 0.296$

$$D_3^+ = \sqrt{\begin{matrix}(1.788 - 0.894)^2 + (1.22 - 0.914)^2 + (1.264 - 0.632)^2 \\ +(1.192 - 1.192)^2 + (0.565 - 0.565)^2\end{matrix}}$$
$= 1.136$

$$D_4^+ = \sqrt{\begin{matrix}(1.788-1.788)^2+(1.22-0.914)^2+(1.264-1.264)^2\\+(1.192-0.596)^2+(0.565-0.424)^2\end{matrix}}$$
$= 0.683$

Formula: $D_i^- = \sqrt{\sum_{j=1}^{n}(y_i^- - y_{ij})^2}$; i=1,2,…,m.

$$D_1^- = \sqrt{\begin{matrix}(0.894-1.341)^2+(0.914-0.914)^2+(0.632-0.632)^2\\+(0.596-1.192)^2+(0.424-0.424)^2\end{matrix}}$$
$= 0.744$

$$D_2^- = \sqrt{\begin{matrix}(0.894-1.788)^2+(0.914-1.22)^2+(0.632-1.264)^2\\+(0.596-0.894)^2+(0.424-0.565)^2\end{matrix}}$$
$= 1.182$

$$D_3^- = \sqrt{\begin{matrix}(0.894-0.894)^2+(0.914-0.914)^2+(0.632-0.632)^2\\+(0.596-1.192)^2+(0.424-0.565)^2\end{matrix}}$$
$= 0.611$

$$D_4^- = \sqrt{\begin{matrix}(0.894-1.788)^2+(0.914-0.914)^2+(0.632-1.264)^2\\+(0.596-0.596)^2+(0.424-0.424)^2\end{matrix}}$$
$= 1.094$

Distance Calculation Results Alternatively From Ideal Positive Solutions and Ideal Negative Solutions, presented in table 13:

Table 13. Distance Calculation Results

Alternative	D^+	D^-
Drexel	0.842	0.744
Voresti	0.296	1.182
Luccaman	1.136	0.611
Eighty Eight	0.683	1.094

d. Calculating Preference Value (V) for Each Alternative.

Formula : $V_i = \frac{D_i^-}{D_i^- + D_i^+}$; Where i = 1,2,3,….,m.

$$V_1 = \frac{0.744}{0.744+0.842} = \frac{0.744}{1.586} = 0.469$$

$$V_2 = \frac{1.182}{1.182+0.298} = \frac{1.182}{1.48} = 0.798$$

$$V_3 = \frac{0.611}{0.611 + 1.136} = \frac{0.611}{1.747} = 0.349$$

$$V_4 = \frac{1.098}{1.098 + 0.683} = \frac{1.098}{1.777} = 0.617$$

e. Rank
The following is a table of the results of each alternative, while the reference in this ranking is based on the highest value of each alternative that will be selected for quality clothing. Here is the alternative ranking of the largest value V to the smallest value V. The alternative with the largest V value is the best quality clothing. Presented in table 14.

Table 14. Rank

Alternatif	Total of Value	Feasibility Status
Voresti	0.798	Verry Good
Eighty Eight	0.617	Good
Drexel	0.469	Not Bad
Luccaman	0.349	Not Bad

4. Conclusion

In analyzing and resolving the problems that occur regarding the assessment of clothing mode using a decision support system with the topis method. Provide knowledge to the general public about the quality of good clothing. Make one of the references in using the topsis method to make elections with different cases.

Limitations

This study has limitations in adding criteria, further research needs to be extended especially from the alternatives used and the selection of criteria used.

Acknowledgment

The researcher thanked my institution, "STMIK Triguna Dharma" Indonesia for providing opportunities and funds to conduct this research.

References

[1] M. Velazquez and P. T. Hester, "An Analysis of Multi-Criteria Decision Making Methods," *Int. J. Oper. Res.*, vol. 10, no. 2, pp. 56–66, 2013.

[2] L. Abdullah and C. W. Adawiyah, "Simple Additive Weighting Methods of Multi criteria Decision Making and Applications: A Decade Review," *Int. J. Inf. Process. Manag.*, vol. 5, no. 1, pp. 39–49, 2014.

[3] E. K. Zavadskas, P. Vainiūnas, Z. Turskis, and J. Tamošaitienė, "Multiple criteria decision support system for assessment of projects managers in construction," *Int. J. Inf. Technol. Decis. Mak.*, vol. 11, no. 2, pp. 501–520, 2012.

[4] D. R. Sari, A. P. Windarto, D. Hartama, and S. Solikhun, "Sistem Pendukung Keputusan untuk Rekomendasi Kelulusan Sidang Skripsi Menggunakan Metode AHP-TOPSIS," *J. Teknol. dan Sist. Komput.*, vol. 6, no. 1, p. 1, 2018.

[5] A. H. Nasyuha, M. Dahria, and T. Tugiono, "SISTEM PENDUKUNG KEPUTUSAN PEMILIHAN MANDOR MENGGUNAKAN METODE TOPSIS," *J. Ilm. Saintikom*, vol. 16, no. 1, pp. 77–86, 2017.

[6] S. Opricovic and G. H. Tzeng, "Compromise solution by MCDM methods: A comparative analysis of VIKOR and TOPSIS," *Eur. J. Oper. Res.*, vol. 156, no. 2, pp. 445–455, 2004.

[7] Darsono Nababan and Robbi Rahim, "Sistem Pendukung Keputusan Reward Bonus Karyawan Dengan Metode Topsis," *Simetris*, vol. 3, no. 6, pp. 2–6, 2018.

[8] A. A. Chamid and A. C. Murti, "KOMBINASI METODE AHP DAN TOPSIS PADA SISTEM PENDUKUNG KEPUTUSAN," 2017, pp. 115–119.

Distance Learning via MOOCs: Improving ESL Learners' Writing Skills

Nurakhma Shabani Jasni[1], Harwati Hashim, Melor Md Yunus*, Siti Fatimah Abd Rahman

[1]Faculty of Education, Universiti Kebangsaan Malaysia (UKM)

*melor@ukm.edu.my

Abstract. Students nowadays are exposed to technology and they are always interested in learning new things. By incorporating technology as a means of learning, the students would be able to develop interest in learning language and improving their ICT skills. Crossing the line of time and distance and applying the saying that learning could be done anywhere and every time, applying the theory of distance learning, this paper reviews the effectiveness using MOOCs to improve students' writing skills. Prior research regarding the use of MOOCs and technology, especially in language education and writing skills, is reviewed. Emphasizing on MOOCs as the platform of improving students' writing skills, this paper explains how MOOCs helps in improving the process of teaching and learning and the importance of self-directed learning.

1. Introduction

Students nowadays are exposed to technology and they are always interested in learning new things [1]–[3]. Traditional setting such as classroom no longer helps in motivating and developing their interest in learning. Thus, by incorporating media and learning MOOCs helps in developing students' interest towards their learning process because it is a new way of learning [4]. In this 21st century, it is extremely important to understand where the students' interest lies especially in order to understand and encourage them. By incorporating technology as a means of learning, the students would be able to develop interest in learning language and improving their ICT skills. One of the most common modes of learning today is via distance education. Distance education is a manner of conveying the process of teaching and learning, often on an individual basis, to students who are not available in the classroom [5]. This is convenient because it helps learning become possible in all kind of manner.

Moreover, [6] stated that distance learning in an educational program is a pedagogical concept applied by using three sets of variables which consists of variables in teaching and in learning and in the interaction of teaching and in learning rather than saying one of the variables are variables in communication and media. By using Massive Online Open Courses (MOOCs), this added the variable of communication and media in distance education. This is because MOOCs is a platform that incorporate media and communication as a teaching method through different set of modules. While distance learning allows students to work at their own pace with themselves being their limit, [6] said that the purpose of distance education is that it is to recap the diverse relationship and strength of relationship among and between the variable that makes up distance learning. Crossing the line of time and distance and applying the saying that learning could be done anywhere and every time, applying the theory of distance learning, this paper reviews the effectiveness of distance learning via MOOCs.

2. Writing Component in ESL

Writing is process of delivering thoughts by jolting it down on a piece of paper. Some people consider it tedious especially when the subject have been decided already and there is no relatable knowledge in regards to the subject. This mostly happens in school. In order for the students to master a language, they have to have a wide range of vocabulary, the ability to form a proper structure of sentence and the ability to understand the language and are able to reply to it fluently.

In Malaysian English context, these components of English in school are assessed from four aspects, reading, writing, speaking and comprehension. Students do not write very often and most of what they write is classroom-bound. Most of their writing are done in classroom or because it is a requirement for their subject. It is hard to choose a topic that the student could relate to and are able to draw out their own feelings and as they went into a higher system of education, the harder it is to comprehend the topic they are writing about. The most important factor in writing exercises is that students need to be personally involved in order to make the learning experience of great value. While encouraging student participation in the exercise, and at the same time enhancing and developing writing skills, compels a definite practical approach [7].

Effective writing, according to [8], consists of central idea, organization, supporting material, expression, word choice and point and the last one is spelling, punctuation and grammar. This means that in order for a student to master writing, they need to have the basic knowledge of the language itself and by mastering writing, they would, indeed, master the language itself. Writing requires not only the knowledge of the language, but also the ability to be able to think critically and the ability to be able to relate to the text so that the idea can be expressed vividly.

3. The Use of Technology in Teaching and Learning Writing

The findings of past studies reported that online learning and instruction have on language learning. For instance, [9] concluded that Internet- based or assisted language learning could help college students in independent language learning and academic writing because these students are enthusiastic and reasonably competent users of Internet-based tools and techniques. This is because the students are able to make their own research and are able to work independently without pressure. Eliminating the pressure of the datelines and being compared of their achievements, the student could work in a more comfortable and stress-free environment.

Technology allows students to work in a more fun and interactive way. They could encourage their friends to work harder from just commenting and giving emoticons. For example, if someone made an error in their writing, a peer could make a comment with "@@" to show they were confused or speechless, or they could follow a comment with ":-D" to show they were happy with the writing or to provide encouragement [10]. Through this learning process, the students had greater chance to observe the mistakes made and mend their past reoccurring mistakes while making connections and interacting with their fellow peers. They were also able to obtain knowledge and skills from the writings, comments, and feedback of others while reflecting on their own. Furthermore, in the process of reviewing and commenting on their peers' work and offering comments and feedback, the students were able to modify their original work and improve its quality [11].

Through technology, the student could reflect on their mistakes not only on a one to one basis, they were able to reflect and offers feedback from their friends without being discouraging. [12] proclaimed that social constructivist theory undertakes that students act and reflect within an environment, and this is then followed by reflecting, abstracting, and increasing experiential knowledge. Thus, through MOOCs as the platform, the students would be able to do all those without having to change and adapt to anyone's pace of learning.

4. Massive Open Online Courses (MOOCs)

Massive Massive Online Open Courses (MOOCs) is an open free access via Internet platform that focuses on enhancing students' skills by providing distance education for students who are unable to come to class. It provides more than 4312 courses across the world online focusing on the concept of open learning. Open learning is an online learning platform that goes beyond content delivery to focus on community, connectedness and student engagement. This platform uses the concept of social media approach allowing people to interact using chats, liking and commenting on other people's post with the intention of engaging the students in a fun social student experience. Student interactions in online discussions can expedite a learner-centred approach to teaching and provide students with an opportunity to practise and learn knowledge and skills in a supportive and encouraging environment [10].

While maintaining the room of improvement for each individual by completing their modules, MOOCs also provide blog pages and built in galleries for further understanding as well as a column for feedback for them to express their thoughts and understanding for each post. At the end of the course, if the student were able to complete the whole module and the course work, the student will be awarded with a certification for their participation. This will give the student a sense of achievement and will encourage the student to study better. This is based on The Hawthorn Effect theory which stated that some people have the tendency to work harder and perform better when they were being observed. At the end of each course there will be a reflection segment which helps the student understand more of their learning method and also helps the student realize what they gained throughout the whole module. The reflection would help both the student and the instructor because the feedback would help in the development of the course itself and the learning platform. Knowledge is something you gained daily and it evolves rapidly. The feedback would maintain the constantly advancing syllabus and helps the instructor to be more aware of the students' need as well.

MOOCs allows the instructors to watch over the students' progress of the work they do and shows how many percent of the task that they had completed. Although MOOCs allows students to have a complete control over what they do and at their own stride, the instructors were not to just sit and do nothing if the students were not doing anything. They can take part of the learning by pushing them more or asking why was the problem. This one to one observation is more effective than teaching in the classroom when the instructor has to focus on everyone in order for each of the student to understand. This is related to Stephen Downes and George Siemens connectivism theory where learning happens not only when they are by themselves but through connections within networks [13]. The theory uses the idea of a network with nodes and connections to define learning. Students recognize and interpret patterns and are influenced by the diversity of networks, strength of ties and their context. This means that their knowledge is continually being acquired. They learn by themselves and make the effort to search for thing they don't know and are constantly creating new networks because the diversity of opinions let them learn something new each time. These networks are also connected by the fact that when fellow students shared their knowledge and posted something and they constantly learn new things because everyone have their way of interpreting things. Favouring the information sharing and duplication, the pedagogical approach embedded in this type of MOOCs is behaviourism [14]. Behaviourism operates on the principle of stimulus-response where all behaviours are cause by the surroundings and the negative and positive reinforcement. When their peers share new info and commented on their post, they feel compelled to reply and feel encouraged to post more. MOOCs makes use of this theory in order to boost the students' confidence to write more.

5. Pedagogical Implications and Conclusion

Applying the concept of distance learning in a classroom will not only help conserve time, it will also be good hands on experience for the students. The students will be able to develop their ICT skills and able to use technology in a good way for technology is a useful servant but a dangerous master. Future teachers need to be able to teach and at the same time possess other skills as well in order to survive and adapt. While attention in selecting the apt hardware and software for the classroom is essential, it is the skill and attitude of the teacher that ascertains the efficacy of technology amalgamation into the curriculum [15]. While the environment affects the ambiance of studying in class, incomplete accommodation discourages the student to study. Thus, it is the responsibility of the school too, to be able to provide the right material and stable internet connection while limiting the internet only for educational purposes. This will accommodate the teachers and students in implementing technology in classroom.

6. Conclusion

This literature review aims to identify and analyze the trends, datasets, methods and frameworks used in the topic of attribute independence assumption assumptions on NB between 2010 and 2018. To conclude, regarding using MOOCs to assist in learning English writing, it would be beneficial for both the students and the instructor. It is time conserving for both part and a self-directed learning could help the student in becoming more independent and helps them in widening the approach of learning. Additionally, it was much easier for them to learn from each other through posting on each of the modules provided with the additional instructions and built in galleries that provides information and examples. Moreover, using MOOCs to do assignments was convenient and reduces stress and environmental impact. This is due to the pressure elimination and a comfortable environment to work at.

Having a social media flow like, MOOCs allows comments and like on the assessments of the user, these is very useful and beneficial in terms of motivating and encouraging the students to work harder. Incorporating media and technology in learning allows student to have more fun while gaining extra knowledge, forming them into a holistic student with the ability to learn in any way and utilising the technology respectably. MOOCs as stated as [14] is the more traditional approach to learning and a source of information and knowledge. This is because the role of the teacher is arbitrated; the class is 'recorded' beforehand as in they provide a full set of resources, exercises and automated test. This show that even with the absence of a teacher, they still facilitate the learning by providing the kind of help the students might need in order to finish the set of questions prepared in advance.

Acknowledgement

This research is supported by Universiti Kebangsaan Malaysia under research scheme no. (i) GG-2018-001; (ii) KRA-2018-044.

References

[1] Hashim H, Yunus MM, and Embi MA, "Factors Influencing Polytechnic English as Second Language (ESL) Learners' Attitude and Intention for Using Mobile Learning," *Asian ESP J.*, vol. 14, no. 7, pp. 195–208, 2018.

[2] Yunus MM, Nordin NM, Salehi H, Sun CH, and Embi MA, "Pros and cons of using ICT in teaching ESL reading and writing," *Int. Educ. Stud.*, vol. 6, no. 7, pp. 119–130, 2013.

[3] Yunus MM, Maimun AL, and Chua PL, "Language learning via ICT: Uses, challenges and issues," *WSEAS Trans. Inf. Sci. Appl.*, vol. 6, no. 9, pp. 1453–1467, 2009.

[4] Yunus MM, Osman WSW, and Ishak NM, "Teacher-student relationship factor affecting motivation and academic achievement in ESL classroom," *Procedia - Soc. Behav. Sci.*, vol. 15, pp. 2637–2641, 2011.

[5] M. S. Abbasi, A. Tarhini, M. Hassouna, and F. Shah, "Social, Organizational, Demography and Individuals' Technology Acceptance Behaviour: A Conceptual Model," *Eur. Sci. J.*, vol. 11, no. 9, pp. 39–68, 2015.

[6] M. G. Moore, "Three types of interaction," *Am. J. Distance Educ.*, vol. 3, no. 2, pp. 1–6, 1989.

[7] D. Adas and A. Bakir, "Writing difficulties and new solutions : Blended learning as an approach to improve writing abilities Writing Difficulties and New Solutions : Blended Learning as an Approach to PhD in teaching English Language Methods , AL-Quds Open

University ," no. September, 2018.
[8] D. S. Wilbers, "Five Elements of Effective Writing," vol. 4, pp. 1–2, 2018.
[9] M. A. Conroy, "Internet tools for language learning: University students taking control of their writing," *Australas. J. Educ. Technol.*, vol. 26, no. 6, pp. 861–882, 2010.
[10] R. Shih, "Manufacturing aluminium and zinc sacrificial marine anodes," *Corros. Mater.*, vol. 37, no. 2, pp. 36–39, 2012.
[11] S. C. Tseng and C. C. Tsai, "On-line peer assessment and the role of the peer feedback: A study of high school computer course," *Comput. Educ.*, vol. 49, no. 4, pp. 1161–1174, 2007.
[12] H. Uzunboylu, N. Cavus, and E. Ercag, "Using mobile learning to increase environmental awareness," *Comput. Educ.*, vol. 52, no. 2, pp. 381–389, 2009.
[13] N. Bailey, Jemimah; Cassidy, Dara; Breakwell, "Keeping Them Clicking: Promoting Student Engagement In MOOC Design," *All Irel. J. Teach. Learn. High. Educ.*, vol. 6, no. 2, p. 1972, 2014.
[14] M. L. R. Renau, "Designing, Developing and Implementing A MOOC: Improve Your Writing Skills," Secundària Obligatòria i Batxillerat Farmacio, 2014.
[15] N. Bitner and J. Bitner, "Integrating technology into the classroom: Eight keys to succes," *J. Technol. Teach. Educ.*, vol. 10, no. 1, pp. 95–100, 2002.

Beyond ESL Classroom: The Use of MOOC in Enhancing ESL Reading Skills

Muhammad Afif Razali, Harwati Hashim, and Melor Md Yunus*

Faculty of Education, Universiti Kebangsaan Malaysia

*melor@ukm.edu.my

Abstract. Technology is changing the landscape of language learning and is viewed to have a huge potential in enhancing the teaching and learning of English language. Due to the high accessibility of the internet, coupled with the availability of ICT tools in Malaysian schools, it is believed that new methods which can shape new pedagogical techniques could be brought into the Malaysian classrooms. The lessons could then be shaped to become more personalized for the learners' needs and allowing them to learn on the move. This creates the potential for significant change in the teaching and learning process of English as a second language (ESL) in Malaysia. By looking at the existence and potentials of the Massive Open and Online Courses (MOOC), this paper offers reflections on the benefits that MOOCs can offer to ESL classrooms; specifically, to the development of the reading skills, as well as the constraints that might be faced by the English as a Second Language (ESL) learners when using MOOC. Above all, it also looks at the changes MOOC might bring to the pedagogical aspect of the ESL classrooms; towards both the students and the teachers themselves.

1. Introduction

We are living in a world where English is the global language. The usages of the English language have expanded through a wide variety of communicative discourses of its native and non-native speakers around the world, such as in inter-country administrative affairs, international business, science, popular culture, and many more. If a person does not have good grasp of the linguistic skills and knowledges of the English language, it will only cause more harm than good to that person's humane development, especially in this globalized world we are currently living in [1]. The English language consists of four main linguistic skills, but none is more important than the reading skill. It is through reading that most learners are mainly able to gain new information, knowledge, and understanding, among others [2]. In Malaysia, the English language is taught as a second language in schools and universities nationwide, as it is not the native language of Malaysians [3]. The term 'second language', or simply known as L2, indicates that the users have lower level of actual or perceived proficiency, compared to the first language (L1). Due to the fact that English is the 'second' or 'weaker' language in Malaysia, special measures must be taken in order to provide the best platform for the learners to fully grasp the language; proposedly through the usage of technology [4] [5] [6] [7].

Throughout the years, educational technology in Malaysia had gone through several phases of evolvement and development. Before the era of educational technology began, the go-to medium of instruction in the 1920s up until the 1950s was the visual instruction [8]. Back then, visual materials such as charts and flash cards were used to support the chalk and talk pedagogical method [9]. Then in the late 1950s, audio materials such as educational radio and television broadcastings, audio recordings and motion pictures started coming into the play, thus creating a new era called the audio-visual (AV) instruction period [10].

The accessibility and the number of computers in Malaysia started growing exponentially from then on, and this growth hugely affected the education quality of Malaysians. The Malaysian Institute of Microelectronic System; better known as MIMOS, collaborated with the Ministry of Education in 1989 to develop the Computer Integrated Education program [11]. Consequently, all schools nationwide received funding to build computer labs, install internet routers, LCDs, and many other ICT education elements [12]. This paved the way for teachers to incorporate a huge variety of elements of the internet into the classroom lessons. Teachers since then started to use web-based tools such as Blogspot, Blendspace, Prezi, and Kahoot, among others, making the classes more fun and interesting for the students, especially the students from the technologically-competent 'Z' generation.

Educating Malaysian learners with English is a tough task, but technology seems to hold the key on facilitating the process. One of the internet-based learning tools that has the biggest potential of doing so is MOOC. An abbreviation for 'Massive Open and Online Course', it is a learning platform that is based on the online world, rather than the physical world where traditional teaching and learning normally happen. Not only it is 'online', MOOC is also 'open', where the learning process in it is built on the interactions of the participants [13] .The following sections of this research will discuss the potential of MOOC to be used as an educational tool to teach English as a second language in a Malaysian context, with the focus is mainly directed to the all-important reading skills.

2. Reading in Malaysian Education Context

One of the most important skill of the English language is the reading skill. It is the main skill used by human beings to recognize written words, deciphering meaning, and fetching information from texts [14]. Without reading, messages cannot be delivered if the listeners and speakers are far away from each other, thus causing a massive burden to human communication as a whole. The Malaysian educational planners certainly realized how important reading skills is for the development of Malaysian students, so it is according that they implement a lot of reading-based programs and initiatives. In 1983, the government introduced the KBSR and the KBSM with the main goal is to implement an education system with a common goal, direction and approach [15]. The main direction of the English KBSR curriculum was to equip students with the three 'R' skills, namely reading, writing and arithmetic. With that in line, the government introduced literature components to develop reading habits and critical thinking for Malaysian students and enhancing their English proficiency along the way. A range of literary texts were formally picked by the ministry from Malaysian, British, Australian, American and African works. Learners were expected to be able to follow a story-line and understand a poem and to give their own response to literary texts.

According to the Curriculum Development Centre of Malaysia (CDCM), by reading literary texts, students could acquire vocabulary and domain knowledge, enhance their critical thinking, and build a background of verbal and factual information that could help them to approach more complex materials. At the same time, there are also aesthetic values from the texts that students can reap, where students may relate their personal experience with the contents of the texts [16]. This shows how wholesome the benefits of reading components implementation into the curriculum to the humane development of the students. The government also implemented a nationwide reading program known as Nadi Ilmu Amalan Membaca (NILAM) in 1999, with the aim of instilling the reading habits among primary and secondary students as well as computer literacy programs [17]. Not only that, the government also introduced a brand-new English SPM paper known as English 122/322 in the second phase of the English Language Teaching curriculum reforms, where it is implemented between the year 1970 to 1990. In this paper, students' communicative skills were more comprehensively tested, with the focus is on the students' reading, writing and speaking (oral assessment) skills. In 1995, the English 122/322 paper is renamed into English 1119 and it is remodeled to mirror the O-level syllabus set by the Cambridge Examination Syndicate of England [18].

3. Massive Open Online Courses (MOOC)

In 2008, two men named Stephen Downes and George Siemens had come up with an idea to develop an online learning platform where a wide variety of participants can interact with each other in order to create a richer learning experience than what learners usually went through with traditional learning tools. They called it MOOC, that stands for Massive Open and Online Course. MOOC is also defined as an online learning tool that integrates the connectivity of social networking, the facilitation of an acknowledged expert in a field of study, and a collection of freely accessible online resources [13]. Defining it by its abbreviation, MOOCs are indeed massive, where a certain MOOC can entertain a huge number of students per lesson. MOOCs are also open, where it is open for any learners interested with the course offered. MOOCs are also based on the online world, which anyone from any part of the world have the ability to join, given they are connected to the internet. There are also a huge variety of courses offered in MOOC, and the learners are free to choose any course that they are interested in. MOOCs are of no financial burden as the majority of MOOC courses are free of charge, there are no fixed time commitment where users are free to enroll MOOC courses whenever they wanted, and MOOCs do not require its users to attend physical classes [19]. The idea of MOOC really took off with the emergence of a number of private initiatives that carried the idea of MOOC, such as Udacity, Coursera, Edx, openlearning, and many more. Due to the MOOC's high tolerant to its users in three different avenues (the wallet, the time and the energy of its users), it is really popular among learners and educators worldwide, and it is time for Malaysian English learners to also enjoy some learning experience on this online platform. Through the usage of the internet-based MOOC, the students can browse online to gain access to related resources and texts that can enhance their knowledge of the text, and relate the idea of the text with their own experiences and relatable world issues. Not only that, the students can also conduct online discussions with their peers about the preliminary ideas of the text through discussion windows on MOOCs.

4. Benefits and Challenges

There are a lot of studies conducted by educational researches about the prospects of using technology in teaching and learning the reading skills. Surely there are benefits of it for both the educator and the learner, along with some risks that the educator must note in order to create the best learning environment possible. According to [20], there are six benefits of implementing technology in the process of teaching and learning English. Firstly, using technology in teaching and learning can help develop students' interest. By using multimedia technology, educators are able to use a variety of medium to convey knowledge, such as through simple images, audio, and visual animations among others, which helps to deliver information across the boundary of time and space. This will make the learning process more interesting for the students thus captivating their interests. A teacher can fully include multiple forms of media in the MOOC to capture students' attention and develop their interest in learning.

Additionally, technology can help promote the communicative capacity of the learners. Through the usage of technology, students can develop their critical thinking about a certain topic. Most importantly, technology can also help to facilitate the students' delivery of their thoughts and ideas about that particular subject. For example, a teacher can open a book-review discussion window in a MOOC. Students' will give provide their views about that book, which helps to develop their productive communicative skills. At the same time, other students can read the views of the other students, thus developing their receptive communicative skills. Third, technology can help to develop students' insights and knowledge of the world. Multimedia teaching coupled by the internet can provide students more information than what textbooks are able to offer. This will help widen their scope of knowledge and deepen their general understanding thus shaping them to become a more versatile type of learner. Fourth, technology can improve the teaching effect. In traditional learning environment, the learning processes are largely teacher-centered, and students does not participate much, apart from affirming to the teachers' orders upon them. On the other hand, technology opens up

a new space for students to express themselves while at the same time gaining valuable knowledges from the teacher, thus increasing their involvement in the teaching and learning process.
A good type of MOOC is interactive for the students and it involve a lot of discussions and feedbacks between the teacher and learner, and between learners themselves [19], thus providing ample windows for the students to shape the learning process according to their needs. This brings to the fifth benefit of technology in teaching and learning, which is improving the interaction of teacher and learner. With multiple interaction windows between students and educators in MOOCs, students can provide more feedback than they possibly can in traditional lessons, thus creating a special bond between the two like never before. Last but not least, technology also creates more flexibility to the structure of the course content. Through the feedback given by the students, the teacher can make changes to the course in a MOOC according to the students' needs and achievement, thus creating a more impactful learning experience.

Along with the positives of implementing technologies in English teaching and learning, there are some risks and negatives of it. [20] highlighted the fact that through MOOC, the development of students' speaking communicative skill is somewhat hindered. By incorporating technologies such as computers and smartphones in education, learning only involves the usage of audio, visual and textual effects. In MOOCs, the learning process involves the interaction of learners with learning materials, the teacher and other learners in the form of texts, videos and audios, and there is no avenue where the students can use their speaking skills to convey their opinions and messages.

5. Conclusion

Using MOOCs to cover the teaching and learning tasks of the English as a second language (ESL) subject is a simple task, yet it is highly-beneficiary for students and teachers alike. However, it is important to note that the efforts to employ MOOCs in Malaysian ESL classrooms need to be done in a highly holistic manner. Teachers should always put the students at the center of the learning process, and the teacher must heavily rely on the usage of ICT and mobile devices in a classroom. Undeniably, mobile devices which are popular among students are motivational tools to be used in education and various learning activities as many researches have shown the positive outcomes. These could be seen as an indicator that the utilization of educational technology components such as MOOCs in teaching and learning of ESL, particularly reading skills; would be an effective approach in overcoming the difficulties faced by educators in motivating Malaysian students to learn the English language, as well as enhancing their learning experience and eventually increase their proficiency. Hence, we should agree that MOOCs are undeniably the way to go for Malaysian English classrooms, and we have to ensure that the Malaysian educational system should include them effectively and productively; so that Malaysian learners can gain invaluable benefits from it.

Acknowledgment

This research is supported by Universiti Kebangsaan Malaysia under research scheme no. (i) GG-2018-001; (ii) KRA-2018-044.

References

[1] M. S. Reddy and P. Mahavidyalaya, "Importance of English Language in Today's World," Int. J. Acad. Res., vol. 3, no. 4(2), pp. 179–184, 2016.
[2] K. Lawton and H. Warren, "The Power of Reading," 2015.
[3] Thirusanku J, Yunus MM. Status of English in Malaysia. Asian Social Science. 2014 Jul 1;10(14):254. 2014.
[4] H. Hashim, Yunus. MM, and Embi, M. A. "Factors influencing polytechnic English as second language (ESL) learners' attitude and intention for using mobile learning," Asian ESP J., vol. 14, no. 7, pp. 195–208, 2018.
[5] Yunus MM, Nordin N, Salehi H, Sun CH, Embi MA. Pros and Cons of Using IICT in Teaching ESL Reading and Writing. International education studies. 6(7):119-30. 2013.

[6] Yunus MM, Lubis MA, Lin CP. Language learning via ICT: Uses, challenges and issues. Wseas transactions on information Science and applications. Sep 1;6(9):1453-67. 2009
[7] Mohamad M, Ghazali N, Hashim H. Secondary School Students' Perceptions on the Use of Google+ towards Improving ESL Writing Skills. International Journal of Emerging Technologies in Learning (iJET). Sep 29;13(09):224-38. 2018.
[8] H. Yusup and M. Razmah, "Instructional technology: Theory and application," 2006.
[9] Dussek and M. Hashim, "Kitab Pedoman Guru," 1953.
[10] J. Nuruddin, "The development of educational media services in Malaysia 1957-1972," 1978.
[11] MIMOS, "MIMOS Milestone," 2017. .
[12] G. Simin and S. Ibrahim, "Effectiveness of ICT Integration in Malaysian Schools: A Quantitative Analysis," Int. Res. J. Qual. Educ., vol. 2(8), no. August 2015, 2015.
[13] McAuley, B. Stewart, G. Siemens, D. Cormier, and C. Commons, "The MOOC model for digital practice," Massive Open Online Courses Digit. ways knowing Learn., pp. 1–64, 2010.
[14] M. Ko3ak, "THE TYPES OF READING," 2011.
[15] Darus, "The Current Situation and Issues of the Teaching of English in Malaysia," 2009.
[16] D. L. Grossman and J. T. Lo, Social Education in Asia: Critical Issues and Multiple Perspectives. 2007.
[17] Bahagian Teknologi Pendidikan, "Laporan pencapaian pelaksanaan Program NILAM 2003.," 2003.
[18] M. N. N. Lee, "Educational Change in Malaysia," 2002.
[19] E. Bourli, "Language Massive Open Online courses in Teaching English as a Foreign Language," 2017.
[20] D. S. Solanki and M. Phil, "'Use of Technology in English Language Teaching and Learning': An Analysis," 2012.

MOOC for Training: How Far It Benefits Employees?

Karmila Rafiqah M. Rafiq[1], Harwati Hashim[1], and Melor Md Yunus[1]*

[1]Faculty of Education, Universiti Kebangsaan Malaysia

* melor@ukm.edu.my

Abstract. As the term 4th Industrial Revolution emerges globally, various industries are moving rapidly towards it. Malaysia, one of the developing countries, is also embracing the 4th Industrial Revolution. However, with the rapid development of industries to catch up to the ever-expanding world, the English proficiency of non-native speakers is worrying, especially in the working field. Certain industries provide training for their employees to improve their English for Occupational Purposes (EOP), but traditional training is time-consuming and could not cater to individuals' needs. This paper aims to identify the benefits of MOOC for employees' training. The findings discussed the benefits of MOOC. MOOC is indeed beneficial for employees in terms of providing the opportunity for self-paced learning, lifelong learning and professional development at zero cost. However, there is also an interesting finding whereby employers do not want their employees to use MOOC for training because the former does not want private and confidential information of the organization to be publicized. Overall, the benefits will be able to guide and provide insight for educators and researchers as it is important to know what motivates learners to use MOOC for training.

1. Introduction

As the term 4th Industrial Revolution emerges globally, various industries are moving rapidly towards it. One of the elements in the revolution is the Internet of Things (IoT), which is more familiarized by cloud-based technology. A diversity of industries are not exempted from this revolution, including the education industry. Undoubtedly, various technologies are introduced to assist lessons in the educational field as well [1]. Being one of the developing countries, Malaysia is also embracing the 4th Industrial Revolution. The focus of the Ministry of Education is to produce a holistic individual, not only in terms of soft skills but also in English language proficiency. As English is the second language in Malaysia, more employers demand graduates with good proficiency in English [2]. This is because employers wish to expand their business globally, which requires employees to communicate with other speakers of other languages using the international language, English [3].

However, with the rapid development of industries to catch up to the ever-expanding world, the English proficiency of non-native speakers is worrying [4]. Certain industries provide training for their employees to improve their English for Occupational Purposes (EOP) for the workplace, but not all traditional training is proven to be effective. This is because some training needs more time and could not cater to all individuals' needs [4]. Some training is too short and expensive, whereby employers are not willing to pay for their employees. Not to mention that some employees also have a bunch of workloads which inhibit their intention to go for traditional training.

In relation to the problem of traditional training for EOP, learning virtually can address to most of the problems mentioned, whereby there is a variety of open learning platforms for learners to join and learn for free. Open learning or Open Educational Resources (OER) as mentioned by [5] provides an essential learning environment, which can assist learners in lifelong learning. Sharing and contributing ideas are some of the elements in open education, which is governed by the Internet, for people around the world to come together [6], [7]. Another element in the OER is in terms of its cost-free nature, which means that courses offered are relatively free to be used and can be modified according to the context and learners' needs, to ensure the effectiveness of a particular course is maximized [5], [8]. Plus, OER allows learners to access the content anywhere, as long as there is an Internet connection because it is an online resource [8]. One of the well-known open learning platforms, which is expanding exponentially is the Massive Open Online Courses (MOOC) [9], [10]. Hence, it is important to look into the benefits of MOOC, especially for training. This paper aims to identify the benefits of MOOC for workplace training.

2. Self-paced Learning Environment

Self-paced learning brings the meaning that learners are given the opportunity to learn by themselves in their own momentum, without feeling obliged or tied to complete a course within a time frame given. One of the most important elements in MOOC, which benefits learners is to allow learners to take their time completing a course without feeling pressured. Self-paced learning is more towards independent learning, whereby minimal interaction is required in learning [11]. Learners who are bound to a working environment has a lesser opportunity to get a formal education. Hence, working learners can gain benefits from MOOC.

A study by [12] claimed that learners' dropout rates from MOOC can be reduced when the content of a particular course is not too lengthy, as voluntary learners are focusing on learning independently. This shows that learners who use MOOC for their own personal gain are more prone to complete a shorter and relevant course because they are learning at their own pace. Concurrently, in another study by [13] argued that MOOC is indeed the best platform for self-paced learning, but it will not be successful if learners are not committed to complete a course. Relating to the working environment, it was also mentioned that learners with working pressure, such as tight working schedules will affect their commitment in taking up and completing a course in MOOC.

Due to that, as suggested by the same researcher, employers play an equally important role in ensuring their employees can be proactive in working and independently taking up online courses. The role of employers in ensuring the success rate of course completion in MOOC is also emphasized in other studies. An employee with the support of his employer tends to learn more skills by themselves to promote their own professional development [8], [14]. In other studies, learners are more engaged towards self-paced learning through MOOC because of its openness and flexibility, which renders to accessibility anywhere and anytime without restrictions [15], [16].

3. Lifelong Learning Opportunities

Learning a skill which can be useful for life is the aim of many learners. As [17] mentioned, MOOCs demolish the barrier of formal learning as learning can also be done informally in an online platform. In the same study, it was mentioned that there are certain circumstances whereby lifelong learning is not emphasized, such as in social and cultural contexts. In terms of the working environment, developing lifelong learning activities is crucial because a more holistic employee is much valued by an employer [6], [14]. A study by [17] reported that some employees took up MOOC to learn a skill such as an entrepreneur skill to suit the current demands of their organizations and countries. With support from employers, these employees will also benefit the organization.

In another study, which reported a similar finding was done by [13]. This study mentioned that MOOC is not only able to provide content-based training, where learners gain knowledge from joining a course, but also able to provide an online-based learning experience. It was stated that effective online instruction can enhance the learning of a learner, especially employees in acquiring a new skill usable in their own organizations. By taking up a course in MOOC, an employee will also be able to apply the skills in the real world. On another note, [16] stated that MOOCs for lifelong learning is dominated by unemployed people as they have more time and motivation to learn a skill in order to be employed. This shows that employed people have either lesser time to use MOOC or not being supported by employers. Nevertheless, MOOC is still a valuable tool for learners to learn regardless of being employed or not. That is one of the benefits of MOOC, which is promoting lifelong learning opportunities for everyone.

4. Professional Development

Undoubtedly, in an organization, professional development activities are carried out to ensure the efficiency of the employees. Employers invest more in traditional training for professional development, which is why resorting to MOOC is a better solution [14]. MOOCs can be used to replace traditional training, to equip employees with necessary skills for development. Due to its open-access nature, MOOC is said to be able to ensure active participation of learners, especially employees who are required to complete a course in MOOC as required by their employers. Similarly, [8] claimed that most employers demand professional employees, whereby the former felt the need to provide essential training to improve the latter. However, in the same study, employers of certain confidential organizations tend to not use MOOC as a training platform because they are afraid that their employees will disseminate private or sensitive information of the company in MOOC, which is an online open platform.

On the other hand, [12] in their study mentioned that employers support the use of MOOC as a tool which is deemed valuable to promote the professional development of their employees. This is because, MOOC is an open-access technology, which is always up-to-date. However, [18] disagreed that MOOC can help to promote professional development skills. This is because, professional skills such as collaboration and teamwork are vital for an organization, in which MOOC could not offer real-time team-building training. The employees stated in their study mentioned that learning and acquiring skills can be done online, but to apply it requires their colleagues' support, which can bring an impact to their working environment.

However [13] claimed that there is a team-based element in MOOC which can foster interactions among learners. In the same study, it was mentioned that MOOCs are also able to train communicative skills such as public speaking, which empowers an organization as well. MOOC offers a variety of courses, which could cater to learners' desires and the courses' theoretical foundations can also be applied practically in the real world. Supporting the statement, [5] agreed that OER such as MOOC will be able to promote an individual's development, which could contribute to the aspirations in the Sustainable Development Goals (SDG).

5. Cost-free Training

The most important element of MOOC, which could bring benefits for both the employees and employers is that it is entirely free of charge. OER is a free to use platform, where users do not have to pay to access the content [5]. Relating to the working environment, a company or an organization needs to provide training for employees to suit the current demands of the world. MOOC is irrefutably a potential platform for training purposes, which can benefit the organization [15]. They also mentioned that organizations want their employees to be able to compete globally, thus training them is essential. Clearly, MOOC can adhere to various challenges in training. This is because, [13] mentioned that most of the MOOCs are free to be accessed, although there are a few MOOCs which charge for certification purposes. Additionally, he emphasized in his study by stating that MOOC is cost-effective in terms of monetary and time-wise. The training can be done massively at zero cost. This benefits the organizations because providing traditional training takes up time and also involves money, which is the total opposite of using MOOC. MOOC is cost-free and also adheres to self-paced and lifelong learning.

6. Conclusion

This paper has discussed the benefits of MOOC for working learners. With the swift advancement of technology and open education, MOOC can be a boon instead of a ban. The open online platform still emits advantages not only for working learners but also to the employers and organizations too, as many organizations are beginning to recognize MOOC as a flexible training platform for lifelong learning and professional development. This paper will be able to provide an insight to researchers and educators in designing and developing a course in MOOC. In order to ensure effective instructional design, it is vital to look into the benefits of MOOC, especially in terms of the length of course, which could motivate learners in self-paced learning. Overall, exploring MOOC for educational purposes can bridge the gap between formal and informal education all over the world.

Acknowledgement

This research is supported by Universiti Kebangsaan Malaysia under research scheme no. (i) GG-2018-001; (ii) KRA-2018-044.

References

[1] H. Hashim, "Application of Technology in the Digital Era Education," *Int. J. Res. Couns. Educ.*, vol. 1, no. 2, pp. 1–5, 2018.
[2] R. R. Rasalingam and M. A. Embi, "The Use of Communicative Language Teaching (CLT) in the Teaching of English Courses among Malaysian Polytechnics Lecturers," *Int. J. Soc. Sci. Humanit. Res.*, vol. 6, no. 2, pp. 254–264, 2018.
[3] N. Nordin and H. Norman, "Mapping the Fourth Industrial Revolution Global Transformations on 21st Century Education on the Context of Sustainable Development," *J. Sustain. Dev. Educ. Res.*, vol. 2, no. 1, pp. 1–7, 2018.
[4] N. Mustafa, N. Nordin, and M. A. Embi, "A Need Analysis for a Communicative English Mobile Learning Module for Healthcare Professionals," *Int. J. E-Learning High. Educ.*, vol. 6, pp. 13–24, 2017.
[5] R. McGreal, "Special report on the role of open educational resources in supporting the sustainable development goal 4: Quality education challenges and opportunities," *Int. Rev. Res. Open Distance Learn.*, vol. 18, no. 7, pp. 292–305, 2017.
[6] A. Nobre, E. M. Mallmann, V. Nobre, and M. D. Mazzardo, "MOOC and OER," in *Emerging Trends, Techniques, and Tools for Massive Open Online Course (MOOC) Management*, IGI Global, 2018, pp. 1–23.
[7] M. Y. Melor, "Innovation in Education and Language Learning in 21st Century," *J. Sustain. Dev. Educ. Res.*, vol. 2, no. 1, pp. 33–34, 2018.
[8] U. Olsson, "Open courses and MOOCs as professional development – is the openness a hindrance?," *Educ. Train.*, vol. 58, no. 2, pp. 229–243, 2016.
[9] A. Anders, "Theories and Applications of Massive Online Open Courses (MOOCs): The Case for Hybrid Design," *Int. Rev. Res. Open Distance Learn.*, vol. 16, no. 6, pp. 39–61, 2015.
[10] R. Godwin-Jones, "Global reach and local practice: The promise of MOOCS," *Lang. Learn. Technol.*, vol. 18, no. 3, pp. 5–15, 2014.
[11] A. G. Picciano, "Theories and Frameworks for Online Education: Seeking an Integrated Model," *Online Learn.*, vol. 21, no. 3, pp. 166–190, 2017.
[12] F. Schwerer and M. Egloffstein, "Participation and Achievement in Enterprise MOOCs for Professional Learning," in *13th International Conference on Cognition and Exploratory Learning in Digital Age (CELDA 2016)*, 2016, pp. 269–276.
[13] S. Karnouskos, "Massive open online courses (MOOCs) as an enabler for competent employees and innovation in industry," *Comput. Ind.*, vol. 91, no. October, pp. 1–10, 2017.
[14] J. Castaño Muñoz, M. Kalz, K. Kreijns, and Y. Punie, "Influence of employer support for professional development on MOOCs enrolment and completion: Results from a cross-course survey," in *Proceedings of the European MOOC Skateholder Summit on experiences and best*

practices in and around MOOCs - EMOOOCS 2016, 2016, pp. 251–263.

[15] D. Petronzi and M. Hadi, "Exploring the Factors Associated with MOOC Engagement, Retention and the Wider Benefits for Learners," *Eur. J. Open, Distance e-Learning*, vol. 19, no. 2, pp. 112–129, 2016.

[16] J. Castaño-Muñoz, K. Kreijns, M. Kalz, and Y. Punie, "Does digital competence and occupational setting influence MOOC participation? Evidence from a cross-course survey," *J. Comput. High. Educ.*, vol. 29, pp. 28–46, 2017.

[17] C. Maitland and E. Obeysekare, "The Creation of CapitalThrough an ICT-based Learning Program: A Case Study of MOOC Camp," *ICTD*, pp. 1–10, 2015.

[18] F. A. C. Junior *et al.*, "Motivation to Learn and Distance Learning Programs: What Brazilian Workers Think about?," *Creat. Educ.*, vol. 07, pp. 2576–2596, 2016.

Development of Teaching Modules Based on Problem Solving

Haris Mahmudi*, Ah.Sulhan Fauzi, Irwan Setyowidodo

Mechanical Engineering, Nusantara PGRI University of Kediri, Kediri, Indonesia

harismahmudi@unpkediri.ac.id

Abstract. This research aims to develop teaching module products based on problem solving and disseminate product results to students as users. This research is a development research that uses three stages, namely: (1) preliminary research, (2) the stage of making a prototype, and (3) the assessment stage. Based on the data analysis of the validation results from the validator and the readability test analysis by the students' assessment of the feasibility of teaching modules based on overall problem solving, the total mean score was 3.39 with the criteria feasible. The assessment of problem solving based teaching modules consists of two aspects, namely aspects of assessing the feasibility of content and aspects of evaluating the feasibility of presentation. Aspects of the feasibility assessment of the contents of problem solving based teaching modules obtain an average value of 3.22 with eligible criteria. The aspect of the feasibility assessment of presentation obtained an average value obtained from 3.55 with the criteria feasible. While the results of the teaching module based learning on problem solving to students viewed from the results of the student satisfaction questionnaire on the use of teaching modules developed were 83% of students very good, while the remaining 17% were good.

1. Introduction

The learning process in the classroom is the most basic activity in the whole process of education. Learning activities will be good if there is reciprocity (interaction) between lecturers and students. There are three kinds of interactions in the learning: student-teacher, students, and student-subject matter [1]. The courses taken by students include theoretical courses and practical subjects that require teaching materials that can help so that learning objectives can be achieved. In this study the teaching material developed is a teaching module based on problem solving, especially the development of Basic Physics practicum modules for mechanical engineering students. The thing that underlies this research is that the Physics practicum is one of the courses taken by mechanical engineering students. Basic skills needed by students in conducting practicum, among others: planning practical activities, carrying out practicums, analyzing the results of the lab and reporting the results of the lab. Besides this this course is an implementation of the theory course so that learning is more fun. While the meaningfulness of learning physics knowledge will be realized if it is done by several methods of scientific methods and accompanied by cognitive reasoning for the data obtained and observed natural symptoms [2]. The problem that many students experience is assuming that theoretical physics courses are difficult subjects because there are too many memorized formulas, understanding the concepts of physics that are still lacking. Then a tool is needed to facilitate students' understanding of theoretical physics courses. The tool is a practical activity that can implement theories taught in theoretical physics courses.

Physical practicum activities required a practicum module that can be used to help the learning process. The learning module is the smallest unit of learning program, which is taught by students themselves individually or taught by students to themselves (self-instructional) [3]. Whereas according to [4], "modules are teaching materials that are packaged intact and systematically, in which contain a set of planned learning experiences and are designed to help students master specific learning goals. From the explanations of the researchers above, a conclusion can be drawn that modules are one of the teaching materials learned by students that are packaged in a complete and systematic manner. With this it is expected that the assistance of the practicum module will help students to understand a Physics concept through practical activities. And make learning more meaningful. Learning means going to be a teacher experiencing what is learned, "the constructivist perspective describes learning as a change in meaning constructed from experience" [5].

The challenges of students in the practicum process arise from problems presented in the practicum module used. The problems raised in the practicum module are understandable problems but there are challenges to answer them and answers to questions must be procedural. The way to solve problems in problem solving cannot be directly understood. Students need to have the ability to solve problems effectively. The problems presented in learning are non-routine and challenging problems where the problem is complex and there is no way or strategy that is ready to be directly used to solve it, so it takes a creativity to solve it [6], [7]. Problems can be called problems or problems, at least have 2 things, namely: (1) Challenging the mind (challenging), (2) Not automatically known how to solve it (non-routine)

2. Methods

The object of this research is students of mechanical engineering study programs who take basic physics practicum courses. The research conducted is development research. The development model used in developing this basic Physics practicum module is a development model [9]. There are three stages in this model, namely preliminary research, the stage of making a prototype, and the assessment phase. The type of data in this development research consists of quantitative data and qualitative data. Quantitative data is in the form of an average questionnaire. Questionnaire data in the form of a checklist with a scale of 1-4. Scale 1: disagree, incorrect, unclear; scale 2: disagree, inaccurate, unclear; scale 3: agree, right, clear; scale 4: very agree, very right, very clear.

While the qualitative data is in the form of suggestions, criticisms, and responses from the validator. Suggestions, criticisms, and responses from the validator are used as considerations in making revisions to the basic physics practicum module based on problem solving. The validation subject (validator) is 3 lecturers. Validators are lecturers who have expertise and experience in writing teaching materials. Product testing was also conducted on mechanical engineering study program students who had taken basic physics practicum courses, this was done to determine the level of readability of the module developed.

The instrument of data collection used in this development research is a questionnaire. The questionnaire used consists of two parts, namely: assessment questionnaire and questionnaire sheet suggestions and comments from the validator. The aspects contained in the practicum module assessment questionnaire are: (1) the feasibility of the content which includes: the suitability of the material description with the RPS, the accuracy of the material, and learning support material; (2) the feasibility of presentation which includes: presentation techniques, learning presentation, and presentation. The questionnaire was submitted to the validator to provide an assessment of the problem solving based Physics practicum module. The results of the assessment are then analyzed using the data analysis techniques specified.

The data analysis technique used to analyze the results of the due diligence data is the average calculation. Determination of the average value analysis technique based on the opinions of [10], which states that to find out the final grade rating on the research questionnaire, the number of values obtained is divided by the number of respondents who answered the assessment questionnaire.

Presentation of data from the development of Basic Physics Based Problem Solving Modules as Supporting Basic Physics Practicum Courses consists of: (1) description of module development, (2) data from validation module development (3) data on student readability of modules. The development description contains a brief description of the contents of the module being developed. The data from the module development validation contains responses, suggestions, criticisms, and validation data from 3 lecturers. Whereas, data from the readability of students contains difficult words and sentences that are not understood by students.

3. Results

3.1 Results of Module Development Validation

The data from the validation of the development of the Problem Solving Basic Physics module were obtained from 3 validators, namely 3 lecturers who were competent in the research field. The data obtained is quantitative and qualitative data. Quantitative data in the form of an assessment questionnaire with a Likert scale, while qualitative data in the form of responses, suggestions, and criticism from the validator. Data from the validation results were analyzed by the analysis of the average value in each aspect. The data from the validation results in each aspect can be seen in Table 1, Table 2 and Table 3, while the responses, the criterion of the validator is presented in Table 4 as follows.

Table 1. Assesment of The Content of Feasibility Aspect in The Module

No	Aspect assesed	Validator			Average	Remarks
		V_1	V_2	V_3		
1	Material Conformity SK and KD	3.67	3.00	3.00	3.22	eligible
2	Accuracy of Material	4.00	3.33	3.00	3.44	eligible
3	Total Average Value	3.00	3.33	2.67	3.22	eligible
	Total Average Value	3.56	3.22	2.89	3.22	eligible

Table 2. Evaluation of Feasibility Aspects Of Module Prsentation

No	Aspects assesed	Validator			Average	Remark
		V_1	V_2	V_3		
1	Presentation Technique	3.67	3.33	3.33	3.44	Eligble
2	Learning Presentation	3.60	3.80	3.60	3.67	Eligible
3	Completeness Presentation	3.33	3.17	3.17	3.22	Eligible
	Total Average value	3.87	3.43	3.37	3.55	Eligible

Table 3. Assessment of the Feasibility Aspects of the module

No	Aspect assesed	Validator			Average	Remarks
		V_1	V_2	V_3		
1	Content Feasibility	3.56	3.22	2.89	3.22	Eligible
2	Feasebility of Prsentation	3.87	3.43	3.37	3.55	Eligible
	Total Average Value	3.71	3.32	3.13	3.39	Eligible

Table 4. Validators' Responses, Suggestions and Critics

No	**Aspects**	**Response, Suggestions, Critics**
	Content Feasability	
1	Conformity of the Student Characteristics with SK and KD	• Material for each module is compatible with what is in RPS
2	Accuracy of the Material	• The teleprint of the module "measuring the level and uncertainty" is very long, however, it is short and clear.
		• In the tools and materials of the "hot type of solid material" number 8: release and stretch of heat to be modified to burn Bunsen spirits and props.

		• It is good, with simple variations in images and in daily life events
3	Learning Support Materials	• Enough to present and hope to be simplified.
		• It is good, to talk about the problems in daily life
Feasibility of Presentation		
4	Presentation Techniques	• Already good, simplistic concepts are presented before more complicated concepts
5	Presentation of Learning	• Use of lights / burners for many women who have been treated with mathematics and medical risk is quite high.
		• For the use of the term that the English are expected to be translated.
6	Completeness of Presentation	• Already adequate
		• Current list of references.

3.2 Data Analysis of Module Feasibility Assessment

Data analysis on the development of problem solving based Basic Physics modules is based on the results of the analysis of the average validation of the Likert scale questionnaire which consists of 3 validators. Based on the data the overall validation results obtained by the average value is 3.39. From the data from the validation results, it can be concluded that the development of problem solving based Basic Physics modules does not require significant content competition. The explanation of the validation data is as follows.

A. Feasibility of Content

Assessment of the feasibility aspects of the contents of the problem-based Basic Physics module has 3 assessment criteria, namely: the suitability of the material description with SK and KD, the accuracy of the material, and learning support material. While the validation of the 3 validators on the three criteria in the aspect of content feasibility shows that the average value of the material aspects of the description of the material with SK and KD is 3.22, the average value of the material accuracy aspects is 3.44, while the supporting material aspects of the learning value are average. obtained from 3 validators of 3.22. So that the average score on the aspect of the feasibility assessment of the contents of the Basic Physics module based on problem solving is 3.22 with the criteria feasible.

B. Feasibility of Presentation

The assessment of the feasibility aspects of the presentation of the Basic Physics module based on problem solving has 3 assessment criteria, namely: Presentation Technique, Learning Presentation, Complete Presentation. While the validation results from 3 validators on the three criteria in the feasibility aspect of the presentation show that the average value of presentation technique aspects is 3.44 , the average value in the aspect of learning presentation is 3.67, while in the completeness aspect the presentation of the average value obtained from 3 validators is 3.22. So that the total average value in the aspect of feasibility assessment is based on problem solving based on 3.55 with eligible criteria.

C. Module socialization

Socialization activities towards teaching modules based on this problem solving process are done to students who have taken basic practice in Physics. The purpose of this socialization is to measure the understanding of the foundation of the modules and the extent to which the teaching modules developed can help students to receive good material on the basic curriculum practice Physics The basics are taken. Assessments in this stage are in the form of satisfaction of the use of teaching modules given to students.

4. Conclusions

Product development results are in the form of problem solving Basic Physics modules. Problem solving based Basic Physics Module results of the development have gone through the revision stage in accordance with the test results strived by 3 validators and the results of student satisfaction questionnaires that have taken the basic Physics practicum.

1. Aspects of assessment of feasibility of modality Physics Based on problem-solving based on the average value of 3.22 with a questionnaire.
2. Evaluation aspects of the feasibility of the presentation of the average value obtained from the validator is 3.55 with a questionnaire.
3. The assessment of the feasibility of Physical modules on the basis of overall problem solving is obtained by the total mean value of 3.39 with a questionnaire score.
4. The results of the student satisfaction series using teaching modules developed were 83% of the students were very good, while 17% were good.

References

[1] Zhang, Dongsong. 2005. Interactive multimedia-Based E-Learning: A study of Effectiveness. The American Journal of Distance Education. London and New York: Lawrence Erlbaum Association, Inc. Vol. 19 (3) 149-162

[2] Wilhelm, J., Thacker, B. & Wilhelm, R. 2007. Creating Constructivist Physics for IntroductoryUniversity Classes. Electronic Journal of Science Education.

[3] Winkel. 2009. Psikologi Pengajaran. Yogyakarta : Media Abadi.

[4] Daryanto. Menyusun Modul. Yogyakarta: Gava Media, 2013.

[5] Newby, T., Stepich, D., Lehman, J. & Russell, J. 2000. Constructivism, Instructional Design, and Technology: Implications for Transforming Distance Learning. Journal of Educational Technology& Society. Vol 3(2) 16-22.

[6] Callejo, M L & Vila, A. 2009. Aprroach to Mathematical Problem Solving and Students' Belief Systems:Two Case Studies.

[7] Rasiman. 2015. Leveling of Students' Critical Ability in Solving Mathematics Problem Based on GenderDifferences. International Journal of Education and Research, 3(4).

[8] Sumardyono.(2007). Pengertian Dasar Problem Solving.(Online).(http://p4tkmatematika.org/file/problemsolving, diakses 11 Januari 2017).

[9] Plomp, T &Nieveen, N. 2010.An Introduction to Educational Design Research.Enschede: Axis Media-ontwerpers

[10] Arikunto, Suharsimi.2006. Prosedur Penelitian Suatu Pendekatan Praktek. Jakarta:Rineka Cipta

Model of Higher Education Quality Improvement in Indonesia: Relationship between HRM and Information Technology Literacy

Santi Susanti[1], Vina Natalia Van Harling[2], Muh. Ardian Kurniawan[3], Rusdin[4], Asdi[5], Aditya Halim Perdana Kusuma Putra[6]

Email :
ssusanti@unj.ac.id[1], vina.natalia@poltekstpaul.ac.id[2], ardian@unimor.ac.id[3], rus.mathology@gmail.com[4], asdi@unismuh.ac.id[5], adityatrojhan@gmail.com[6]

[1] Faculty of Economics, Universitas Negeri Jakarta, Jakarta, 13220, Indonesia
[2] Faculty of Engineering, Politeknik Katolik Saint Paul, Kota Sorong, 98414, Indonesia
[3] Faculty of Education Science, Universitas Timor, Timor Tengah Utara, 85613, Indonesia
[4] Department of Tarbiyah, STAIN Sorong, Kota Sorong, 98414, Indonesia
[5] Faculty of Economics, Universitas Muhammadiyah Makassar, Makassar, 92161, Indonesia
[6] Faculty of Economics, Universitas Muslim Indonesia, Makassar, 90231, Indonesia

Abstract. The purpose of this study was to produce a model for improving lecturer performance by revitalizing the components of competence, motivation, and information technology literacy. The object of this study took a sample of 160 lecturers from various private universities in the city of Makassar whose scientific fields concentrated on informatics and computer management. The use of mediation models is done using AMOS. Results of this study also state that the high competence of lecturers has not been able to contribute significantly to the improvement of information technology literacy. Low motivation does not contribute to improving information technology literacy. Competence and motivation, as well as information technology literacy, proved to be able to help significantly in improving the performance of lecturers. Besides that, information technology literacy is not able to show its role in explaining the influence of competence and motivation on the performance of lecturers. The novelty of this research is the addition of information technology literacy components as a model to improve lecturers' performance in Higher Education.

Keywords: Higher Education, Human Resources Management, Information Technology Literacy

1. Introduction

One of the novelties of this research is the use of information technology in supporting lecturer performance conducted using modeling to be analyzed using structural equation modeling for mediation and moderation models [1]. The intended modeling is to test whether the mediation model of Information and Communication Technologies has a higher contribution compared to the mediation model in analyzing the performance of lecturers. Changes that have taken place in all sectors are due to a revolution in technology and information [2]. The education sector is one of the social segments in which the global market has been formed, there is more intensive cross-country, and cross-regional interactions and the contribution of information and communication technology has a significant role in the implementation of work [3]. In the world of education, the quality of professionalism is a determining point of the quality of learning [4]. The rapid development of information technology and computers has received positive response in the community. Various public services have implemented ICT [5]. In the future the duties and functions of a lecturer are increasingly complex, this is due to the rapid development of science, and also the demand to use E-learning or Information Technology cannot be avoided. This situation requires competency to adapt according to the needs of higher education development [6].

For this reason, the readiness of a lecturer to improve the quality and intensity of his academic activities is absolutely an awareness and planned comprehensively and integrally. This is mainly related to the implementation of the higher education tri dharma namely education/teaching, research and community service.

These three activities are added with coaching activities and are likely to be followed by structural tasks which are one measure of academic performance [7].

Furthermore, the professionalism of lecturers in Indonesia has not shown encouraging results. Data held by the Ministry of National Education Research and Development shows that of the 120,000 permanent lecturers of a private university (PTS) and public university (PTN) in Indonesia, there is still 50.65 percent or around 60,000 of them have not had a master's degree or have just completed their undergraduate study. This fact is very ironic considering that one of the great aspirations of higher education in Indonesia is to become an international class university (world class university). With 50% of lecturers who are still qualified undergraduate (S1), it is very difficult shortly to reach these goals. Especially in the midst of such conditions, there was no significant effort from the lecturers to improve their professionalism as a central element of the university. Some of them are even less aware that the teaching profession, like other professions, is also related to the dimensions of knowledge, expertise, and ethics that need to be continually developed [8], [9]

With the various forms of responsibilities mentioned above, it is appropriate for a lecturer to have a loud and clear motivation, namely the existence of a high level of dedication to achievement. With the frequency of lecturers motivated to do their jobs well, they will improve the quality and job satisfaction they want because the strength of their motivation or work motivation will determine the size of job satisfaction [10]. Understanding lecturer competencies are still low; it is necessary to socialize the improvement of lecturer competencies through the provision of opportunities to follow higher levels of education for each lecturer, need to be including in various professional education and training in their fields. And actualizing the competency theory that is an underlying characteristic of an individual which is causally related to criterion-referenced effective or superior performance in a job or situation [11]. The theory explains that competency has significance for lecturers. In increasing satisfaction and performance, this condition explains that lecturers as professionals are paid by the State according to their pedagogical, professional, personal and social abilities to achieve organizational goals. The lecturers as educational staff have a very strategic position, which has a direct influence on the learning process, graduate quality, and competitive output patterns. This is means that the quality of their teaching staff determines the quality of students as the central element of education staff [12]. motivation, and utilization of information technology through job satisfaction on the performance of lecturers.

Human Resources Management

The lecturers competencies in Indonesia explained in Government Regulation Number 19 of 2005 concerning National Education Standards, namely pedagogy, professional, personality, and social skills which are four competencies that need to be carried out by lecturers. Competence becomes an essential element in an individual's human resources to achieve his goals. Included in this case are the lecturer competencies needed in the progress of higher education [13]. Because the teaching profession is different from other businesses because specific characteristics must be implemented so that it can direct students to the four pillars of education issued by UNESCO (learning to know, learning to do, learning to be and learning to live together). In this case, the lecturer characteristics needed are: (1) Understanding the teaching profession as a genuineness call, (2) during the teaching process always looking for positive rewards, so that students can do self-esteem, (3) the attitude of the lecturers is not only sympathetic but also needs to be firm, (4) realizing that as a lecturer in the global era it should have the ability to be a learner (lifelong learning) and not just an ambivalent profession.

Ideally, the ability and expertise of the lecturer must continue to be honed and developed by universities from time to time. Improving the quality of lecturers in higher education can be done by various methods and methods, including providing continuous education and training programs, improving lecturer development methods and strategies through fulfilling competencies in their fields based on definite knowledge, skills and work culture [14]. Apart from that, the drivers that can trigger an increase or decrease in a lecturer's performance are mainly influencing by both intrinsic and extrinsic motivational factors. Motivation plays a vital role in changing worker behavior. This also applies to lecturers as part of human resources in this case labor. In other words, improvement in how to motivate lecturers will result in increased quality of lecturer performance. The best-known motivation theory is the hierarchy of needs of Abraham Maslow. Maslow's view of human relations movement focuses on the requirement as a key foundation. Maslow's motivational theory rests on two main principles. First, the policy of deficiencies, namely satisfied needs is not a behavioral motivator [15].

Improving the quality of higher education in Indonesia can be realized if there is an improvement in the quality of the performance of the lecture itself. Lecturer performance is the ability to carry out work or tasks that the lecturer has in completing a job. Lecturers who have performance are lecturers who carry out their duties responsibly. The nature of responsibility characterizes qualified lecturers and have attitudes and personalities in carrying out Tri Darma and mastery of technical expertise, being able to teach students and master concepts, processes, and philosophical foundations. In other words, a qualified lecturer is a professional lecturer in the sense that he can make decisions professionally, responsibly and provide services to the community. Therefore the performance of lecturers has specific criteria that are by their competencies as educational educative staff [10].

Information Technology Literacy as Supporting the Quality of Higher Education in Indonesia

Technology is seen as a tool used by individuals in helping carry out tasks in the evaluation. In this context, information technology includes computers (mainframes, mini, micro), software (databases), networks (internet, intranet), electronic commerce, and other types of technology [16]. Because the experience of using technology will affect users, whether the technology has a better or worse impact on performance, it will affect future performance [17]. Empirically, it was founded the fact that to receive information technology it is not necessary to think first whether the information technology provides benefits or not but is a necessity and necessity so that information technology must be considered easy to use. Information systems and information technology can be used effectively to contribute to performance, so members of the organization must be able to use the technology properly. At present, information technology is considered natural, because training has often been carried out aimed at improving the skills of using information technology by lecturers.

Linkages between Human Resources Components and Information Technology Literacy

Information technology and job satisfaction significantly affect workers to work productively because job satisfaction can support the excellent behavior of the workers themselves [18], that information systems and information technology can be utilized effectively to contribute to performance, so members of the organization must be able to use the technology properly. The ability is an essential factor in improving work performance capabilities related to the knowledge and skills that a person has. [19] state that the use of computers provides excellent benefits for companies, primarily related to efficiency and effectiveness. The process of changing manual work methods into a computerized system is expected to make the work can be done quickly and accurately so that the overall performance of the organization can increase. the potential benefits of computer media that can be used to improve the effectiveness of the learning process to be able to create a continuous learning process. Apart from that social factors and conditions that facilitate positive and non-significant effects on the use of information technology, while effect, complexity, and suitability of tasks have a negative and no significant impact on the use of information technology.

2. Research Methods

This research was conducted on the basis of the development of several models and theoretical frameworks that have been built such as the model for lecturer performance (WL) in Indonesia which has been previously examined by Kusuma, et.al [7] which contains lecturers' main tasks in Indonesia which include Tri Dharma Higher Education (Education, Teaching, and Community Service). Then the competency model (COM) which is measured by the dimensions of Pedagogic competence, Professional Competence, Personality Competence, and social competence, motivation (MOV) and job satisfaction (WS) and Education quality are measured by dimensions of responsibility, progress, achievement, recognition, and specifications work refers to the Mc Clelland theory. The use of technology (IT) proposed by Robbins [20] measured by dimensions of measurement of hardware, software, and network data management. Dimensions and variables were estimated based on the Likert scale (1 = Strongly disagree - 5 Strongly Agree). The sampling method used proportionate stratified random sampling with the criteria of the population of the number of lecturers at the private tertiary education sector (STMIK) in Makassar - Indonesia as many as 285 lecturers. Samples were 39 samples (STIMIK Handayani), 74 samples (STIMIK Dipanegara), 18 samples (STIMIK Kharisma), 20 samples (STIMIK Akba), 15 samples (STIMIK Professional). So that the total sample total is 166 samples. Modeling tests and data analysis methods are carried out with different programs and models. The use of mediation models using Structural Equation Model with AMOS [21], [22], [23]

3. Results and Discussion

3.1 Characteristics of Respondents

The characteristic in question is the identity of the lecturer at each university which consists of; (1) gender, (2) age, (3) rank and class, (4) years of service and (5) level of education. Briefly, the characteristics of respondents can be seen in (table 1):

Table 1. Features of Lecturer

No	Criteria	Classification	Total Amount	
			Frequency	%
1	Gender	Men	91	56,87
		Women	69	43,13
2	Age (Year)	30 <	43	27,18
		31 - 40	57	34,96
		41 – 50	41	25,72
		> 50	19	12,14
3 .	Rank and group	Asisten Ahli	65	40,62
		Lektor	55	34,46
		Lektor Kepala	40	25,24
4	Working Time (Year)	5 >	60	37,57
		6 – 10	64	39,81
		11 – 20	24	15,04
		< 21	12	7,28
5	Education Status	Magister	147	91,87
		Doctoral	23	8,13

The description of the characteristics of the respondents in table 1 shows that the lecturers who taught at STMIK in the city of Makassar were mostly male, 91 people or 56.87%, respondents had the age of 31-40 years as many as 57 people or 34.96%, rank and the class of expert assistant respondents as many as 65 respondents or 40.62%, the working period of respondents between 6-10 years as many as 64 people or 39.81%, and education of respondents is magister 147 people or 91.87% and the rest who have doctoral training have just reached 8.13% or around 23 people.

3.2 *Model Analysis of improving lecturer performance with AMOS*

The results of confirmatory factor analysis (CFA) show that the overall value of the latent construct variable or loading factor the required(cut off point) ^{3} 0.60 or so it can be stated that the indicator of variable formation is seen to be of the same dimension with other indicator variables to explain a variable. Competency variable contributions (X1) and work motivation (X2) towards information technology literacy (Y1) of R2 = 0.005 while the contribution of Competency variables (X1), Work motivation (X2) and Information technology Utilization (Y1) on lecturer performance by R2 = 0.794 or 79.40%. The results of testing the feasibility of structural models through the path diagram are presented in (Fig. 2)

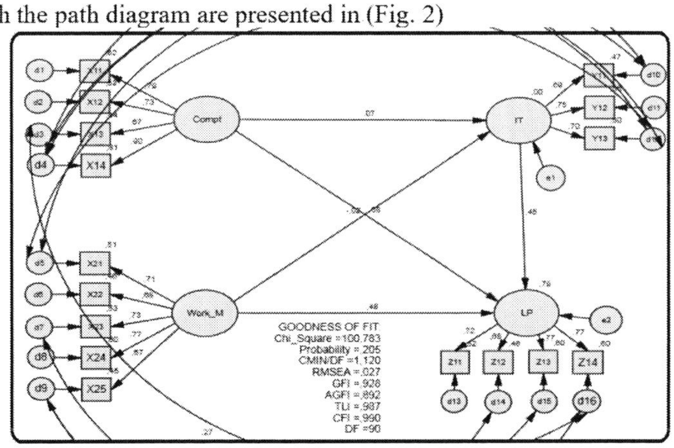

Figure 1. Full Path Diagram of AMOS Structural Model Testing Model

Furthermore (Fig. 1) also shows the results of testing the full model path diagram of the structural model, which is the basis for conducting a comparison of the index of the suitability of the model with the cut-off value presented in (Table 3) Goodness of fit model.

Table 2. Fit-Index

The goodness of fit index	Cut-off Value	Result Model*	Description
Chi_Square (DF = 90)	expected small	100,783 ≤ (0,05 ;90 = 113,145)	Good
Probability	≥ 0,05	0,205	Good
CMIN/DF	≤ 2,00	1,120	Good
RMSEA	≤ 0,08	0,027	Good
GFI	≥ 0,90	0,928	Good
AGFI	≥ 0,90	0,892	Marginal
TLI	≥ 0,94	0,987	Good
CFI	≥ 0,94	0,990	Good

Based on Figure 1, it is shown that the results of the conformity analysis of the structural model above are built as a basis for analyzing the relationship between latent variables and hypothesis testing with the aim of knowing the relationship between hypothesized latent variables and the level of significance of causality between the following:

Table 3. Hypothesys Test

	Hypothesis	P-Value	Direct Effect	Indirect Effect	Total Effect	Info
H_1	Competence ➔ Literacy of Information Technology	0,502	0,066	0,000	0,066	Rejected
H_2	Motivation ➔ Literacy of Information Technology	0,832	-0,021	0,000	-0,021	Rejected
H_3	Competence ➔ Lecturer Performance	0,000	0,579	0,030	0,609	Accepted
H_4	Motivation ➔ Lecturer Performance	0,000	0,477	-0,009	0,468	Accepted
H_5	Literacy of information Technology ➔ Lecturer performance	0,000	0,453	0,000	0,453	Accepted

3.3 Discussion

The results of the analysis show that the essential or dominant factor that reflects the lecturer competency variable is the indicator of social competence which means that the lecturer must have the ability to communicate and interact effectively and efficiently with students, fellow lecturers, and the surrounding community so that social competence can be formed. In reflecting the competence of lecturers. The results of this study confirm that the lowest competency indicators based on respondents' responses and current loading factor values are indicated by indicators — personality competencies that show that lecturers have not been fully able to maintain authority and have not been fully able to become role models for students, so it is important to be improved. Apart from that the high competence of lecturers can lead to an increase in information technology literacy. The results of hypothesis testing prove that lecturer competency has no significant effect on information technology literacy. The statement can be interpreted that the high competence of lecturers has not been able to provide a real meaningful contribution to the improvement of information technology literacy. This is because the lecturer has not been able to manage himself in carrying out daily tasks and Tri Dharma, that lecturers have not fully possessed the maximum mastery of teaching materials and have limitations in regulating learning and the evaluation process that is carried out in the teaching and learning process is not optimal. so that it has an impact on the low use of Hardware or hardware in supporting the smoothness of the lecture process and other Tri dharma University. So that mastery of technology is considered important to be improved in order to improve performance and quality Education in college.

In the aspect of Human Resources component, the dominant factor that reflects the lecturer motivation variable is the Indicator of Award Needs, which means that the lecturer in carrying out his activities is encouraged because of the desire in the lecturer to be appreciate including feelings, beliefs, competencies, achievements, mastery, independence and academic freedom. The results of this study confirm that the Self Actualization currently indicates the lowest indicator of motivation Needs sign, which states that the lecturer has not entirely been able to prove and show himself to others, the lecturer has not been fully able to develop as much as possible the potential he has. Apart from that the low motivation of lecturers resulted in a decrease in the mastery of information technology literacy. The results of hypothesis testing prove that lecturer motivation has a negative and not significant effect on information technology literacy. Of course, low motivation will influence interest in enriching technology literacy, because information technology literacy possessed by lecturers is now the primary and mandatory requirement to support the performance of the lecturer and as a benchmark of the quality of education in higher education

3.4 *Conclusion*

High lecturer competencies have not been able to provide a real (meaningful) contribution to the improvement of information technology literacy. This is because the lecturer has not been able to manage himself in carrying out daily tasks and Tri Dharma College, that lecturers have not fully possessed the maximum mastery of teaching materials and have limitations in regulating learning and the evaluation process that is carried out in the teaching and learning process is not optimal so that it has an impact on the low use of Hardware or hardware in supporting the smoothness of the lecture process and other Tri dharma (Lecturer Workload). The low motivation of lecturers has not been able to make a real (meaningful) contribution to the improvement of information technology literacy. This is because the lecturer has not fully developed all the potential that is contained in him so that it has an impact on the not yet optimal use and utilization of available network or internet connections for lecturers to support the implementation of the tri dharma of higher education. High lecturer competencies are proven to be able to make a real contribution (meaningful) to the improvement of lecturer performance. This is because the lecturer has a high understanding and knowledge of the lecture material and the lecturer can evaluate the teaching and learning process thoroughly and correctly and can develop the potential of the student so that it has an impact on the ability of the lecturer to carry out the Tridarma for support activities.

High lecturer competencies are proven to be meaningful for improvement of lecturer performance. This is because the lecturer has a high understanding and knowledge of the lecture material and the lecturer can evaluate the teaching and learning process thoroughly and correctly and can develop the potential of the student so that it has an impact on the ability of the lecturer to carry out the Lecturer Workload for support activities. High information literacy is proven to be able to make a real contribution to the improvement of lecturers' performance. This is because lecturers are highly capable of operating software, that in general, the implementation of Lecturer Workload tertiary institutions requires software support in its application and these conditions have been implemented well so that it can provide a real influence on the implementation of community service activities and Lecturer Workload for support activities other colleges. Information technology literacy is not able to show its role in explaining the impact of competence and motivation on the performance of lecturers. It is necessary to change the paradigm of thinking and severe efforts of university leaders and lecturers in the city of Makassar to improve information technology literacy, because in testing the model it was found that information technology literacy support was not able to enhance competency and motivation towards lecturer performance. It is necessary to change the paradigm of thinking and severe efforts of university leaders and lecturers to improve the competence of lecturers 'personalities as a representation of lecturers' competence through increasing the authority and ability of lecturers to become role models for students. It is necessary to change the paradigm of thinking and severe efforts from the leaders and lecturers in particular and Indonesia in general to increase the motivation of lecturers through increasing needs Actualization by increasing efforts from within the lecturer to prove and show themselves to others through increasing all potential owned by the lecturer. It is necessary to change the paradigm of thinking and severe efforts of university leaders and lecturers to improve the mastery and ability of lecturers on hardware because the use of information technology is always up to date. It is necessary to change the paradigm of thinking and severe efforts of university leaders and to improve the performance of lecturers through increasing the quantity and quality of research and development of scientific work that has an impact on the quality of education in higher education.

References

[1] S. Aryee, F. O. Walumbwa, Q. Zhou, and C. A. Hartnell, "Transformational leadership, innovative behavior, and task performance: Test of mediation and moderation processes," *Hum. Perform.*, vol. 25, no. 1, pp. 1–25, 2012.

[2] I. Hsu and R. Sabherwal, "Relationship between intellectual capital and knowledge management: an empirical investigation," *Decis. Sci.*, vol. 43, no. 3, pp. 489–524, 2012.

[3] S. Marginson, "Rethinking academic work in the global era," *J. High. Educ. policy Manag.*, vol. 22, no. 1, pp. 23–35, 2000.

[4] E. Kleinhenz and L. Ingvarson*, "Teacher accountability in Australia: Current policies and practices and their relation to the improvement of teaching and learning," *Res. Pap. Educ.*, vol. 19, no. 1, pp. 31–49, 2004.

[5] F. Bannister and R. Connolly, "ICT, public values and transformative government: A framework and programme for research," *Gov. Inf. Q.*, vol. 31, no. 1, pp. 119–128, 2014.

[6] S. C. Eze, H. O. Awa, J. C. Okoye, B. C. Emecheta, and R. O. Anazodo, "Determinant factors of information communication technology (ICT) adoption by government-owned universities in Nigeria: A qualitative approach," *J. Enterp. Inf. Manag.*, vol. 26, no. 4, pp. 427–443, 2013.

[7] A. H. P. Kusuma, R. Rina, and A. H. Syam, "The Main Role of Locus of Control and Professional Ethics on Lecturer's Performance (Indonesian Lecturer Empirical Study)," *Int. Rev. Manag. Mark.*, vol. 8, no. 5, pp. 9–17, 2018.

[8] D. H. Clark and M. Oey-Gardiner, "How Indonesian Lecturers have Adjusted to Civil Service Compensation," *Bull. Indones. Econ. Stud.*, vol. 27, no. 3, pp. 129–141, Dec. 1991.

[9] I. S. Anggraini, "Motivasi Belajar dan Faktor-Faktor yang Berpengaruh: Sebuah Kajian Pada Interaksi Pembelajaran Mahasiswa," *Prem. Educ. J. Pendidik. Dasar dan Pembelajaran*, vol. 1, no. 02, 2016.

[10] D. S. Sukirno and S. Siengthai, "Does participative decision making affect lecturer performance in higher education?," *Int. J. Educ. Manag.*, vol. 25, no. 5, pp. 494–508, 2011.

[11] L. M. Spencer and P. S. M. Spencer, *Competence at Work models for superior performance.* John Wiley & Sons, 2008.

[12] M. Fang, T. Nastiti, and C. H. V. Chen, "The tug of work and family: A study of the sources of the work family conflict among Indonesian lecturers," *Int. J. Manag. Enterp. Dev.*, vol. 11, no. 2–4, pp. 127–141, 2011.

[13] B. E. Kaufman, "Market competition, HRM, and firm performance: The conventional paradigm critiqued and reformulated," *Hum. Resour. Manag. Rev.*, vol. 25, no. 1, pp. 107–125, 2015.

[14] A. Carmeli and A. Tishler, "The relationships between intangible organizational elements and organizational performance," *Strateg. Manag. J.*, vol. 25, no. 13, pp. 1257–1278, 2004.

[15] C. Toban and H. Sjahruddin, "The antecedent and consequence of Organizational Commitment and Job Satisfaction," *J. Bus. Manag. Sci.*, vol. 4, no. 2, pp. 26–33, 2016.

[16] R. Taylor and H. Lee, "Occupational therapists' perception of usage of information and communication technology (ICT) in Western Australia and the association of availability of ICT

on recruitment and retention of therapists working in rural areas," *Aust. Occup. Ther. J.*, vol. 52, no. 1, pp. 51–56, 2005.

[17] R. L. Thompson, C. A. Higgins, and J. M. Howell, "Personal computing: toward a conceptual model of utilization," *MIS Q.*, pp. 125–143, 1991.

[18] L. Evans, "Professionalism, professionality and the development of education professionals," *Br. J. Educ. Stud.*, vol. 56, no. 1, pp. 20–38, 2008.

[19] A. Herdiyanti, A. N. Adityaputri, and H. M. Astuti, "Understanding the Quality Gap of Information Technology Services from the Perspective of Service Provider and Consumer," *Procedia Comput. Sci.*, vol. 124, pp. 601–607, 2017.

[20] S. P. Robbins and T. A. Judge, "Perilaku organisasi edisi ke-12," *Jakarta: Salemba Empat*, 2008.

[21] M. Haming, I. Murdifin, A. Zulfikar Syaiful, and H. P. K. P. Aditya, "The Application of SERVQUAL Distribution In Measuring Customer Satisfaction of Retails Company," *J. Distrib. Sci.*, vol. 17, no. 2, p. 25, 2019.

[22] A. Arfah and A. H. P. K. Putra, "Analysis of Productivity and Distribution of Female Workers in {FB}'s Industries," *J. Distrib. Sci.*, vol. 17, no. 3, pp. 31–39, Mar. 2019.

[23] I. Murdifin *et al.*, "Environmental disclosure as corporate social responsibility: Evidence from the biggest nickel mining in Indonesia," *Int. J. Energy Econ. Policy*, vol. 9, no. 1, 2019.

The review of utilization in natural environment by the teacher as a source for learning science at elementary school in low class

N Astimar[1], Y Helsa[1], A K Kenedi[2]
[1]Universitas Negeri Padang Jln Prof. Hamka, Air Tawar, Padang, Indonesia
[2]Universitas Samudra, Jl. Meurandeh, Langsa, Indonesia

E-mail: yullys@fip.unp.ac.id

Abstract. Teacher is the one of the component which is determine of successful for students. Based on observations , The utilization in natural environment was still lack for the teacher as a source as learning science at the begalung sub-distric padang . The study aimed at paving the review as well as as particularize how much the utilization of natural environment can enhance by the teacher as a source of learning science in low class at elementary school. This study employed descriptive, the subject of the study was the 44 teachers joining in elementary school. The data were gotten from observation which later analyzed. The result of study showed that there has lack of utilization in natural environment of the teacher as the source science which amount 46.48% of teachers have used to be natural environment as a source for learning science.

1. Introduction

The successful of students is established by a lot of factors. As one of the factor is learning process for teachers. The teacher have a role as influencing, foster and enhance intelligence and skills of the students. Teachers should have able to comprehend and implement in learning process significantly. Based on the constitution of RI No. 20 year 2003 concerning the system of national education on article 1 stated that the learning process is the communication between students toward teachers and sources of learning process [1].

Sciences became one of the subjects that related with natural environment [2] [3]. The natural environment can be used by teachers as a learning resource [4]. That will be expected to help the successful of students in the learning process. Basically, students of low class in primary school was on stage of developing intellectual. As the result , the experience is required in field, that will make student to be understand about science easily. Furthermore, Science is not material to be memorize, but Science is giving opportunities for students to conduct of all kind of observation and practice to develop of thinking logically and reasonably [5]. In primary school, Students attained to understand about science so that the students will be easier know about concept of science immediately and logical way [6].

In science learning needs of supporting books. Primary school doesn't have of laboratory particularly in science otherwise media to support of the lesson. Therefore, the utilize in natural

[1] To whom any correspondence should be addressed.
[2] To whom any correspondence should be addressed.

environment as source learning is relevant both of them because biology as science that study about living things, objects and nature of the symptoms and the problem of natural environment

The utilize of natural environment for student learning is a real object [7] [8]. Learning source means effort to be used for the benefit of the process or learning activities either directly or indirectly. The student prepared by themselves while learning [9]. The natural environment become a closest learning sources that can be utilized by teachers in the process of learning to their students.

Using the natural environment is necessary as a source of learning that have impact for students will be active, motivate , enthusiasm and interest of learning. Using the natural environment will help student to understand the concept of living things and their environment as well as the characteristics of a good environment, because living beings is itself often was shown and utilized in everyday life. Therefore, it would make it easier for teachers to deliver of science for students.

In addition, students are expected to be able to love and preserve the natural environment in their life. The natural environment as source of learning, either students will be appreciate, protect , and preserve the natural environment [10].

Based on observation the researcher discussion towards teachers in the primary school that there a few teachers who utilize the natural environment particularly in science. Furthermore, the ability of students to achieve in science that has yet significant. In addition, teachers as a source concern about the utilization natural environment so that will obtain the result clearly.

2. Method

This study employed descriptive. The subject of the study was the 44 teachers joining in elementary school who are active in low class of begalung district .This study used primary data, means data is required by the respondent. This study used of variable those are living of plants and animals and human.

The instrument this study used questionnaire which is giving by the researcher for respondent. The questionnaire arranged based on the purpose of this research. Respondent choose one alternative answer suitable on research guideline. Statement arranged by positive sentence. There are consist of category always (S), often (O), rarely (R.), and never (N).

The instruments should tested than it can be called as a valid test. The result of validity required by questionnaire divided into live things, animal, plants. That was arranged 38 items and after that was tested then gotten 15 items. The instruments have qualified of reliability. The test was conducted to 30 primary teacher. The data was collected and was put the data using of tabulation.

3. Results and discussion

3.1. Utilization of natural environment as learning sources the concept of a live things

Utilization of natural environment as a learning source for the concept of a live things consists of two indicators, The characteristic of live things , and the maintenance of live things.

The characteristics of live being was elaborated that follows: 1) Comparing plants which is watered everyday and plants is not watered 2) Observing the features of animal to move 3) Observing the growth of the sprouts movement 4) Observing of the increase the heavy animals; 5) Measurement the growth of the sprouts; 6) Observing the way grafting of plants; 7) planted -seed the plants 8) students to inhale and exhale. The result of the data can be shown in figure 1

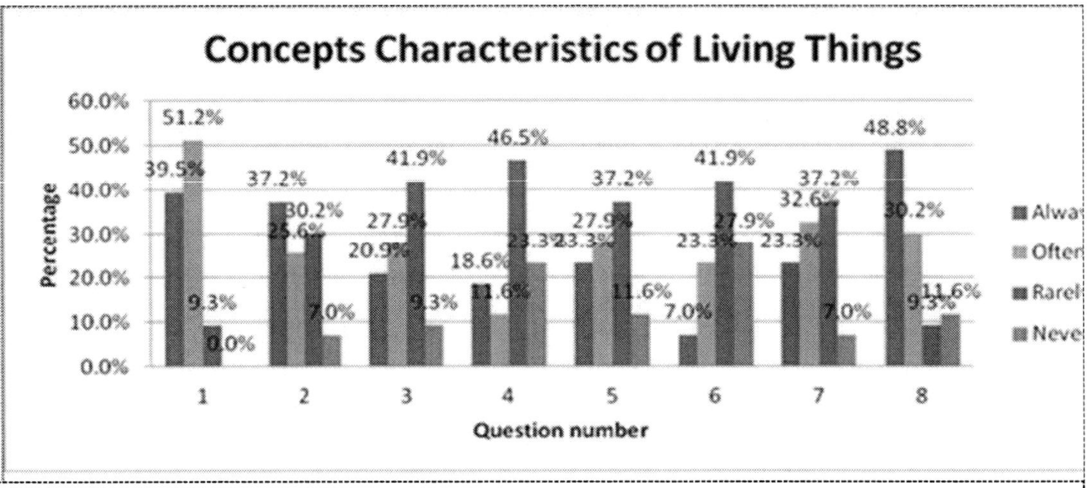

Figure 1. The Percentage teachers Assigning to students about the concept of characteristics of live being

Based on the figure 1. Generally, most of the teacher assigning for students to utilize the natural environment in understanding of the characteristics of live things. The first statement most teachers (51.2%) often assigning students to compare the plants are watered every day and plants is not watered. In the second statement about 37.2%) Observing the features of animal to move. On the third statement approximately 41.9% who rarely assigning their students to observe the growth of the sprouts movement. fourth most teachers (46.5%) rarely assigning their students to observe increasing the heavy animals. fifth statement about 37.2% of teachers rarely assigning students to measure of the growth of the sprouts. sixth statement about 41.9% of teachers rarely assigning students to observe the way grafting of plants. Seventh statement about 37.2% of teachers rarely assigning to students planted-seed of plants. In a statement the eighth most teachers (48.8%) always assigning students to inhale and exhale.

Utilization of natural environment as a learning resource for the maintenance concepts living was elaborated as follows: 1) Brings of students to the location of the farm around the school 2) Observing the ways of living animals in the environment. The results of the data analysis can be shown in figure 2.

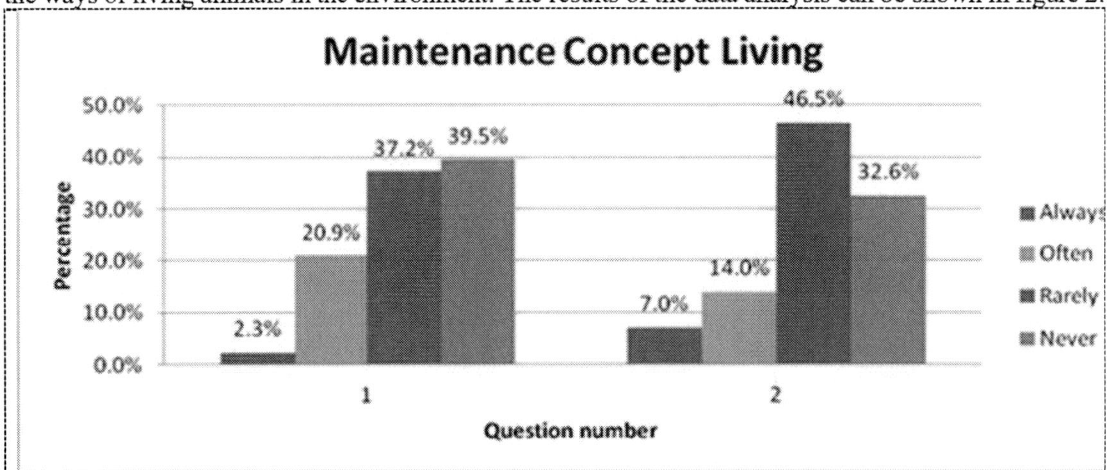

Figure 2. The Percentage of teachers Assigning to their students about maintenance concept living things

Based on the figure 2 that can be shown generally, Most of teacher who never and rarely assign students to utilize the natural surroundings in understanding the concept of the maintenance of living beings. the first statement about 39.5% of the teachers never bring students to the location of the farm around the school. the second statement most of the teachers (46.5%) rarely assign students to observe the ways of living of the animals in the environment.

Teachers should use the natural environment around student learning as a source. The selection of the natural environment which as a learning resource of science is precisely because the concept of living things to discuss about plants, animals and their environment. The selection of learning resources can improve of motivations, so that role students can be active in the learning process. [11] [12] [13]. Environment provides stimulation toward individuals [14] [15]. On the contrary, the individual giving response of the environment so that will be meaningful and that will be reference to develop of concept of learning and teaching.

3.2. Utilization of natural environment as a learning source for plant concept

The concept of the plant consists of two indicators, as follow : the concept of plants have specific part and the benefits of plants. The concept that plants have specific part was elaborated on statement as follow :1) Classify the type of root plants 2) Classify the type of plant stems 3) Classify the type of leaves of plants; 4) counting the number of Crown flowers; 5) observing the type of grain; 6) observing the part of fruits. The results of the data analysis can be shown in figure 3

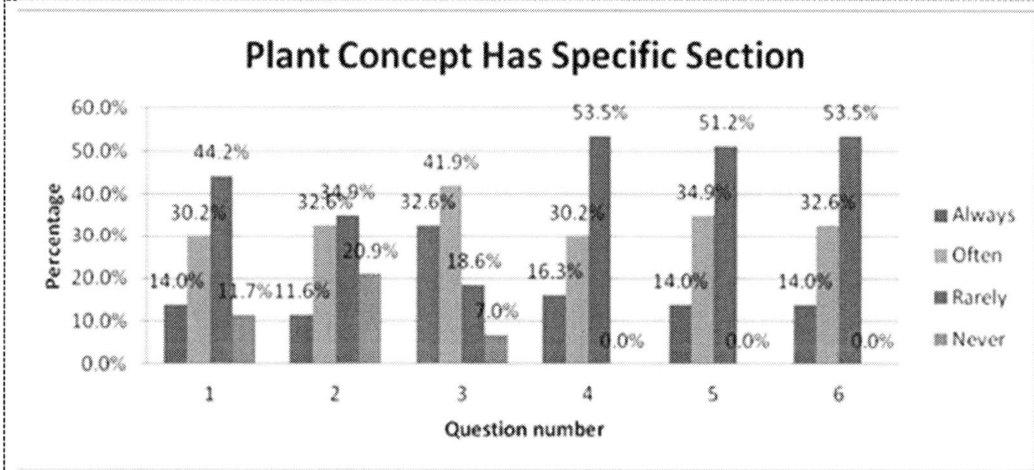

Figure 3. Percentage of teacher assigning to their students about concept of plants have specific part

Based on the graph 3 above it can be shown that generally, most of the teachers rarely assigning their students to understand about the concept of plants have specific parts. first statement about 44.2% of teachers rarely assigning for students to classify of plant roots. second statement about 34.9% teachers rarely assigning for students to classify of plant stems. third statement 41.9% often assigning for students to classify leaves of plants. fourth , Most of teachers (53.5%) seldom to count of Crown flowers; fifth , Most of teachers (51.2%) rarely assigning for students to observe of grain. sixth Most of teachers (53.5%) rarely assigning for students to observe of the fruits.

learning source for concept of benefit plants was elaborated statement as follows: 1) Note taking about plants as a food ingredients; 2) Note taking about plants which are utilized as an ingredient of drugs. s; 3) note taking plants which are used as a flavor enhancer.

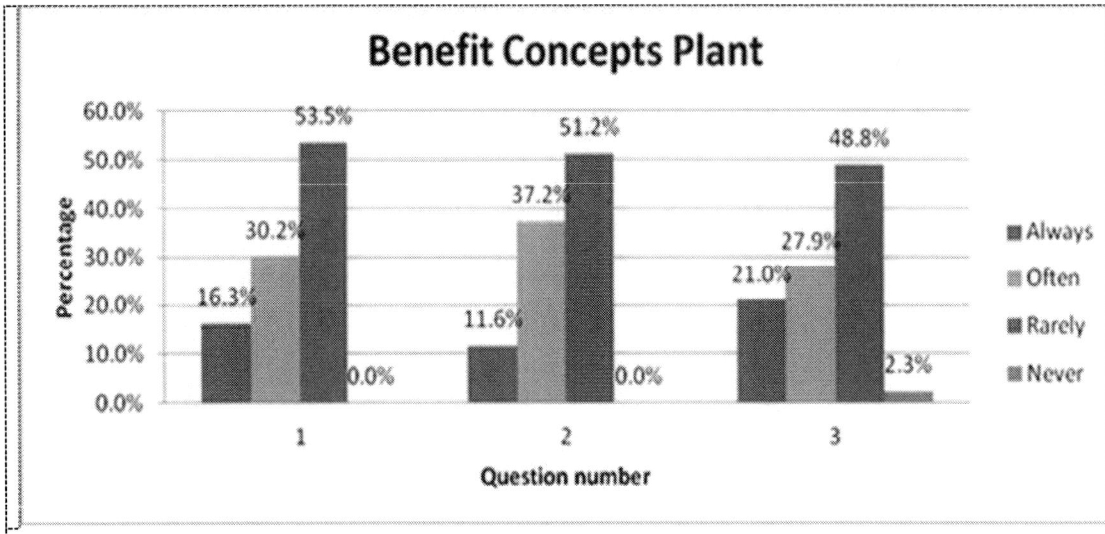

Figure 4. The Percentage of teachers assigning to their student about benefit concepts plants

Based on graph 4 that can be show generally, Most of teachers rarely assigning for students to utilize the natural environment especially the benefits of the plant. first statement most teachers (53.5%) rarely assigning for students in note taking about types of plants as food ingredients. second statement some teachers (51.2%) rarely assigning for students in note taking plants which are utilized as an ingredient of drugs. The third statement most of the teachers (48.8%) rarely assign students in note taking plants which are used as a flavor enhancer.

Teachers should be able to assign students , they observe the type of plants that is used in daily life, so that students will be familiar with the environment. Finally, that will give impact on the students to love their environment [16] [17]. the utilization natural environment as learning process that will be autonomous learner, creative, and they will be able to adapt to face the problems.

The teacher must be able to instill the concept of plants from the environment itself. The environment provides direct plants as objects of student learning. This will be more beneficial because the environment is one of contextual learning and can present learning directly to students [18] [19] [20]. Learning plants that are learned directly in the environment will be able to make elementary school students easier to understand. Students can get to know plants in a real and complete way.

3.3. Utilization of natural environment as a learning source for the concept of animals

The concept consists of two indicators, as follow : the concept of pets and wild animals and the concept are ovipar and vivipar

concept of pets and wild animals was elaborated as statement as follows: 1) Classify types of wild animals; 2) observing the way of life of the animal 3) Observing/showing an example of a food which descend of pets. The results of the data analysis can be show in figure 5.

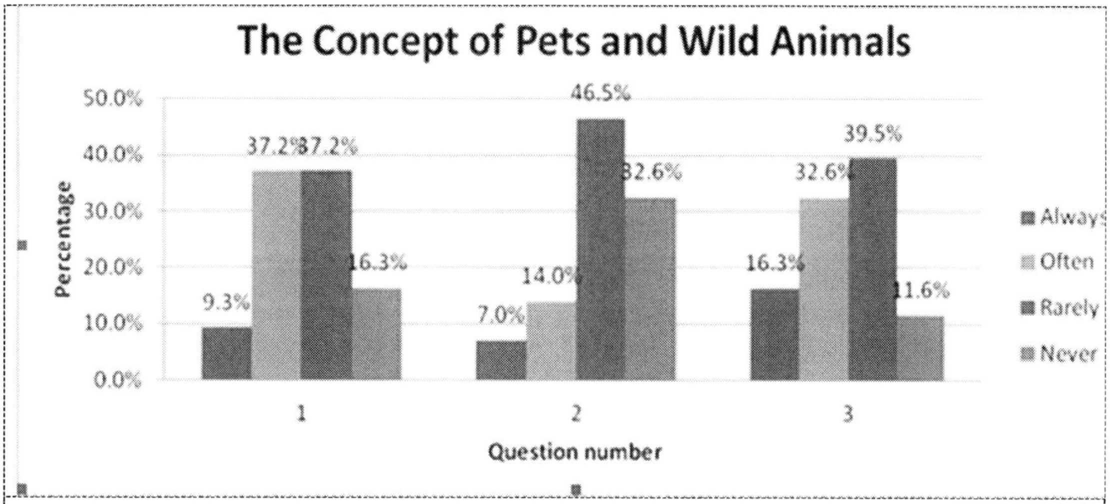

Figure 5. Percentage teacher assigning their student about the concept of pets and wild animals

Based on graph 5 that can be shown , Generally, most of the teachers rarely assigning students to understand about the concept of pets and wild animals .first statement about 37.2% of teachers who often and rarely assigning students in clustering about kind of wild animals. second statement most of the teachers rarely assigning students to observe the way of life of the animals. third statement about 39.5% teachers rarely assigning to observe/show an example of a food which descend of pets

The concepts of ovipar and vivipar that was elaborated at statement as follows: 1) visiting chicken farm to get information about animal breeding as lay their eggs 2) Observing the mammals characteristic. The results of the data analysis can be shown in figure 6.

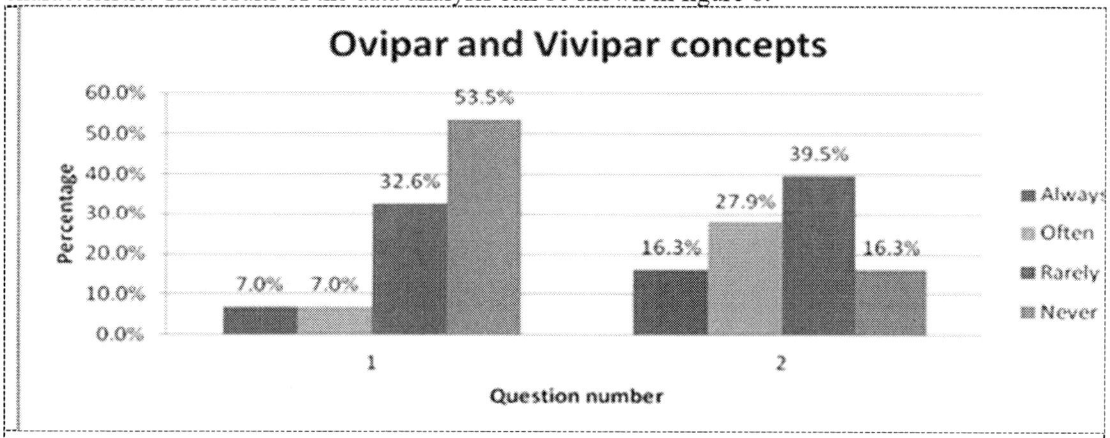

Figure 6. The Percentage of teachers Assigning to their students about maintenance concept living things

Based on the graph 6 that can be show that generally, most of teacher never and rarely assigning to their to understand about concepts vivipar and ovipar. first statement most of teachers (53.5%) never invite the students to visit a chicken farm. second statement about 39.5% teachers rarely assigning their students to observe of the mammals characteristic

In order that students can observe directly about pets and breeding , the teacher have to utilize the farms as learning source. Achievement as result, it can not be separated from the existence which supporting on learning process. Teacher is an important role to achieve of successful for students [21] [22]. Learning source will be meaningful for students or teachers, if learning source would be organize through a plan that probably to utilize in learning.

The environment provides a variety of information that is highly related to the contents of nature. One of them is animal. Animals are a review of natural science learning. Therefore animal studies are very suitable to be taught directly in the environment

The environment is a habitat for animals [23]. Students will more easily understand animals directly. The environment can easily object to animal studies for elementary school students. Elementary school students are students who are in the concrete operational phase [24] [25]. Therefore the need for real learning that can make students easy to understand information.

4. Conclusion

Based on finding research, that can be conclude as follows: teacher have been already utilized was still lack percentage, source of learning science about 46.48% which consist of concept living things of 49.60%, plants of 48.95% of the plants and animals of 34.88%.

Acknowledgment

Thank you to Universitas Negeri Padang for permitting collaborative research. Thank you to the Universitas Samudra that permitted research collaboration.

References

[1] Taniredja T, Afandi M and Faridli E M 2012 *EDUCARE* **5** 107
[2] Liu T Y, Tan, T H and Chu Y L 2009 *Journal of Educational Technology & Society* **12** 161
[3] Mirrahimi S, Tawil, N M, Abdullah N A G, Surat, M and Usman I M S 2011 *Procedia Engineering* **20** 389
[4] Laurillard D *et al* 2013 *Journal of Computer Assisted Learning* **29** 15
[5] Posner G *J et al* 1982 *Science education* 66 211
[6] Astimar N *2014 Scientific journal Science Education* **16** 98
[7] Shabiralyani G, Hasan K S, Hamad N and Iqbal N 2015 *Journal of Education and Practice* **6** 226
[8] Cheryan S, Ziegler S A, Plaut V C and Meltzoff A N 2014 *Policy Insights from the Behavioral and Brain Sciences* **1** 12.
[9] Spiritually A 2010 *Manajemen Pembelajaran* (Jakarta: Cipta Rineka)
[10] Sagala S 2010 *Konsep Dan Makna Pembelajaran* (Bandung: alfabeta)
[11] Blumenfeld P C, Soloway E, Marx R W, Krajcik J S, Guzdial M and Palincsar A 1991 *Educational psychologist* **26** 369
[12] Hamalik 2010 *The Process Of Teaching And Learning* (Jakarta: Aksara Earth)
[13] Lepper M R and Chabay R W 1985 *Educational Psychologist* 20 217
[14] Mattila A S and Wirtz J 2008 *Journal Of Services Marketing* 22 562
[15] Vlek C and Steg L 2007 *Journal of social issues* **63** 19.
[16] Sanders D L 2007 *International Journal of Science Education* **29** 1209
[17] Syamsudduha S and Tidy M 2012 *Journal Of PMIPA* **15** 18
[18] Byriel M G and Burchett M 2011 *Interior Plantscape* **37** 5
[19] Jennings P A, Frank J L, Snowberg K E, Coccia M A and Greenberg M T 2013 *School Psychology Quarterly* **28** 374.
[20] Oliver-Hoyo, M T, Allen D, Hunt, W F, Hutson J and Pitts A 2004 *Journal of Chemical Education* **81** 441.
[21] A Majid 2009 *Perencanaan Pembelajaran* (Bandung : Rosdakarya)
[22] Danijela M 2018 *International Journal of Cognitive Research in Science, Engineering and Education* **6** 33
[23] Batool S and Hussain M 2016 *Journal of Entomology and Zoology Studies* **4** 508

[24] Kenedi A K, Helsa Y, Ariani Y, Zainil M and Hendri S 2019 *Journal on Mathematics Education* 10 69
[25] Kiswanto A 2017 *9th International Conference for Science Educators and Teachers (ICSET 2017)* vol 118 (Semarang/Atlantis Press)

Problem based learning: a way to improve critical thinking ability of elementary school students on science learning

Zuryanty[1], A K Kenedi[2], R Chandra[1], Hamimah[1], Y Fitria[1]
[1]Universitas Negeri Padang, Jl. Prof Dr. Hamka Air Tawar, Padang, Indonesia
[2]Universitas Samudra, Jl. Meurandeh, Langsa, Indonesia

E-mail: zuryanty@gmail.com

Abstract. The background of this research is to find information on learning models that can improve the thinking skills of elementary school students in science learning. The purpose of this study was to determine the effectiveness of the problem-based learning model to enhance the critical thinking skills of elementary school students in science learning. This study was a true experimental design with a type of post-test group. The results showed that the problem-based learning model could improve the critical thinking skills of elementary school students in science learning. This study implies that elementary school teachers can use a problem-based learning model in enhancing students' critical thinking skills in science learning in elementary schools.

1. Introduction

The learning process at school is still dominated by traditional learning methods so that elementary school learning is still one-way (teacher centred) which results in weak student thinking skills [1]. The one-way learning model will cause information that is not useful to students and becomes useless [2]. Teachers often require students to learn but do not explain how to learn well and solve problems in the learning process [3], included in the process of learning science in elementary school.

Science learning is learning that is integrated with other learning in elementary schools. Science learning in elementary schools is learning that contains natural phenomena, identifying and formulating natural phenomena based on observations, and finding solutions to problems [4]. Science learning in elementary schools has the goal that students can have the ability to learn, be responsible and can implement their knowledge in social life [5]. Science learning can also improve the critical thinking skills of elementary school students [6]. This fact is in line with the objective of science learning that science used in making decisions on sociocultural issues in democratic societies so that it requires strong critical abilities [7]. Therefore science learning must be carried out properly.

Critical thinking ability is a high level of skill to solve problems and make decisions [8]. Critical thinking ability is an intellectual ability that can make concepts, apply concepts, analyse ideas, synthesise concepts and evaluate information collected from observation, experience, reflection, reasoning, or communication [9] [10]. Dell`olio states that the ability to think critically is the ability of students to analyse, focus, observe, hypothesis, assume, conclude review, and reflection [11]. Critical thinking is an ability that must be possessed in the 21st century because this ability requires individuals

1 To whom any correspondence should be addressed.
2 To whom any correspondence should be addressed.

to be able to use their thinking skills in everyday life [12]. Therefore this ability must be developed since elementary school age.

Analysis of science learning that is still one-way and the importance of developing the thinking skills of elementary school students requires the effort to improve the quality of science learning to enhance the critical thinking skills of elementary school students. One attempt to overcome this problem is to use a learning model. The suitable learning model used based on these problems is a problem-based learning model.

Problem-based learning is a process that can create existing relationships between students and students, using problems related to students' real lives, ill-structured and open-ended [13]. In PBL students are given the opportunity to solve problems together and make the learning experience the basis for learning [14]. Problem-based learning is learning that focuses on learning from students' direct experience or learning from real problems faced by students [15]. Problem-based learning can improve students' skills, improve student communication, cooperate, increase motivation, and increase the convenience of collaborating in solving problems [16]. Based on the opinion above, the need for proof that states problem-based learning can improve the critical thinking skills of elementary school students in science learning.

2. Method

The study was a True experimental design that uses simple random sampling. This study consists of two variables, namely the independent variable and the dependent variable. The independent variable in this study is the problem-based learning model, and the dependent variable is the thinking ability of the elementary school students. The study design uses the PPost test group. Data collection methods include documentation methods (name and semester grades) and test methods to measure critical thinking skills of elementary school students.

3. Formatting the title, authors and affiliations

Analysis of critical thinking in elementary school students in the experimental class and control class can see in table 1 below:

Table 1. the results of research on critical thinking skills

Component	Post Test	
	Control	Experimen
The number of Students	35	35
Average	73,5	82,5
% Mastery Learning	65,71	94,28
Complete Learning	12	2

Based on the t-test obtained t count equal to 3, 983 with t table equal to 1.690. This result states that $t_{count} > t_{table}$. The point is the existence of differences in critical thinking skills of elementary school students who use problem-based learning between the control class and the experimental class.

The results of critical thinking skills of elementary school students experienced a significant increase between classes in traditional learning (the method used by the teacher is the lecture method) with the class learning using problem-based learning.

The increase that occurred in the experimental class was due to changes in the learning model that could train students' critical thinking skills. Thomas stated that the problem-based learning model requires students to be active in the learning process [17]. This fact is supported by the stages of problem-based learning that requires active students in the learning process. Arends states that there are five steps of problem-based learning namely problem orientation, learning organisation, individual and group investigation, development and presentation of the results of problem-solving, analysis and evaluation of problem-solving processes [18]. While in the control class using the conventional method (lecture method). In the lecture method, students only listen to the teacher's explanation so that students become passive [19]. This result in students being only able to remember the information provided by the teacher. The student active is essential in the learning process [20]. But in learning using the lecture method of student activity is low because the learning process carried out is still a student centre. This result in the critical thinking skills of students not being trained so that students' critical thinking skills are low. Therefore the critical thinking ability of students in the control class is lower than the experimental class. This fact is by the statement that learning that uses problems in the real world students will be able to train the ability of critical thinking skills and knowledge to solve the issues and can obtain necessary information in the learning process [21].

This statement support by research conducted by Sihaloho which states that problem-based learning can train students' thinking skills such as critical thinking skills, problem-solving, and creative thinking skills [22]. Problem-based learning that requires students to be able to find solutions to problems from problems that occur in the real world automatically both individually and in groups [23]. The problem-based learning model is the flow of constructivism that can develop the critical thinking skills of students who can analyse problems in the real world [24]. Also, problem-based learning can train critical thinking skills that are embedded in a long time [25].

The problem-based learning model has the characteristics of problems that used as the basis for learning. The student uses the issue for finding information and solving problems in groups. In this process, students collaborate with other students in solving problems. Problem Based Learning allows students to exchange information and solve problems so that students' critical thinking skills automatically increase. This case is contrary to the conventional method (lecture method), students do not give a problem, students are only asked to listen to what is conveyed by the teacher so that students record what is sent by the teacher. This result in the critical thinking skills of untrained/low students.

Every elementary school student can think critically. This ability better train from elementary school age through a learning process such as the science learning process. This study proves that the science learning process that uses a problem-based learning model can improve the critical thinking skills of elementary school students.

4. Conclusion

The results of the calculation of the t-test show that t is equal to 3, 983 with t table of 1.690 with this result stating that $t_{count} > t_{table}$. This study shows that the problem-based learning model can improve the critical thinking skills of elementary school students. This model can be an effort to enhance the thinking skills of elementary school students in science learning.

Acknowledgment

Thank you to Universitas Negeri Padang for permitting collaborative research. Thank you to the Universitas Samudra that permitted research collaboration.

References

[1] Stošić, L 2015 *Journal of Cognitive Research in Science, Engineering and Education* **3** 1
[2] M Lak *et al Journal of Advances in English Language Teaching* **5** 1
[3] D A M Lidinillah 2008 *Jurnal Pendidikan Dasar* **10** 1
[4] B Karyadi *et a*l 2018 *J. Phys.: Conf. Ser.* **1013** 012074
[5] N Fazriyah *et al* 2017 *J. Phys.: Conf. Ser.* **812** 012014
[6] Suryanti *et al* 2018 *J. Phys.: Conf. Ser.* **1108** 012128
[7] H A Yacoubian 2015 *Canadian Journal of Science, Mathematics and Technology Education* **15** 248
[8] Al-Mubaid H 2014 *International Journal of Advanced Corporate Learning* **7** 34
[9] N Sumarna *et al* 2017 *J. Phys.: Conf. Ser.* **812** 012067
[10] L Yuliati *et al* 2018 *J. Phys.: Conf. Ser.* **1013** 012025
[11] A S Vong 2017 *Cambodia Kasetsart Journal of Social Sciences* **38** 2017
[12] I Puspita *et al* 2017 *J. Phys.: Conf. Ser.* **895** 012100

[13] A Kiswanto 2017 *Advances in Social Science, Education and Humanities Research* **118** 1040
[14] E H J Yew & Karen G 2016 *Health Professions Education* **2** 75
[15] V F C Servant-Miklo 2018 *Health Professions Education* **1** 1
[16] A Alrahlah 2016 *The Saudi Dental Journal* **28** 155
[17] A Gurses *et al* 2015 *Procedia - Social and Behavioral Sciences* **197** 2390
[18] R Mushlihuddin *et al* 2018 J. *Phys.: Conf. Ser.* **948** 012028
[19] S Sukaesih and Sutrisno 2017 *J. Phys.: Conf. Ser.* **824** 012070
[20] I Emaliana 2017 *Jurnal Sosial Humaniora* **10** 59
[21] M Kek *et al* 2011 *Higher Education Research and Development* **30** 329
[22] R R Sihaloho *et al* 2017 *IOSR Journal of Research & Method in Education* **7** 11
[23] Z A Putra and M Dewi 2018 *IOP Conf. Ser.: Mater. Sci. Eng.* **306** 012140
[24] N Ragilia *et al* *IOSR Journal of Research & Method in Education* **8** 1
[25] B Birgili 2015 *Journal of Gifted Education and Creativity* **2** 71

Development of Teaching Material of Final Assignment Subject Based on PBL Using ICT

Radeswandri[1], Afiq Budiawan[2], and Rian Vebrianto[3*]
[1]Universitas Terbuka, UPBJJ Pekanbaru, Riau, Indonesia
[2]STAI H.M Lukman Edy Pekanbaru, Riau Indonesia
[3]Universitas Islam Negeri Sultan Syarif Kasim Riau, Pekanbaru, Indonesia

*rian.vebrianto@uin-suska.ac.id

Abstract. Open universities provide educational opportunities for workers, especially teachers who have not met the academic qualification requirements. Final Assignment (FA) is a subject aims to measure advanced thinking skills comprehensively on theory and practice, which is designed based on problem-solving. Therefore, this study designed and developed PBL-integrated FA teaching materials to meet the 2021 Open University Vision and Missions. This was a research and development study of FA teaching materials using PBL-based ICT with the ADDIE model. The study samples were 17 open universities students who were taking the FA subject. The instrument used was a media assessment questionnaire consisting of three aspects, namely the aspect of appearance, ease of use, and efficiency. The Alpha Cronbach value of the questionnaire was 0.82. The results show that the aspect of appearance had min = 3.65; SD = 0.316, ease of use had min = 3.63; SD = 0.247, and efficiency had min = 3.66; SD = 0.292. This means that PBL-based Online Tutorial Materials are excellent to help all students and tutors of the FA subject to develop a good and effective learning process.

1. Introduction

Today we often hear various concerns addressed to scholars, low levels of reasoning and creativity. The theme of this study, for 2018, is in line with the Open University (UT) Vision and Mission 2021 as well as developing strategic issues in UT, namely UT being a world-class PTTJJ (open distance higher education) institution that implies that the implementation of UT has met PTTJJ's best implementation standards recognized, both by the community and the institutions or associations of the PTTJJ world institutions. As a consequence and in accordance with the principles of the PTTJJ system, UT also produces a variety of high-quality higher education products that are standardized. The definition of higher education products includes academic products that substantially follow the development of science and technology (Science and Technology) and maintain their relevance to the development of society and national development; and graduates who are globally competitive. The definition of standard is the quality of products guaranteed by standards at each PTTJJ location by UT [1].

In addition, according to Government Regulation of the Republic of Indonesia Number 19 of 2005 concerning Education National Standards and Number 32 of 2013 concerning Amendments to National Education Standards article 1 paragraph 8, explicitly confirms that educators and grooming staff must meet the criteria and feasibility of pre-job education and position.

The teacher is a professional educator with the main task of educating, teaching, guiding, directing, training, evaluating, and evaluating students in early childhood education in formal education, basic education, and secondary education. Professional teachers will produce quality education processes and results in order to realize intelligent and competitive Indonesians, namely human beings who believe and fear God Almighty, are noble, healthy, knowledgeable, capable, creative, independent, and become citizens democratic and responsible.

To become a professional teacher, you must have the ability to develop competency aspects that exist in him and think comprehensively and at a high level or Higher Order thinking Skills (HOTS). Mastering science with professionals means being able to think comprehensively in solving problems from various integrated scientific aspects where this skill is an objective of learning from the Final Assignment (FA) [1].

Education is the key to all quality progress and development in a country because with education, humans can realize all the potential that exists in themselves both as individuals and as citizens of society [2]. The need for alternative ways, one way to improve and develop teacher competencies is to provide a good understanding of various sciences into a comprehensive unit [2][3].

Professionalism defines the commitment of an individual to improve his professional skills and develop strategies to perform his duty admirably. A professional should manage everything appropriately, particularly aspects related to his profession[4].

The professional development is a process improving an individual or a group involving various methods, such as reflection, renewing, and improving commitment to achieve better performance. The methods than realized as an effort to improve knowledge, skills, and morals to develop a better idea, design, and practice [5].

The success of a learning process in the university not only influenced by the lectures, but also the administrative staffs. OECD (2005) reported that the human resources of Indonesia would not provide any benefit without the improvement of their quality. Data from ASEAN Productivity Organization (APO) shows that only 4.3% of 1000 Indonesian labors were skilfull while the Philippines, Malaysia, and Singapore reached 8.3%, 32.6%, and 34.7%, respectively [6].

With this data, the responsibility of a teacher is getting bigger and must be followed up immediately by providing understanding and fertilization of various concepts of good high-level thinking skills, for that. The researcher wants to develop material of final assignment subject based on PBL using ICT.

1.1 Research Methodology

The study was conducted in the open universities from September to december 2018. The research objects were the student in the Faculty at Open universities in Riau. This was a research and development study of FA teaching materials using PBL-based ICT with the ADDIE model [7].The study samples were 17 open universities students who were taking the FA subject. The instrument used was a media assessment questionnaire consisting of three aspects, namely the aspect of appearance, ease of use, and efficiency. The Alpha Cronbach value of the questionnaire was 0.82.which proved the quality of the instruments.

2. Result and Discussions

2.1 The procces of ADDIE model

The basic design and stages of learning development refer to the ADDIE model instructional design (Analyze Design Develop Implement Evaluate) [7]. The ADDIE model is chosen because it is based on a system that involves fostering teaching materials for a course or a curriculum. This model is usually awakened by a research in producing products. Which were FA teaching materials using PBL-based ICT with the ADDIE model.

2.2 Descriptive analysis

The assessment results of FA teaching materials using PBL-based ICT with the ADDIE model are implemented during the tutorial process who were taking the FA subject. In this assessment assessed from the aspect of appearance, ease of use and efficiency. The quality of the instrument is essential in ensuring the obtained data are consistent and relevant. The analysis results of the students' perception of FA teaching materials using PBL-based ICT with the ADDIE model are shown in Tables 1 to 3.

Table 1. The students' perception of FA teaching materials using PBL-based ICT with the ADDIE model from appearance aspect

No	Appearance Aspect	Descriptive Analysis Result			
		N	Min	SD	Category
1	The appearance based on ICT	17	3,53	0,514	excellent
2.	The content is clearly	17	3,35	0,606	excellent
3.	Animated graphics based on ICT	17	3,24	0,437	good
4.	language and Language Effects of use	17	3,65	0,493	excellent
5	Design of the operation of teaching materials	17	3,41	0,618	excellent
6	General effects teaching materials	17	3,53	0,514	excellent
Appearance Aspect		17	3,53	0,514	excellent

Table 1 displays that in the aspect of appearance, FA teaching materials using PBL-based ICT with the ADDIE model in All items of appearance aspect were good and excellent category which is satisfying at the mean of 3.24 which is in the Animated graphics based on ICT, while other aspects are excellent category with the highest mean, namely the aspects of language and Language Effects of use (mean = 3.65). So that it can be concluded that according to users in the implementation of this media, it is known that from the aspect of appearance of FA teaching materials using PBL-based ICT with the ADDIE model from appearance aspect (mean = 3.45) were excellent category. Then for the aspect of ease of use can be seen in Table 2.

Table 2. The students' perception of FA teaching materials using PBL-based ICT with the ADDIE model from the aspect of ease of use

No	Ease of use	Descriptive Analysis Result			
		N	Min	SD	Category
1	Easy to use because it uses instructions	17	3,71	0,470	excellent
2.	Using language that is easy to understand	17	3,71	0,470	excellent
3.	Arranged systematically	17	3,65	0,493	excellent
4.	Can be used for independent learning	17	3,76	0,437	excellent
5	Helps in understanding concepts	17	3,59	0,507	excellent
6	The menus on media teaching materials are very suitable	17	3,35	0,606	excellent
Ease of use Aspect		17	3,63	0,246	excellent

As can be seen in Table 2, in the aspect of ease of use consists of six indicators. in the six indicators, the results of the assessment in the excellent category are low, namely the aspects of the The menus on media teaching materials are very suitable, namely the mean of 3.35, while the highest in the aspect can be used for independent learning (mean = 3.76). So it can be concluded that from the ease of use aspect of FA teaching materials using PBL-based ICT with the ADDIE model from the ease of use aspect (mean = 3.63) were excellent category. Then for the aspect of efficiency can be seen in Table 3.

Table 3. The students' perception of FA teaching materials using PBL-based ICT with the ADDIE model from the aspect of efficiency.

No	**Efficiency Aspect**	**Descriptive Analysis Result**			
		N	**Min**	**SD**	**Category**
1	Learning Becomes Efficient	17	3,65	0,606	excellent
2.	workable anywhere	17	3,88	0,332	excellent
3.	Examples of media Questions greatly help understanding	17	3,47	0,624	excellent
4.	Use of Media Reduces Tutor Expenses	17	3,65	0,493	excellent
Efficiency Aspect		17	3,66	0,292	excellent

Furthermore, it can be seen from Table 3 it is known that for the efficiency aspects of the assessment results from users of FA teaching materials using PBL-based ICT with the ADDIE model were excellent category seen from the four indicators in sequence, namely: Learning Becomes Efficient (mean = 3.65), workable anywhere (mean = 3.88), the Examples of media Questions greatly help understanding (mean = 3.47) and the use of Media Reduces Tutor Expenses (mean = 3.65). This means, that the total mean value for the aspect of efficiency is in the excellent category with a mean = 3.66.

From the explanation of Tables 1 to.3, it can be seen that the FA teaching materials using PBL-based ICT with the ADDIE model that has been developed by researchers can be a good and effective learning resource because it has been assessed from the aspect of appearance, ease of use and efficiency in a excellent category. So this media can be reproduced and used in a good learning process online and helps all students who were taking the FA subject.

To strengthen this study, judging from the results of the participants' responses to the media, it is known that with ICT, it is very helpful for students and lecturers in learning. I really like this smartphone-based lesson, can facilitate students in learning, only I beg in the provision of material can be adjusted so that it suits me who is old. Furthermore, other students said that it was very helpful with ICT-based materials, lots of information and could be used by everyone. It is a very important thing to have functions such as processing data, looking for material, is a necessity for anyone who is the general dominant in life.

So it can be concluded that the FA teaching materials using PBL-based ICT with the ADDIE model were very feasible to be applied and implemented in open university learning because it can be studied and used anywhere, anytime online [8].

3. Conclusion

It can be concluded that the FA teaching materials using PBL-based ICT with the ADDIE model that has been developed by researchers can be a good and effective learning resource because it has been assessed from the aspect of appearance, ease of use and efficiency in a very satisfying category. So this media can be reproduced and used in a good learning process online and helps all students who are taking TAP courses. Then for advice can be a rule model in producing interesting learning media according to the vision and mission of an open university that can carry out learning anywhere, anytime and in time that can be adjusted online.

References

[1] Universitas Terbuka. (2013). *Tugas Akhir Program*. Banten: Penerbit.

[2] OOn Seng Tan. (2003). *Problem-Based Learning Innovation:Using Problem to Power Learning in the 21st Century*. Singapore: GALE Cengage Learning.

[3] R. R. K. O. Rian Vebrianto. (2016). "Biomind Portal For Developing 21st Century Skills And Overcoming Students' Misconception In Biology Subject," *Int. J. Distance Educ. Technol.* 14 (4): 55–67.
[4] L. S. Christopher Day. (2009). "The effects of reform: Have teachers really lost their sense of professionalism?," *J. Educ. Chang.* 10 (2–3): 141–157.
[5] A. K. A. Muhammad Faizal A. Ghani. (2015). "Model Program Perkembangan Profesionalisme Guru Malaysia Satu Kajian Analisis Keperluan di Sekolah Berprestasi Tinggi dan Sekolah Berprestasi Rendah," *JuPiDi J. Kep. Pendidik.* 2 (2): 1–16.
[6] O. for E. C. and Development. (2005). *The definition and selection of key competencies: Executive summary.* Paris: France: OECD.
[7] Y. P. Chuan. (2009). *Statistik Penyelidikan Lanjutan: Ujian Regresi, Analisis Faktor dan Analisis SEM.. II.* Mc Graw-Hill.
[8] K. O. Rian Vebrtianto. (2013) ."Fostering Science Process Skills and Improving Achievement Through The Use Of Multiple Media," *J. Balt. Sci. Educ.* 12 (2): 191–204.

Locking Target on Missile System

Syafruddin R[1]*, Givy Devira Ramady[1], Rahmad Hidayat[1], Hetty Fadriani[1], Decy Nataliana[2]

[1]Sekolah Tinggi Teknologi Mandala, Bandung, Indonesia
[2]Institut Teknologi Nasional, Bandung, Indonesia

*syafruddinr@yahoo.com

Abstract.
The ability to create a locking target on missile system is very limited. For this reason, researchers tried to examine the title Locking Target On Missile System. As a researcher, his goals and hopes from this simple one can become a prototype, the endurance test, then it can be used as mass production. Where the stages of the research method locking target on missile system are through image processing. Here the researchers used Visual Studio 2010 software to create a software application. By processing the image BMP from the camera. In signal processing, there are many ways and languages that can be used, from the results of the author's experience, first with the C++ language the, for that the author uses the C# language. As for the validation of the signal and the format results obtained are quite slow, then the C# language, the results are quite fast, from all that info sent, the signal reception simulation test is performed with the same application software. Overall research on locking target on missile system which is a real-time control signal application software is then sent via radio.

1. Introduction

1.1 Background

The background of this research is in order to achieve the development of locking target[1,2] on missile systems. The long-term goal of this research activity is to design and develop locking targets on missile systems, with the hope of being able to develop their own locking targets on missile systems. The need for locking targets on missile systems is one of the obligations in the development of science and technology, while for developed countries is a daily requirement for the industry, but it does not hurt us to start from now, even though in the world today, for the industry segment locking target on missile systems already very advanced, it's hard for us to compete in this field. This research topic includes high technology in the fields of informatics, electronics, and other supporting technologies which are not included in this study such as chemistry for rocket motors, mechanics for manufacturing processes and its aerodynamic calculations. Actually, for now, it is independent in locking the target on the missile system, becoming very urgent and including promising industrial products.

1.2 Formulation

In the case of special objectives researchers limit themselves only the Locking Target On Missile System relating to information technology, electronics, including the procurement of actuators in the form of geared dc servomotor and control casing.

1.3 Identification

The object of the target locking missile system consists of:
1. Software application
2. The remote system uses a pair of radio transmitters
3. Booster.

1.4 Research Purposes And Objectives

Journal of Physics: Conference Series **1424** (2019) 012039 doi:10.1088/1742-6596/1424/1/012039

The findings targeted in this study are software application control signals sent via radio with the title Locking Target On Missile System.

1.5 Use Of Research

The purpose of locking target on missile system research through image processing[3,4] is expected to be useful to make a fundamental contribution to the field of control system science in supporting the development of science and technology. Locking target on missile system is a weapon control system with a high degree of accuracy because it is equipped with a target locking as guidance. Locking the target on missile system which is also called in other words the fin control is an important part of the missile system, with this fin control we can direct the missile that is sliding to the target to be destroyed. Developing your own locking target on missile system is much more possible to do, because the data needed is only the philosophy of the system, we don't need to know the data processing and wiring system of the existing system, just read the brochure about other missile system products, watch how to use it, then translate it into our language and imagination, then pour it into our desired concept, the components needed are adapted to existing components and can be purchased on the free market, so that our design is not constrained by limited supply of components and sources of references and information. The benefit of the final result of this activity is to increase the capability of locking target on missile system design technology. But in this study we limit ourselves to examine the Locking Target On Missile System, henceforth in the field of developing Servo Control For Missile System is another title, continued this research, so that when combined into a missile control system known as fin control. The scope of this research develops its own locking target on the missile system, including 1. camera, 2. BMP, 3. convert BMP color to gray bmp[5,6,7,8], 4. Image processing, 5. send out data in the form signal of radio [9,10,11]. Expect from the results of this research to accelerate the achievement of long-term plans for mastering control system technology in the Locking Target On Missile System. Activity description: The description of activities in this explanation is assisted by Fig. 1 up to Fig. 3, where locking targets on missile systems is an important part of missile systems.

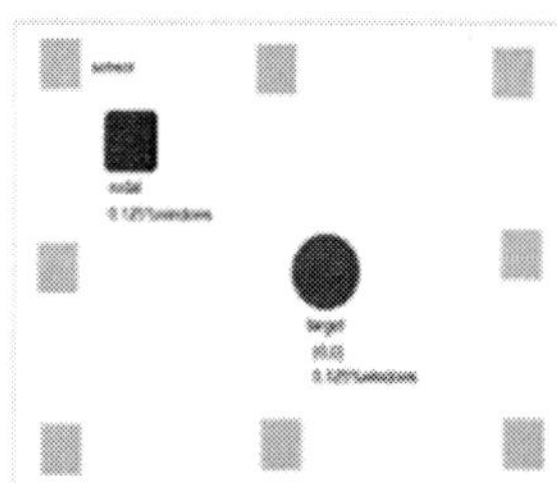

Figure 1. Image calculation.

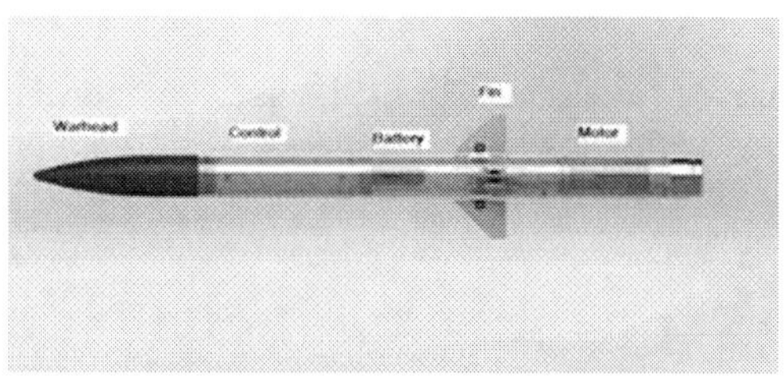

Figure 2. Control missile.

Journal of Physics: Conference Series **1424** (2019) 012039 doi:10.1088/1742-6596/1424/1/012039

Figure 3. Existing control application.

In the case of Locking Target On Missile System has the following limitations: Fast-moving objects of order 2 mach = 600 m / s. The controlled also moved quickly to order 2 mach = 600 m / s. For that to be controlled and the object can overlap, of course, the track must have a certain heading angle, if the angle of 0° certainly will not be able to overlap because the speed is controlled and the target object is the same. In connection with the above restrictions, the Locking Target On Missile System must also be faster than the target, in this case only image processing is used by people as sensor devices for making high-speed control signals.

2. Research Methods

The research method used is the design and manufacture of prototypes, prototype testing, prototype test data retrieval and analysis of test results. The test results are said to be successful if the sent signal can be received according to the format [12] sent, for that as a testbench the same application software is used as the recipient pair, and be checked in the result, Fig.4 as an illustration of this research method. For this reason, the researchers try to make a research method in the form of signal processing flowcharts as follows:

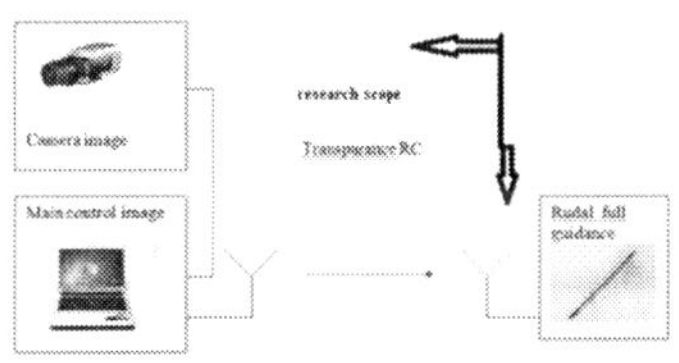

Figure 4. Research scope.

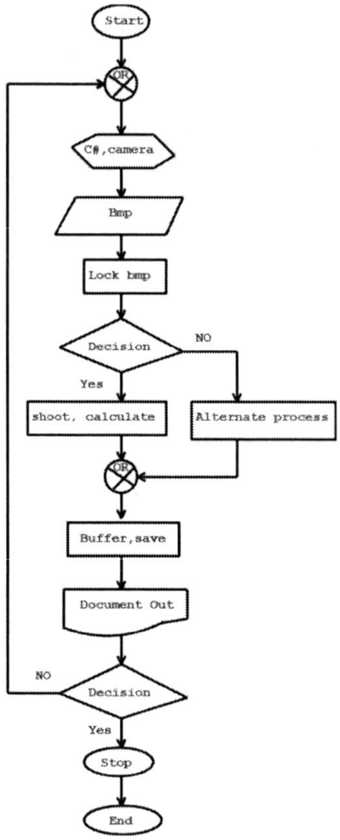

Figure 5. Signal processing flowchart.

3. Results And Discussion

3.1 Results

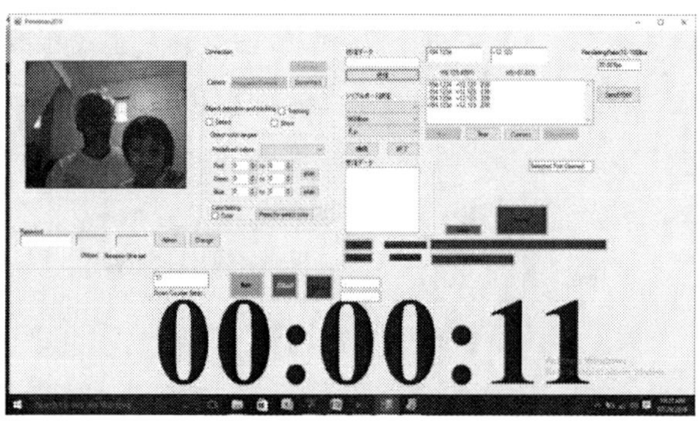

Figure 6. Dashboard control.

Table 1. Measurement chart.

No.	Tx	Rx	According to √	Not according to X
1	0x12	0x12	√	
2	0x18	0x18	√	
3	0x20	0x20	√	
4	0x22	0x22	√	
5	0x24	0x24	√	
6	0x26	0x26	√	
7	0x28	0x28	√	

3.2 Discussion

The result of locking the target on missile system sent is received in the recipient's locking target on missile system the same value as expected. This research can be concluded successfully according to the design plan.

4. Conclusions and Suggestions

4.1 Conclusions

From the results of the test: The reliability of the dashboard is turned on 7x24 hours successfully without problems, Intermittent data transmission within 7x24 hours was successful without problems, The intermittent button functions within 7x24 hours succeed without problems, The run button to the intermittent exit button within 7x24 hours was successful without problems, From these results it can be concluded that the research was successful,

4.2 Suggestions

The results of the locking target on missile system research are ready to be integrated with the servo control for missile system research which is another title, continued this research.

References

[1] Erik, C. (2005) Kalman filter for vision tracking, Av. Revolucion No. 1500, C.P. 44430, Guadalajara, Jal., Mexico, Technical Report B 05-12

[2] Simone, M. (2009) Metadata Extraction from PDF Papers for Digital Library Ingest, 10th International Conference on Document Analysis and Recognition, Dipartimento at Sistemi e Informatica - Universit'a in Florence Via S. Marta, 3 -50139- Florence - Italy

[3] Gaikwad, L. A. (2016) Cancer detection using digital Image processing On CT scan Images, International Research Journal of Engineering and Technology (IRJET), vol.3, issue 04

[4] Tian, L. (2016) Research on image segmentation based on clustering algorithm, Jilin University College of Geoexploration Science and Technology, Changchun 130012, China, International Journal of Signal Processing, Image Processing and Pattern Recognition, vol. 9, no. 2

[5] Christophoros, N. (2008) Digital Image Processing Basic Methods for Image Segmentation, Computer Vision course by Svetlana Lazebnik, the University of North Carolina at Chapel Hill

[6] Bernd, J. (2005) Digital Image Processing, Interdisciplinary Center for Scientific Computing of the University of Heidelberg Im Neuenheimer Feld 368 69120 Heidelberg Germany, 6th revised and extended edition

[7] Rafael, C. G. (2001) Digital Image Processing, Prentice-Hall Upper Saddle River, New Jersey 07458, Second Edition

[8] Muzamil, B. (2014) Digital image processing, International Journal of Scientific & Technology Research, vol. 3, no. 1

[9] Ramady, G. D. (2018) Bluetooth beacon battery monitoring system using LabView. Online Journal of Mandala College of Technology, 13 (2), pp. 45-50

[10] Syafruddin (2018) Analysis of harmonized content in 3 phase ac motor 0.12 kW controlled by 3 phase inverters, Online Journal of Mandala College of Technology, 13 (1), pp. 32-38

[11] Syafruddin (2018) Analysis of total harmonic distortion in ACPDB panels due to non-linear loads, Online Journal of Mandala College of Technology, 13 (2), pp. 33-44

[12] Hidayat, R. (2018) Smart key implementation for BTS gate door based on the internet of things, The 1st International Conference on Computer Science and Engineering Technology, Muria Kudus University, vol.1, no. 1.

Servo Control for Missile System

Decy Nataliana[1]*, Syafruddin R[2], Givy Devira Ramady[2], Yakob Liklikwatil[2], Andrew Ghea Mahardika[2]

[1]Institut Teknologi Nasional, Bandung, Indonesia
[2]Sekolah Tinggi Teknologi Mandala, Bandung, Indonesia

*decyhari@gmail.com

Abstract.
The ability to make control systems is very limited, especially high-speed two-dimensional motion controls such as missile controls, for that reason researchers try to examine the title Servo Control For Missile System. As a researcher, his goals and hopes from this simple one can become a prototype, the endurance test and then can be used as mass production. Where the stages of this research method make Servo Control For Missile System, the control signal is received through the radio to be processed. Here the researchers used codevision software to create a software application, the signal from the radio then combined with the optical encoder position signal, to get smoother and more precise motion, stopping at a position that is has been ordered then as feedback is input from the encoder servo motor, then the signal is amplified to drive motor power amplifier driver circuit. In this case, the signal processing uses a microcontroller, with several considerations, including a fairly small physical dimension where the outer diameter of the casing is 120mm, the inner diameter is 116mm. Overall Servo Control For Missile System which is a real-time software and hardware control signal application to drive fin.

1. Introduction.

1.1 Background.

The background of this research is in order to achieve the long-term design of the development of a servo control[1] for the missile system. The long-term goal of this research activity is to design the servo control for the missile system, with the hope of being able to develop servo control for the missile system. The need servo control for missile system for the development of knowledge in college institutions is a must, for that it does not hurt us to start from now, even though in the world today, for the industrial servo control for missile system segment is very advanced, it is difficult to compete in this field. Developing servo control for missile system is the rationale for researchers to conduct this research, although this research topic includes high technology in the fields of informatics, electronics, and other supporting technologies which are not included in this study such as chemistry for rocket motors, mechanics for manufacturing processes and its aerodynamic calculations. Actually, for now, it is independent in servo control for the missile system, becoming very urgent and including promising industrial products.

1.2 Formulation.

In this paper, researchers limit themselves to researching the development of the prototype servo control for the missile system, and to further develop the prototype locking target on missile system is another title, continued this research, so that when combined into a missile control system known as fin control.

1.3 Identification.

Servo Control For Missile System research objects consist of: receiver, software, hardware driver control, printed circuit board, Kalman filter[2,3,4,5,6,7] driver software for geared dc servo motor drivers with rotary encoder so that the movement is smoother and more precisely where it stops in accordance with the commands given by the signal command, power electronic driver, drivers for two geared dc servo motors with rotary encoder.

1.4 Research Purpose and Objectives.

In the case of special objectives, researchers limit themselves only to the Servo Control For Missile System related to the microcontroller, electronics, electrotechnology programs, including mechanical design of the casing and the procurement of control actuators [8,9,10,11].

1.5 Use of research.

Making application software, hardware with the title Servo Control For Missile System is expected to be useful to make a fundamental contribution to the field of control system science in supporting the development of science and technology. Servo control for missile system is a part of a weapon system with a high degree of accuracy because it is equipped with controls as guidance. Servo control for missile system which is also called in other words fin control is an important part of the missile system, with this fin control we can direct the missile that is sliding to the target [12] that would be destroyed. The price of missile systems is expensive because of the high price of Servo control for the missile system. Stressing in this research and paper: developing a servo control for missile system is much more possible because it does not require large costs. The description of activities in this explanation is assisted by Fig. 1 up to Fig. 3 In this study, there are limitations as follows: 1. Fast-moving objects of order 2 mach = 600 m/s. 2. The controlled also moves quickly to order 2 mach = 600 m/s. For that to be controlled and the object can overlap, of course, the track must have a certain heading angle, if the angle of 0° certainly will not be able to overlap because the speed is controlled and the target object is the same.

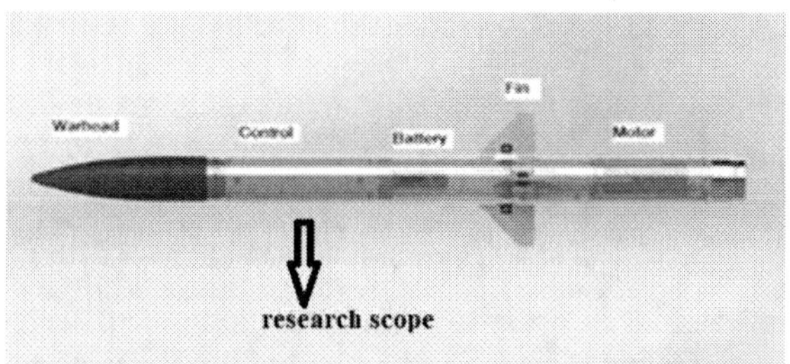

Figure 1. Servo control for missile system.

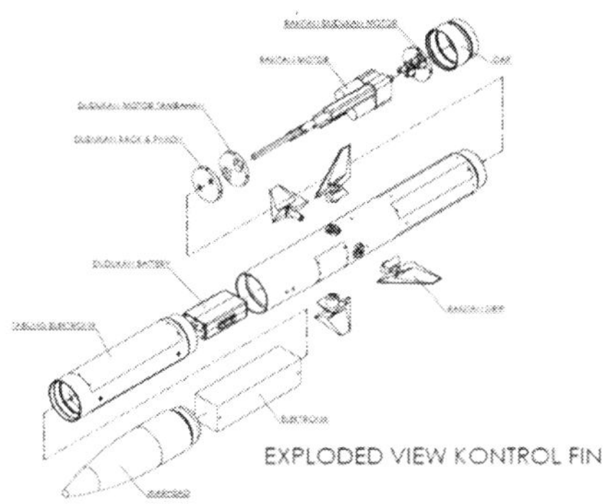

Figure 2. Exploded servo control for missile system on the market.

Figure 3. Servo control for missile system applications that are already on the market.

2. Research Methods.

The research method is the method of design and manufacture of prototypes, prototype testing, prototype test data retrieval and analysis of test results. Figure 4 illustrates this research method.

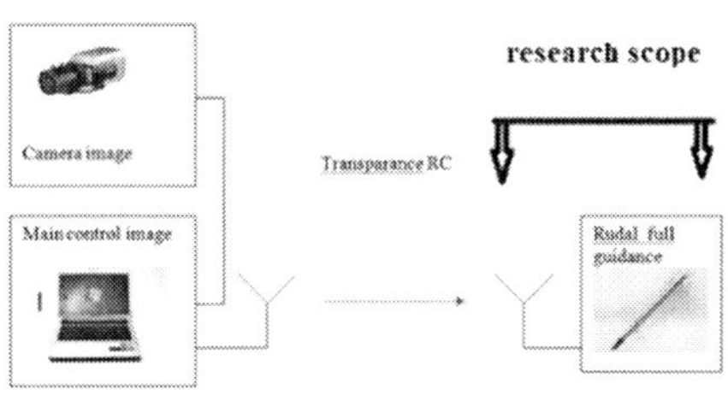

Figure 4. Research scope.

For this reason, researchers try to make a research method in the form of signal processing flowcharts as follows:

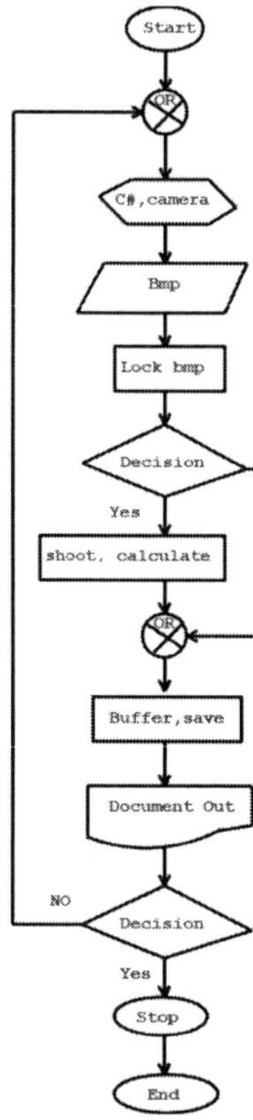

Figure 5. Signal processing flowchart.

3. Results And Discussion.

3.1 Results.

Figure 6. Ready-made PCB.

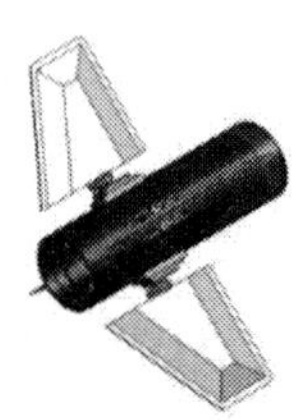

Figure 7. Outer diameter 120mm control housing.

Figure 8. PCB has been assembled.

3.2 Discussion.

First receive the signal via radio, using ccodevision software to create a software application in its processing. In this case signal processing uses a microcontroller, with some consideration, the physical dimensions are quite small where the outer diameter of the casing is 120mm, the inner diameter is 116mm, the PCB diameter must be 110mm or the square dimension of the PCB is 80mm, the I/O pin is sufficient, available other features needed such as the number of interrupts, the number of timers/ PWM pulse wide modulation, including the clock microcontroller which is sufficient 8MHz and above.

4. Conclusion.

From the results of this study, aesthetics, performance, reliability, response time and other functions went well. servo control can receive radio signals and scan according to the format [13,14] Morse, frequency, rs232c formula that has been determined and is shown by the left pin motion if it receives left orders and right if it receives right orders. This Servo Control For Missile System is in accordance with the expected results and is ready to be integrated with the research on locking target on missile system which is another title, continued this research.

Reference

[1] Aulia, Y.R., 2017. Design and implementation of hardware in the loop simulation for electric ducted fan rocket control systems using an 8-bit microcontroller and real-time open-source middleware, *Journal of Mechatronics, Electrical Power, and Vehicular Technologies*, pp.60-69

[2] Pei, Y., 2017. An Elementary Introduction to Kalman Filtering, the University of Texas at Austin

[3] Syafruddin, 2018. Analysis of total harmonic distortion in ACPDB panels due to non-linear loads, *J. Online Sekolah Tinggi Teknologi Mandala*, vol.13, no.2, pp.33-44

[4] Auger, F., 2013. Industrial applications of the Kalman filter: a review, *IEEE Transactions on Industrial Electronics*, vol. 60, no. 12

[5] Valade, A., 2017. A Study about Kalman Filters Applied to Embedded Sensors, LAAS-CNRS, Université de Toulouse, CNRS, INSA, 31031 Toulouse, France

[6] Ramady, G.D. and Wowiling, R. G., 2017. Analysis of vehicle rate prediction using linear regression method as a rate level indicator, *J. Online Sekolah Tinggi Teknologi Mandala*, vol. 12, no. 2, pp. 22-28

[7] Syafruddin, 2018. Analysis of harmonized content in 3 phase AC motor 0.12 kW controlled by 3 phase inverters, *J. Online Sekolah Tinggi Teknologi Mandala*, vol. 13, no.1, pp. 32-38

[8] Johan, A. K., 2018. Advances in PID Control, Department of Automatic Control, Lund University
[9] Dimitry, G., 2005. Integrated Diagnostics of Rocket Flight Control, EEE Aerospace Conference, Big Sky, MT
[10] Antoine, R., 2017. A flexible real-time simulation platform dedicated to embedded rocket engine control systems development and testing, 7th European Conference for Aeronautics and Space Sciences (EUCASS)
[11] Bindu, R., 2012. Tuning of PID controller for DC servo motors using genetic algorithms, *International Journal of Emerging Technology and Advanced Engineering*, vol.2, issue 3
[12] Jeffrey, T., 2017. Designing a variable drag rocket control system for target altitude, Worcester Polytechnic Institute, Rocketry ISP, Final Report
[13] Hidayat, R., 2018. Smart Key Implementation for BTS Gate Door Based on the Internet of Things, The 1st International Conference on Computer Science and Engineering Technology, Muria Kudus University
[14] Ramady, G.D., 2018. Bluetooth beacon battery monitoring system using Labview, *J. Online Sekolah Tinggi Teknologi Mandala*, vol.13, no.2, pp. 45-50

Simulation of Robot Arm for Diabetes Mellitus Patients

Subekti[1], Aan Budi Setiawan[2], and Abdul Hammid[1]
[1]Mechanical Engineering Vibration Laboratory, Universitas Mercu Buana, Meruya Selatan, Kembangan, Jakarta Barat, 11650, Indonesia
[2] Student of Mechanical Engineering Vibration Laboratory, Universitas Mercu Buana, Meruya Selatan, Kembangan, Jakarta Barat, 11650, Indonesia

[1]subekti@mercubuana.ac.id, [2]setiawan.aanbudi@gmail.com,
[3]dr.hamid.abdul@mercubuana.ac.id

Abstract. Diabetes is caused by metabolic disorders. The treatment is to control blood sugar levels which can be done with non-pharmacological therapy, namely reflexology. The intensity of therapy continuously carried out will speed up the healing process of diabetes. On this basis, we simulated a robotic arm for reflexology treatment for diabetics. The robotic arm system used was five degree of freedom, which used 5 servo motors in the design. The simulation results show that $\sigma_{material} \geq \sigma_{structure}$, so it can be concluded that the material used is rigid and strong to withstand the loads acting on the robot arm structure. While the magnitude of M1 servo motor torque is 2502 Nmm, M2 is 7492 Nmm, M3 is 7492 Nmm, M4 is 6934 Nmm and M5 is 4351 Nmm. In designing the robot arm, 1,2,3,4 servo motors used 8000 Nmm of torque, and 5 servo motors used 4500 Nmm of torque, so the servo motor used was safe to hold the torque on the robot arm.

1. Introduction

Diabetes mellitus (DM) is a chronic metabolic disorder caused by the pancreas not producing enough insulin or the body cannot use the insulin produced effectively, and in Indonesia, this disease is more commonly called sugar disease [1]. Efforts to handle diabetes mellitus must be done properly, one of the treatments for this disease is to regulate blood sugar levels within normal limits. According to Apriyanti [2], controlling blood sugar levels is the best way that can be done to avoid complications in people with diabetes mellitus. Blood sugar level control can be done in various ways, including pharmacological therapy and non-pharmacological therapy [3]. Pharmacological therapy is done by giving medicines to patients with diabetes mellitus. It could be by insulin therapy or oral hypoglycaemic drug therapy, or a combination of both. While non-pharmacological therapy is with treatment without drugs. One of them is reflexology, according to Kamaluddin (2010), non-pharmacological therapy is considered to have fewer side effects and more economical [4].

According to a study conducted by Robiul (2014), there were significant differences between blood sugar levels before and after reflection [5]. His results showed the median blood sugar after reflection was 150.50 mg/dl and before reflection was 181 mg/dl. While the average blood glucose level of patients with DM before reflexology was 199.76 mg/dl while in the control group was 183.18 mg/dl [6]. The average blood glucose level of patients with DM after reflexology is 159.14 mg / dl while in the control group was 170.43 mg/dl. In addition to lowering blood sugar levels, according to McRitchie (2013), reflexology can improve respiratory function, improve blood circulation, reduce tingling, and reduce stress in diabetes mellitus [7].

The key success of reflexology is the intensity of therapy carried out continuously [8]. But the limited time and a limited number of therapists who are able to do foot therapy cause the routine exercise of the patient becoming hampered. To allow people with diabetes to do foot therapy without a therapist, it is necessary to provide tools so that people with diabetes can still practice wherever and whenever without dependence on the therapist's schedule. This can increase the recovery ratio for diabetics. Based on the above considerations, technology foot reflection is needed. One of the technologies used in the treatment of diabetes is robotic technology. Interactive, high-intensity and repetitive foot therapy robots can perform foot therapy without a therapist. The existence of this robot makes the budget for therapists trimmed. Robot therapy will overcome the problem of lack of therapists for people with diabetes.

A robot is a multi-function manipulator that can be programmed to do certain tasks. In general, there are only two types of robots, namely robots that move, and robots that do not move (robot arm). Their use depends on the needs. The robot arm has a working principle adopted from the human arm. The components in the robot arm consist of three parts, namely, mechanical construction, actuator, and control system. The mechanical structure of the robot is very important in building a robot arm. According to Buchori (2019), designing a robot not only illustrates it but also calculates the mechanical strength of the robot arm structure [9]. To obtain quality and balanced mechanical structure, the calculation of each component must be accurate. Based on the movement of robots, Adriansyah (2013) implements communication on multi-robot systems using Bluetooth technology [10].

In this study, we will do a dynamic simulation that produces the material stress and torque of the servo motor needed by the robot arm. The results of the calculation of this robot arm can be used as a reference for selecting the material needed by the robot arm. In order to obtain high-accuracy analysis, we use Solid works 2018 software, considering that this software has been equipped with a finite element analysis program that allows users to analyse their design results. The author will conduct material simulations and servo motors used on robotic arms to carry out reflexology treatment for diabetics.

2. LITERATURE REVIEW

Manipulators are a mechanical structure consisting of several rigid bodies (links) see in Figure 1, which are connected with joints. To determine the thickness of the robot arm material structure and the servo motor used, it is very important to do the calculation at the beginning. This calculation is done to find the minimum value needed by the material structure and servo motor, which aims to enable the robot arm to withstand the loads acting on the robot arm [11]. The specifications of motor 1 and motor 2 can be calculated using the equation as below [12]:

$$T1 : W1*L1 + W2*L2 + W3*L3 + W4*L4 \quad (1)$$

$$T2 : W3*L5 + W4*L6 \quad (2)$$

Information:

T : Torque (Nmm)

W : Weight (N)

L : Length (mm)

The selection of motor specifications is determined by the amount of torque that occurs in the mechanical structure of the robot arm [13].

$$T_{motSer} \geq T_{Structure} \quad (3)$$

The required robotic arm structure thickness can be calculated using the equation below:

$$\sigma_{mat} : M/Z \quad (4)$$

Information:

σ_{mat} : Material permit voltage (N/mm^2)

M : Bending moment (Nmm)

Z : Cross section modulus (mm^3)

In the case of calculating the robot arm structure, bending moments that occur in the robot arm structure is equal to the rate of torque that occurs in the robot arm. So that it can also be written as:

$$T = M \quad (5)$$

Material is declared safe with conditions [14].

$$\sigma_{material} \geq \sigma_{Struktur} \quad (6)$$

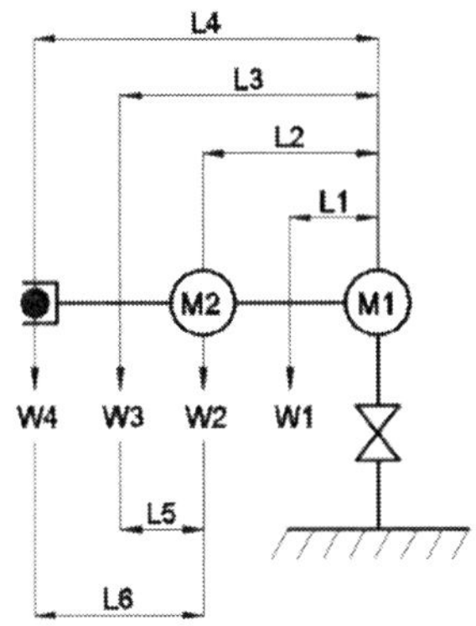

Figure 1. Robot Arm

3. MATERIALS AND METHOD

In designing robot arms, we need to use materials that are cheap and have good strength. This also applies to the servo motor that was used. S45C material is medium carbon steel which has a carbon content of 0.3% - 0.5%, so it has excellent workmanship and strength properties [15]. The thickness of the material used in this study was 3 mm and the safety factor were 1.2. To find out the properties of S45C material see Table 1.

Table 1. Material Properties S45C

Index	Unit
Yields strength	3.43×10^8 N/m^2
Tensile strength	5.69×10^8 N/m^2
Allowable stress	2.86×10^8 N/m^2
Thickness	3 mm

Source: (Aida, 2017)

In this paper, the author used a servo motor as the most important component in the robot arm. The servo motor has a function to move the joint and base arm of the robot arm. The selection of servo motors was strongly influenced by the torque that occurred in the robot arm structure caused by the weight of the robot arm components as well as the forces acting on the robot arm. Servo motor specifications used in the study can be seen in Tables 2 and 3. The robot arm is designed with the same function as the human arm. The end effector is designed to resemble a pressure device commonly used by therapists to perform reflexology on the feet, which is commonly used for the treatment of diabetics.

Table 2. Specifications for servo motors 1-4

Description	Unit
Torque	8000 Nmm
Power	6.5V~13V

Source: (Feetech, 2019)

In terms of the robot arm structure, the arm length has been adjusted to the area of the foot. Thus, all the reflection points on the soles of the feet can be reached using this robot arm. In addition, this robot is designed with attention to its aesthetics and beauty. The parts on the robot arm can be seen in Figure 2. The robotic arm components can be seen in Table 4. In general, the element method was used to run the simulation, so that the material voltage could be obtained. The robot arm design that had been completed was then simulated. First, the robot arm components were given gravitational force. Giving gravity was intended so that each component of the robot arm had gravity. Then, second, we added a motor to each robot arm connection. This motor drove all robot arm components during the simulation. The servo motor movement on the robot arm was limited to 180°. Next, we added a normal force that was perpendicular to the pressure component, which was 25N. This style resulted from the action-reaction that arose between the suppressor component and the sole of the foot.

Table 3. Specifications for servo motors 5

Description	Unit
Torque	4500 Nmm
Power	6.0~7.4 V

Source: (Hsservo, 2019)

The third step, we selected the components to be simulated by selecting the von Mises or deformation results. This step was to determine the value of the voltage that occurred in the robot arm component. Finally, we displayed the simulation results of the torque that occurred in each servo motor, resulting from the loads that had been added to the previous steps. For more details, the simulation of the robot arm can be seen in Figure 3.

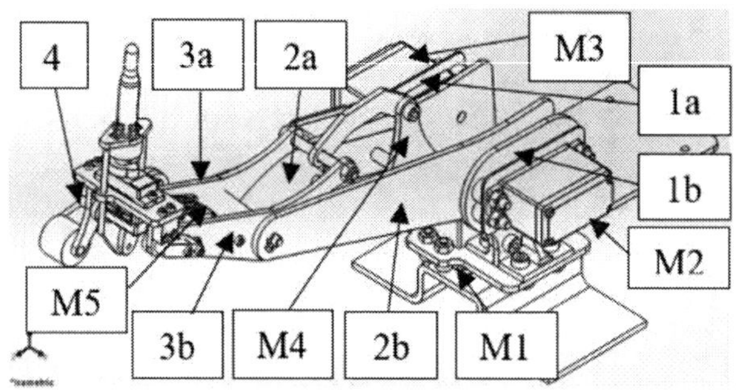

Figure 2. Robot arm design

Table 4. Description of robot arm components

NO	Remark
1a	Shoulder-arm 1a
1b	Shoulder-arm 2b
2a	Arm 2a
2b	Arm 2b
3a	Arm 3a
3b	Arm 3b
4	Arm support end effector
M1	Servo motor 1
M2	Servo motor 2
M3	Servo motor 3
M4	Servo motor 4
M5	Servo motor 5

4. RESULTS AND DISCUSSION

Motion Analysis Results

The simulation results of motion analysis on structures 1a and 1b based on Von Mises showed in Figure 5 (a) show that the maximum stress that occurred in the structure was 2.777×10^7 N/m^2 & 1.698×10^7 N/m^2. The lowest voltage was 8.156×10^3 N / m^2. The simulation results of motion analysis on structures 2a and 2b based on Von Mises showed in Figure 5(b) show that the maximum stress that occurred in structures 2a and 2b was 2.568×10^4 N/m^2 & 2.342×10^8 N/m^2. The lowest voltage was $1,464 \times 10^4$ N/m^2. The simulation result of motion analysis on structures 3a and 3b based on Von Mises showed in Figure 6 (a) indicate that the maximum stress that occurred in structures 3a and 3b was 5,38 x 10^7 N/m^2 and 3.698 x 10^7 N/m^2. The lowest voltage was 2.023 x 10^4 N/m^2.

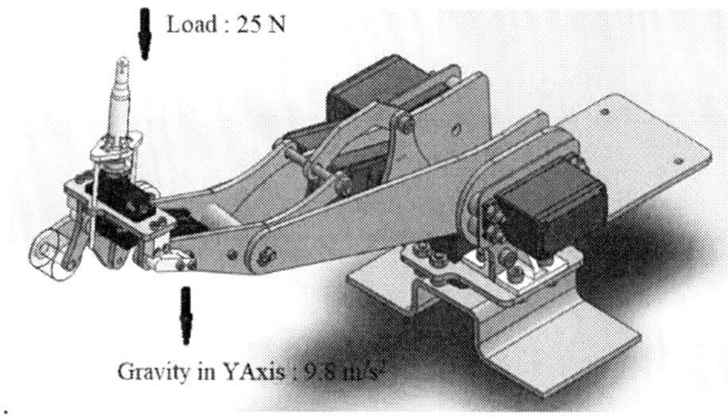

Figure 3. Simulation of the robot arm

The motion analysis simulation on structure 4 based on the Von Mises shown in Figure 6 (b) shows that the maximum stress that occurred in structure 4 was 1.033 x 106 N/m^2. The lowest voltage was 6.7 x 10^3 N/m^2. Referring to equation (5), a structure is declared safe if the $\sigma_{material} \geq \sigma_{structure}$. The simulation results above show that the highest stress occurred in structure 2b was 2.342 x 10^8 N/m^2. The material used in this design was S45C with a thickness of 3mm, and the permitted voltage was 2.86 N/m^2. Based on the above equation, we can conclude that the S45C material with a thickness of 3 mm is safe to withstand the loads acting on the robot arm structure.

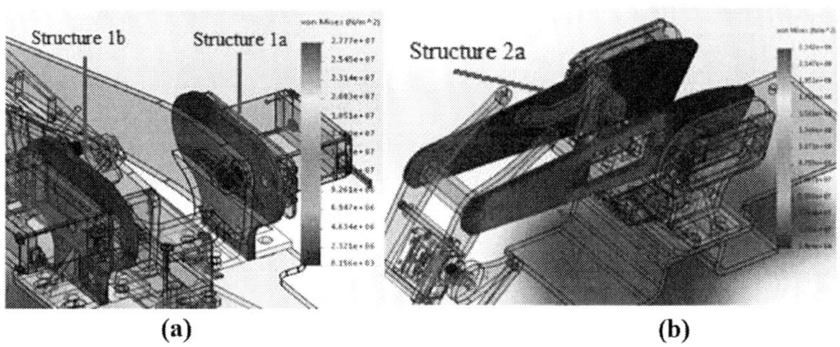

Figure 4. Motion analysis on (a) structure 1a & 1b (Von Mises) and (b) structure 2a & 2b (Von Mises)

Motor Torque Analysis

Figure 7 (a) shows a graph showing the need for an M1 servo motor. The graph shows that the torque value was 2504 Nmm. Figure 7 (b) shows a graph showing the need for M2 & M3 servo motors. From the graph, we can know that the torque value was 14983 Nmm, because the servo motors used were 2, the torque could be divided by 2, each M2 servo motor and M3 receive a 7492 Nmm torque load.

Figure 8 (a) shows a graph indicating the need for an M4 servo motor. From the graph, we identify that the torque value was 6934 Nmm. From the calculation of the torque as presented above, the highest torque was found on 2 & 3 servo motors, which was 7492 Nmm. Referring to equation (2), the servo motor is declared safe if the $\sigma_{material} \geq \sigma_{structure}$. The torque of the servo motor used was 8000 Nmm, while the 5 servo motor had a torque of 4351 Nmm, the servo motor capacity used was 4500 Nmm, it can be concluded that the two motors servo used could withstand the loads acting on the robot arm structure. Figure 8 (b) shows a graph showing the need for an M5 servo motor. The graph shows a torque value of 4351 Nmm.

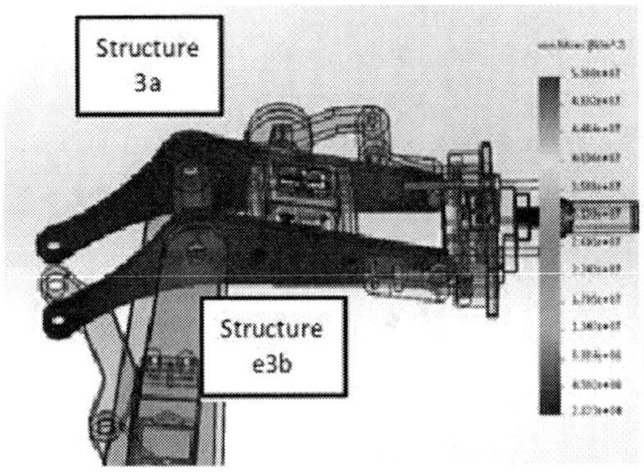

(a) structure 3a & 3b (Von Mises)

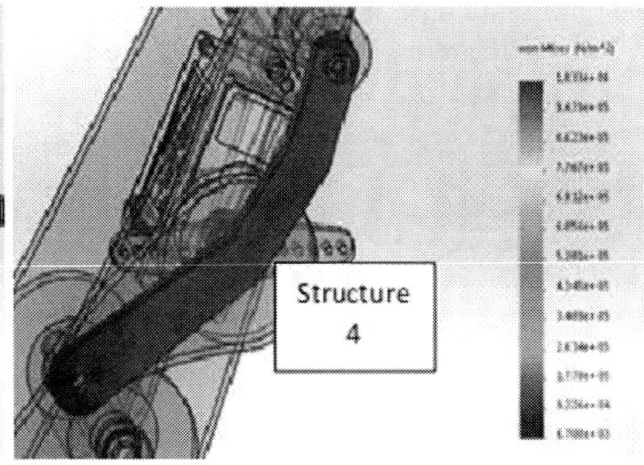

(b) structure 4 (Von Mises)

Figure 5. Motion analysis on structure

CONCLUSION

From the above analysis, we could obtain the following results:

1. The material voltage of each robot arm structure was Table 5.

Table 5. Simulation results of robot arm voltage

No	Component	The highest voltage N/m^2	The lowest voltage N/m^2
1	Structure 1a	2.777×10^7	8.156×10^3
2	Structure 1b	1.698×10^7	8.156×10^3
3	Structure 2a	2.568×10^4	1.464×10^4
4	Structure 2b	2.342×10^8	1.464×10^4
5	Structure 3a	5.38×10^7	2.023×10^4
6	Structure 3b	3.698×10^7	2.023×10^4
7	Structure 4	1.033×10^6	6.7×10^3

From the simulation, we identify that the highest voltage was in the 2b structure with material stress of 2.342×10^8 N/m^2, the material used was S45C with Allowable stress of 2.86×10^8 N/m^2. By referring to equation (6) we can conclude that the material used by the robot arm was safe to accept the load on the robot arm.

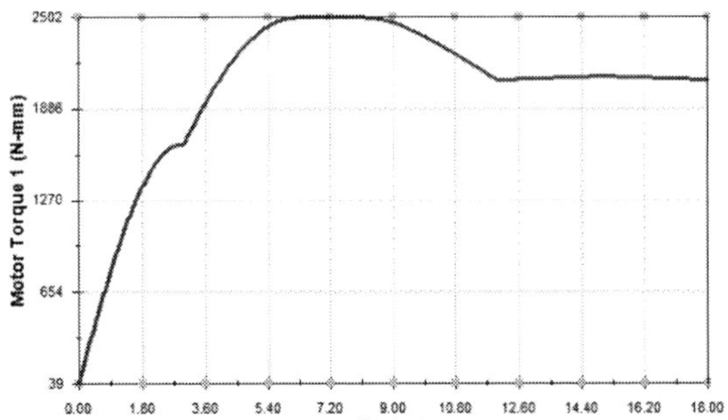

(a) Motor 1 (M1)

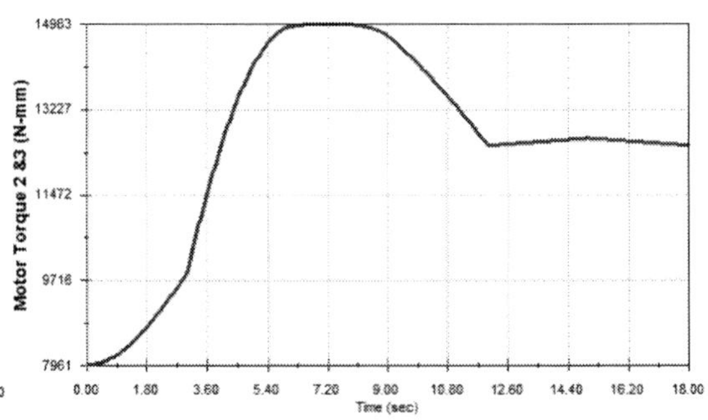

(b) Motor 2 (M2) and motor 3 (M3)

Figure 6. Motor torque requirement

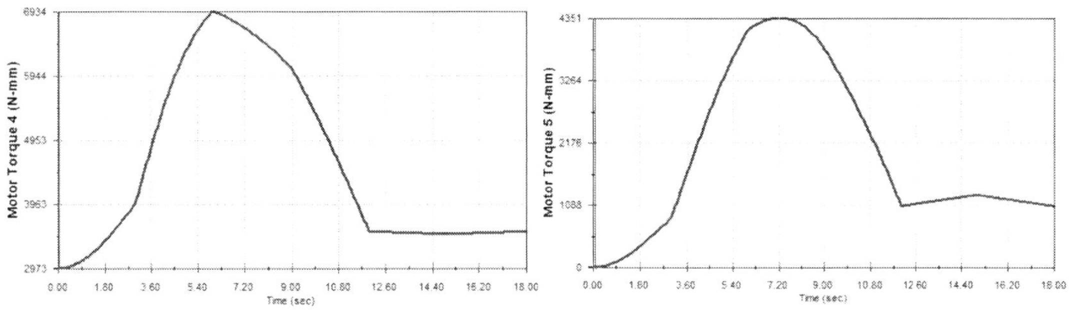

(a) Motor 4 (M4) (b) Motor 5 (M5)

Figure 7. Motor torque requirement

2. The torque that occurred in each servo motor was Table 6

Table 6. Simulation results of Torque

No	Component	The Highest torque (Nmm)
1	Motor 1	2502
2	Motor 2	7492
3	Motor 3	7492
4	Motor 4	6934
5	Motor 5	4351

From the simulation results from 1 servo motor to 4 servo motor, it was found that the highest torque occurred in 2 & 3 servo motors with the torque of 7492 Nmm. The specification of the servo motor used was 8000 Nmm. While the 5 servo motor has a torque of 4351 Nmm, the servo motor used has 4500 Nmm of torque. Referring to equation (2) the 1,2,3,4,5 servo motor is declared safe to hold the load and move the robot arm.

REFERENCES

[1] Infodatin (Pusat Data dan Informasi Kementrian Kesehatan RI), "Situasi dan Analisis Diabetes," *American Journal of Medical Genetics, Part A*. 2013.
[2] Apriyanti and R. F. Nuwarda, "Diagnosis dan pengobatan diabetes melitus tipe 2 : artikel review," *Farmaka*, 2017.
[3] R. WULANDARI, "Perbedaan Kadar Gula Darah Setelah Terapi Bekam Basah Dan Pijat Refleksi Pada Penderita Diabetes Mellitus Di Karangmalang," 2015.
[4] A. Komplementer, B. Di, K. Banyumas, and U. J. Soedirman, "Ipi10481," vol. 5, no. 2, pp. 95–104, 2010.
[5] R. F. Masithoh, H. Ropi, and T. Kurniawan, "PENGARUH TERAPI AKUPRESUR TERHADAP KADAR GULA DARAH PADA PASIEN DIABETES MELITUS TIPE II DI POLIKLINIK PENYAKIT DALAM RS Tk II dr. SOEDJONO MAGELANG," *J. Holist. Nurs. Sci.*, 2016.
[6] M. Musiana, T. Astuti, and R. Dewi, "EFEKTIVITAS PIJAT REFLEKSI TERHADAP PENGENDALIAN KADAR GLUKOSA DARAH PENDERITA DIABETES MELLITUS," *J. Ilm. Keperawatan*, 2017.
[7] W. I. S. Diabetes and C. Studies, "SHIATSU & DIABETES," 2013.
[8] M. Hablul Barri, A. Ryandika, A. Cesario, and A. Widyotriatmo, "Desain dan Kontrol Posisi dari Arm Manipulator Robot Sebagai Alat Rehabilitasi Pasien Pasca Stroke," *J. Otomasi Kontrol dan Instrumentasi*, 2017.

[9] "Rancang Bangun Miniatur Robot Lengan Menggunakan Mikrokontroler Atmega 8535," *J. Tek. Elektro*, vol. 6, no. 1, p. 7, 2014.
[10] A. J. D. L. Rezky Ardiansyah, "in System Programming Avr Menggunakan Koneksi Bluetooth," *Widya Tek.*, vol. 10, no. 1, pp. 54–67, 2011.
[11] K. S. O. E. Nwe and W. A. I. P. Maung, "Dynamic Simulation and Motion Load Analysis of Six," no. March, pp. 24–28, 2018.
[12] S. Pachaiyappan, M. M. Balraj, and T. Sridhar, "Design and Analysis of an Articulated Robot Arm for Various Industrial Applications," *IOSR J. Mech. Civ. Eng.*, 2014.
[13] D. Arhur, Y. Perrot, C. Bidard, J. P. Friconneau, J. D. Palmer, and L. Semeraro, "ITER Articulated Inspection Arm (AIA): Geometric calibration issues of a long-reach flexible robot," *Fusion Eng. Des.*, 2005.
[14] S. S. Duicu and L. Popa, "Modeling and simulation of the 5-axis robot arm trajectory 3 Robot Arm Modeling," pp. 154–159.
[15] A. Mahmudah, G. Kiswanto, and D. Priadi, "Fabrication of punch and die of micro-blanking tool," in *IOP Conference Series: Materials Science and Engineering*, 2017.

The Technology Of Sterilizing Brackish Water Become Drinking Water In The Society

Muhammad Helmi*, **and Rostian Nafery**

University of Tridinanti Palembang, Indonesia

*deltahelmi@gmail.com

Abstract. Banyu Urip Village, Hamlet III, Tanjung Lago Sub-district, Banyuasin Regency, South Sumatra. The distance from the Palembang city is ± 40 km. Lack of awareness of the importance of maintaining the personal and environmental health is also seen from the absence of a place for taking clean water. People still take water which is red, smelly, sticky and salty from the river in the dry and rainy season. The results obtained from the technology of sterilizing brackish water into drinking water for the society have succeeded well. The final results of measuring treated water are Ph 7.7 and TDS / TS 794 mg / liter. 1.36 scale turbidity of NTU. TSS 0.2 mg / liter. Specific gravity of 0.1 mg, very clear and odorless. It is expexted that the Science and Technology for society that has been implemented can be a role model for people in other villages, and can form cooperation villager's community through partners that have been formed.

Keywords: Technology, Sterilization, Brackish Water, Society.

1. Situation Analysis

Banyu Urip Village, Hamlet III, Tanjung Lago Sub-district, Banyuasin Regency, South Sumatra is located between 1,30° - 4,0° South Latitude (LS). The average rainfall is 1700 m per year, the topography is flat and bumpy, the average temperature ranges from 24^0C-29^0C with the type of *Clay Humus* soil. In tidal and swamp areas, people have difficulties to obtain clean water for household needs, especially drinking water. This is caused by the peat water in that area is brown and acidic, and the results; 4.66 soil pH measurement is smelly and sticky. The acidic nature of peat water is caused by the presence of clay containing sulfide, which is oxidized to sulfuric acid.

2. Partners Problems

Based on the situation analysis, can be identified several problems that faced by the partners, namely:

- Water condition in Banyu Urip Village is not good for their daily activities because the water's quality is not qualifying. The well water in this area is red, smelly and salty while the water from the river in this area is clear but sticky and salty.
- Lack of awareness of the villagers about the importance of maintaining the personal and environmental health.
- The low level of formal education is possessed by villagers.
- Mastery of technology, especially water treatment (taking bath, washing, drinking) is still very low.

- The role of the Regional Government, especially Banyuasin Regency is still very limited.

Example of river water

Counseling on water purification methods

Figure 1. Example and Counseling

3. Solution

- The results of the meeting of related elements, the problem that is faced by the society today is the needs of clean water (for taking bath and washing) and drinking water for daily activity needs. Built clean water installations that are suitable for health (drinking water, taking bath, and washing).
- Banyu Urip villagers are aware of their personal and environmental health.
- Helping the government of Banyuasin Regency build clean water installations with technology that has been applied to the Banyu Urip village to other nearby villages.

4. The Implementation Method

In the implementation of IbM activities involved 2 partners, namely the Pokja Group 1 and Pokja Group 2. The implementation methods included:

- The first stage is to provide information on PKM activities that will be carried out on referring to the Regulation of the Minister of Health of the Republic of Indonesia Number 492/MENKES/PER/IV/2010, which is meant by drinking water through processing that is qualifying of healt requirements and can be drunk directly.
- The second step is to build clean water sterilization technology and prepare the necessary materials, such as Poly Aluminum Chloride (PAC), which is $Al_m(OH)_n\ Cl_{(3m-n)}$.. as a coagulant or flocculant to decompose cloudy solutions and agglomerate particles, it is possible to separate from the solution medium[1]
- The implementation of assisted society groups that carry out the provision of drinking water by providing training through assisted groups about clean water supply systems and processing water for drinking.

Meeting with society

Intake tub and Pre-sedimentation tub

Tedmond on the WTP Steel Frame

8 Inch Tube Filter Media

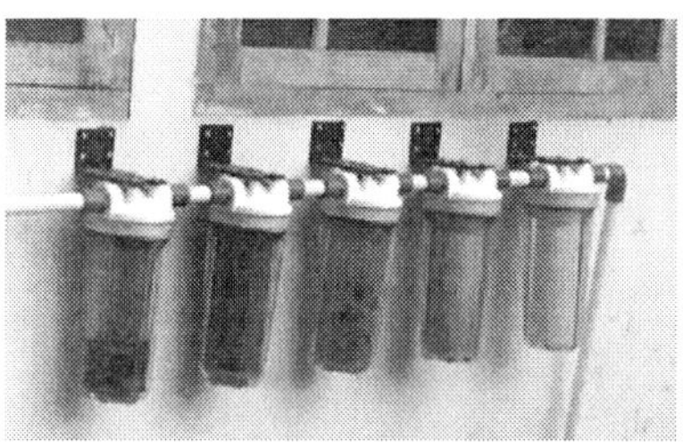
10 Inch Filter Housing

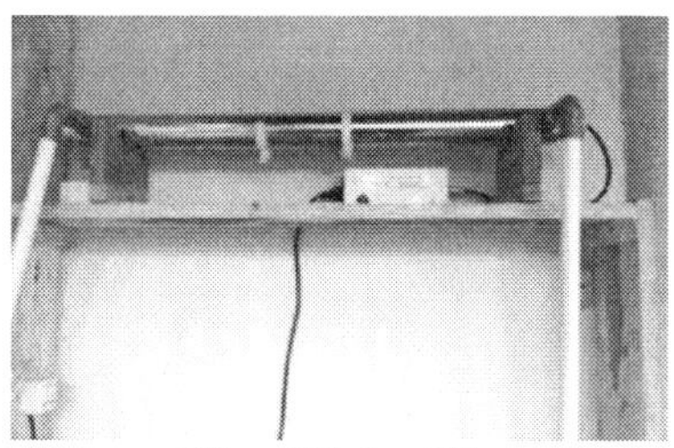
Ultra Violet Filter

Three bottles of water sample (UV filter, Presidation, River)

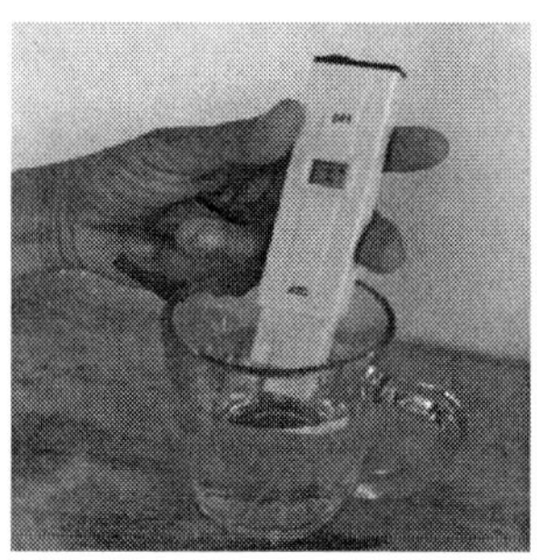
Water pH measurement

5. Results And Discussion

1) Intake tub, the coagulant used in processing river water through pump 1 is PAC as a coagulant or flocculant to decompose turbid solutions and agglomerate particles, making it possible to separate

from the medium solution for 4 hours. The deposited clots will be visible, then flow through the drain faucet.

2) Pre-sedimentation tub accommodates water through a pipe faucet from the processing tub. Remaining particles that clot will precipitate to the bottom to be released through the mixing faucet. Water in a reservoir that has been clear, not too clear, still contains: Fe CO3 (Ferro Carbonate); Mn CO2 (Manganese Carbonate); Zn CO3 (Zinc Carbonate)
3) Processed water from the settling basin is transferred to the reservoir tank via pump 2. Water from the tedmon can be directly flowed or pumped into the manganese zeolite filter.

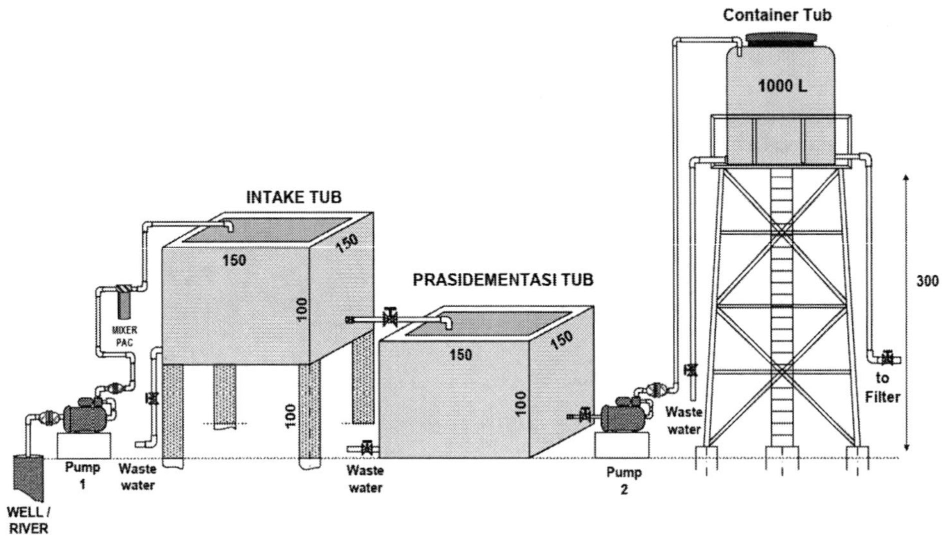

Initial process from the intake tub, pre-sedimentation tub and reservoir

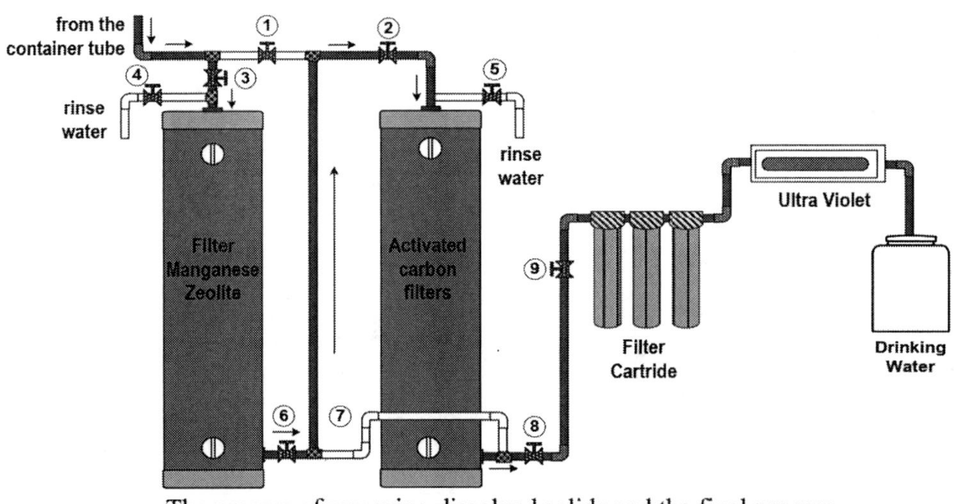

The process of removing dissolved solids and the final process

4) Filter manganese zeolite to filter or remove substances that are still attached to the water and remove suspended solids: Mn CO3 (Manganese Carbonate), and Zn CO3 (Zinc Carbonate).
5) Activated carbon filters to remove organic matter, odors, flavors and other micro pollutants.
6) Then the water is flowed into the cartridge filter as a filter filter which can remove dissolved solids.

7) Water that has passed through the filter cartridge is very clear and categorized as clean water is collected in the reservoir. The water can already be used for bathing and washing washing for various household needs.
8) Ultra violet sterilizer, its function is to kill pathogenic microorganisms in water. This process without requiring large energy because it works with a gravity system and only requires 30 Watts of electrical energy.

Final Results of Processed Water Inspections

No	Examination	Content Allowed	Results examination	Unit
1	TDS/TS	1000	794	mg/l
2	Turbidity	25	1,30	Scale mtu
3	TSS		0,20	Mg/l
4	Density		0,10	Mg
5	pH	6,5 – 8,8	7,7	-

From the final results of examination of processed water in Brackish Water Sterilization Technology to drinking water for the Banyu Urip Village Village 3, Tanjung Lago Sub-district, Banyuasin Regency, South Sumatra has succeeded well, where:
1) TDS / TS (Total Dissolve Solid) or often said the amount of dissolved solids is 794 mg / liter.
2) Turbidity is a measure that uses light effects as a basis for measuring the state of raw water with the NTU scale (Nephelo metrix turbidity unit). Turbidity caused by the presence of mixed objects or colloidal objects in the 1.30 scale water.
3) TSS (Total Suspended Solid) is the residue and total solids held by a filter with a particle size of 0.20 mg / l.
4) Weight The type of solution is the measurement of mass per unit volume of objects with a measuring instrument picnometer 0.10 mg.
5) Ph water measurement results 7.7 are alkaline, odorless and tasteless, meaning processed water is safe for consumption.

6. Conclusion

The results obtained from the technology of brackish water sterilization into drinking water for the people of Banyu Urip Village, Hamlet 3, Tanjung Lago Sub-district, Banyuasin Regency, South Sumatra have succeeded well. The final result of measurement of TDS / TS processed water is 794 mg / liter. 1.36 scale turbidity of NTU. TSS 0.2 mg / liter. Specific gravity of 0.1 mg. The value of Ph 7.7 and very clear does not smell. Technology transfer, was given a demonstration of how to purify water into clean water and how it was subsequently processed until the water could be drunk and the sustainability of the program, carried out monitoring and evaluation programs.

References

[1] Ayatlangit, Penggunaan-Polyaluminium-Chloride (PAC), https://bisakimia. com/ 2013/12/01.
[2] Anonim, 1991. Kimia Tanah, Direktorat Jenderal Pendidikan Tinggi, Departemen Pendidikan dan Kebudayaan.
[3] Anonim, 2016, Buku Manual air gambut & pengolahannya, https:// ipehijau.files.wordpress.com/2016/03.
[4] Alfian Putra, dkk. Rancang Bangun Alat Penjernih Air Gambut Menjadi Air Bersih, Jurusan Teknik Kimia Politeknik Negeri, 17 Mei 2016.
[5] Anonim, Instalasi Pengolahan air gambut untuk penyediaan air bersih, Pusat Penelitian dan Pengembangan Sumber Daya Air Kementrian Pekerjaan Umum, http://www.pusair-pu.go.id, 1 Mei 2016.
[6] Darmawi. 2011. Penelitian mandiri potensi energi air pada pintu air saluran irigasi sekunder desa Bangun Sari Telang II Banyuasin.

[7] Dyah Agustiningsih dkk, 2012. Analisis Kualitas Air dan Strategi Pengendalian Pencemaran Air Sungai Blukar Kabuaten Kendal, Program Magister Ilmu Lingkungan Universitas Diponegoro, Jurnal PRESIPITASI, Vol. 9 No.2 September 2012, ISSN 1907-187X.
[8] Farhatush Shoalihat, 2015. Iptek bagi masyarakat (IbM) Dusun Lebari dan Dusun Krajaan untuk pengelolaan air buangan rumah tangga, Program Studi Biologi, Fakultas Saint dan Teknologi, UIN Sunan Kalijaga, Yogyakarta .
[9] Kepmenkes, 2010. Peraturan Menteri Kesehatan Republik Indonesia Nomor 492/ Menkes/Per/IV/2010. Tentang Persyaratan Kualitas Air Minum
[10] Peraturan Pemerintah Republik Indonesia Nomor 122 tahun 2015, Tentang Sistem Penyediaan Air Minum. www. hukumonline.com.

Socio-Cultural and Settlement Factors That Influence Pro Environmental Behavior on Rural-Urban Citizen

Haris Mukti Amrullah[1], **Herdis Herdiansyah**[1*]

[1]School of Environmental Science, Universitas Indonesia, Salemba, Indonesia

*Email: herdis@ui.ac.id

Abstract. Environmental behavior in the community has a vital role in structuring the city environment — identification of factors that influence the development of pro-environmental behavior. Various models of the formation of environmental behavior theories show various internal and external factors that determine the development of pro-environmental behavior in a person. Previous studies had explained how socio-cultural, and place attachment are the external factors that contribute to shaping pro-environmental behavior. However, further studies are necessary to determine the most appropriate solution can be applied in emerging cities like Jakarta, based on the grading systems of socio-cultural, and place attachment determinants. This study evaluated the proposed solutions based on those determinants, and calculates the rank based on the implementation criteria, namely: effectiveness, cost, duration, and sustainability using AHP as a tool in the decision-making process in solving the problems with high complexity. It is determined that the prioritized alternative to be implemented in the urban/rural area in Jakarta is Place Attachment, as it is the most sustainable and effective alternative in developing the pro-environmental behavior within the community.

Keywords: Environmental Behavior, Socio-cultural, Settlement, Rural-Urban, Citizen

1. Introduction

Pro-environmental behavior in the community determines the quality of the environment. Establishing pro-environment behavior as a daily behavior of the community is the most important thing to achieve a sustainable environment [1]. A good environment will be difficult to achieve if only done by someone or a group but must involve all elements and all people who have interests in the environment. Pro-environment behavior in the community can also support global actions to create a sustainable environment [2],[3]. So, a good environmental behavior in society is needed to form a better environment. Environmental behavior will be preferred by the community to do if there are no barriers that prevent them from doing so [4]. These barriers can be in the form of lack of supporting facilities, more expensive costs or concerns about their safety. If there are no obstacles in pro-environment actions, then it will be able to motivate the community to implement pro-community behavior as their daily behavior. It is easier for environmental behavior to be carried out by the community, so it will be more desirable to do.

Indonesia as a developing country that has socio-cultural wealth in it makes its people have different views about their environment. This difference of opinions will ultimately affect the differences in attitudes, behaviors, and ways in which environmental management is carried out, even though the characteristics of the environment are the same [5]. Based on the 2013 Environmental Care Behavior Survey (SPPLH) conducted by the Central Statistics Agency (BPS), the Indonesian people have an environmental awareness index at 5.7. It means that the environmental concerns of the Indonesians are in the moderate category [6].

From the results of the literature study that has been conducted, it was found that pro-environment behavior is influenced by at least three fundamental factors, namely internal factors such as knowledge, concern, and perception [7],[8] external factors such as socio culture status [4],[9] and psychological factors [4],[8]. These three factors have an attachment to each other in shaping one's environmental behavior. It is known that socio -cultural factors and conditions of settlements are external factors that influence pro-environment behavior in a person. The socio-cultural factors discussed in this article include income, education, and variable housing conditions because these external factors are more dynamic to be used as alternative program choices rather than referring to static internal factors, such as gender or group age.

The socio-cultural factors in this study focus on community income and its correlation with pro-environment behavior. Revenue determines the behavior of an individual. People who live in developed countries with higher average income levels than people living in developing countries have better environmental behavior and attention to the environment than developing countries [8]. Pro-environment behavior in the form of consumptive behavior and resource use and waste behavior in high-income communities is not good when compared with low-income communities [10]. The level of income in households is also able to predict the impact on the environment caused by the behavior of the community [11]. Based on studies of the results of research that have been discussed [8],[10],[11] the relationship between income levels with different environmental behaviors in low-income communities, compared to high-income communities. In high-income communities, consumptive behavior is generally worse than low-income people, but they have a better environmental concern.
In contrast to this, economic constraints make low-income people more efficient in using resources, but their motivation to do so is not from environmental factors. In this case, other factors such as education that will provide knowledge and increase environmental awareness can influence the relationship between income and environmental behavior. This result is also supported by the results from previous studies [12],[13].

The socio-cultural factors discussed in this study are the level of education. Awareness of the importance of environment management in accordance with the actual problems that are occurring will encourage correct pro-environmental behavior. Environmental education is also one of the effective ways to shape environmental behavior in society [14]. In urban communities, environmental education will provide an overview of how to manage the environment well and solve various environmental problems that are so complex in urban areas.

Settlement factors discussed in this study are how settlements as artificial environments and human daily environments have their own role in shaping human behavior. In better settlements such as luxury housing or apartments, supporting facilities for environmental management are generally complete, so the people who live there are not hard to behave environmentally [15]. The presence or absence of supporting facilities will determine how pro-environment behavior in someone. In addition, personal attachments to his residence can also determine how a person's environment behaves. A person's attachment to his environment, especially the housing environment, the environment in the workplace and the city environment can affect the behavior of his daily environment [16]. A person's attachment in a place will be stronger along with the length of time and activities carried out in that place. Juneman and Rufaedah (2013) in their study concluded that place attachments at the level of settlements, offices and cities were able to influence a person's daily behavior in a person. In general, place attachment is divided into a natural place attachment, namely one's attachment to the natural environment and civic place attachment, namely one's attachment to an artificial environment [16]. Halpenny (2006) also states that place attachment has a strong influence in predicting environmental behavior in a specific place [17].

From the description above, it is observed that the development of pro-environment behavior is a fairly complex problem because it is related to 3 aspects of sustainable development, namely: environmental and social. To solve this problem, all of these 3 aspects must be considered, not by looking from one aspect only. Therefore, this research focuses on solving the problems that exist in the community to achieve the main objective, namely to develop pro-environment behavior in rural-urban environments such as the City of Jakarta.

Analytical Hierarchy Process (AHP) one of the multiple decision-making methods that model decision-making processes mathematically and are used to solve complex problems [18]. For this reason, the AHP method is chosen to describe the existing problem as the main objective, using selection criteria that make an activity manageable and sustainable, namely: program effectiveness (duration), duration of the program (duration), costs to be incurred during implementation program (cost), and an estimate of whether the positive impact of the program will continue or only succeed for a short period of time (sustainability).

2. Methodology

This research uses the combination of the literature study and for self-judgment using the Analytical Hierarchy Process (AHP) method. The literature study aimed to determine the most important factors that influence the development of pro-environmental behavior in the rural-urban community. Next step is generating the AHP model, on which the selection criteria from the study are utilized to benchmark the offered alternative solutions. Finally, the scoring and grading are done with AHP Calculator Software, while maintaining the Consistency Ratio (CR%) under 10%. The obtained priority ranking helps to decide which solution is the best to achieve the study objective.

3. Result and Discussion

The model of Analytical Hierarchy Process (AHP) method in determining the main objective, selection criteria, and alternative solutions in Jakarta rural-urban area is presented in Figure 1. The main objective in this study becomes the aim of model, and the criteria and alternative solutions are generated based on literature studies. Next, each alternative will be benchmarked on priority, according to the selected criteria, thus, determining the decision that is manageable, and most important to be implemented in Jakarta.

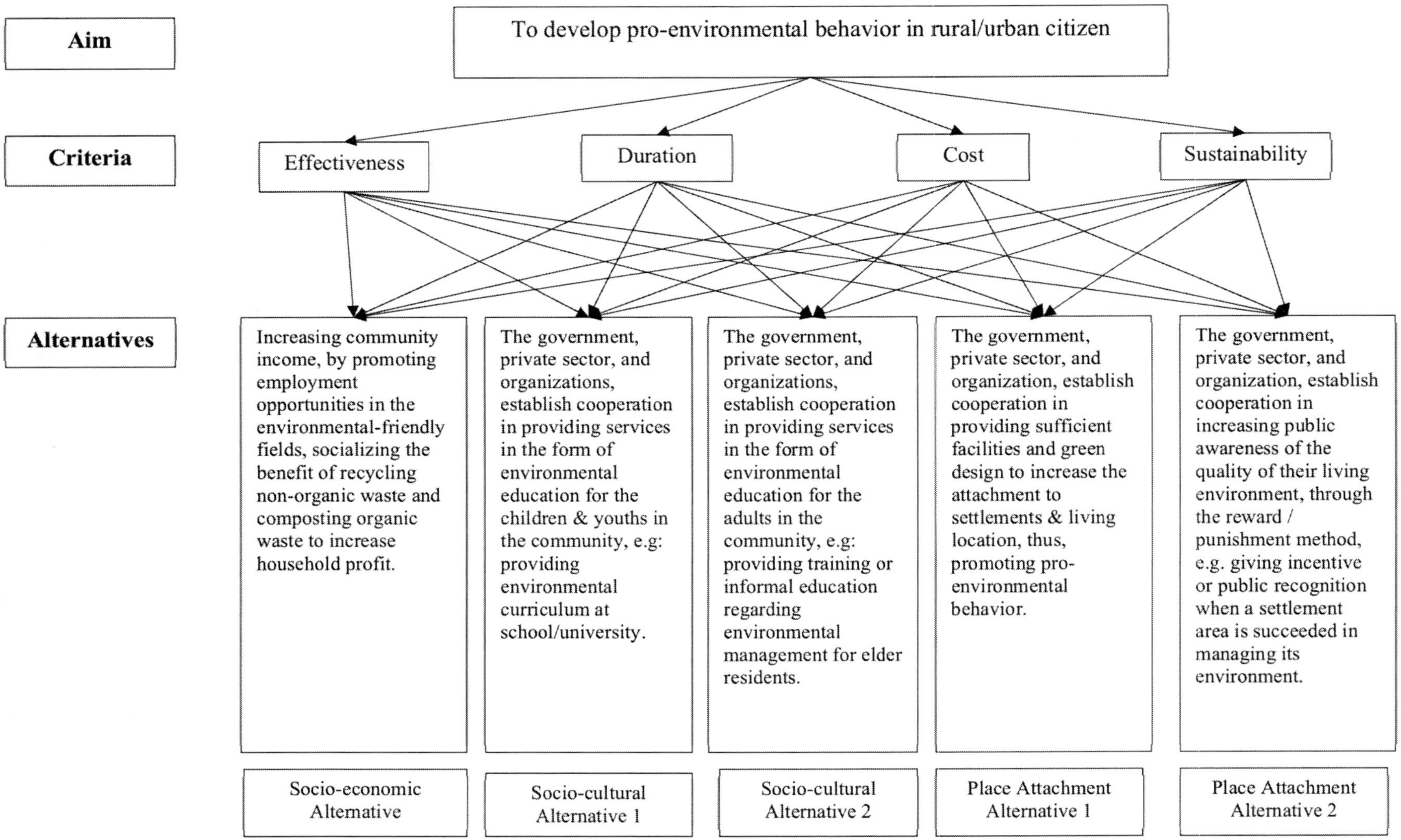

Figure 1. AHP model to determine of the main objective, selection criteria, and alternative solutions

The Analytical Hierarchial Process (AHP) calculation is conducted using software to ensure that the Consistency Ratio (CR%) is ranged under 10% to maintain the consistency of ranking the criteria and alternatives. The calculation result of the priority of the criteria is presented in Table 1, where sustainability of the alternative activities becomes the first priority of all 4 criteria (=0.434).

Table 1. Result of AHP Calculation on Criteria Priority (CR% = 3.8%)

	Effectiveness	Duration	Cost	Sustainabilty	Priority
Effectiveness	1.00	7.00	1.00	1.00	0.307
Duration	0.14	1.00	0.20	0.11	0.044
Cost	1.00	5.00	1.00	0.33	0.216
Sustainabilty	1.00	9.00	3.00	1.00	**0.434**
TOTAL					**1.000**

Based from the calculation, sequentially ranked criteria from the first to last priority are: sustainability, effectiveness, cost, and then duration. This ranking method is accepted due to the perspective that the results of activities related to environmental management is mostly visible in long term, not in short span of time, e.g.: the development of open green space, the benefits that can be obtained from environmental education and training, and the benefit of maintaining the hygiene of the residential settlements. All of these activities are not beneficial if they are only done once at a time. Hence, if maintaining the environment becomes habitual and conducted in daily activities, the result can be obtained in long term.

Table 2. Result of AHP Calculation on Priority Towards Sustainability (CR% = 2%)

	SE	SC_1	SC_2	PA_1	PA_2	Priority
SE	1.00	0.20	1.00	1.00	0.20	0.073
SC_1	5.00	1.00	7.00	5.00	5.00	0.350
SC_2	1.00	0.14	1.00	1.00	0.20	0.069
PA_1	1.00	0.20	1.00	1.00	0.20	0.074
PA_2	5.00	2.00	5.00	5.00	1.00	**0.434**
TOTAL						**1.000**

Table 3. Result of AHP Calculation on Priority Towards Effectiveness (CR% = 6.7%)

	SE	SC_1	SC_2	PA_1	PA_2	Priority
SE	1.00	0.14	0.20	0.33	0.11	0.033
SC_1	7.00	1.00	5.00	7.00	0.50	0.318
SC_2	5.00	0.20	1.00	1.00	0.11	0.082
PA_1	3.00	0.14	1.00	1.00	0.14	0.068
PA_2	9.00	2.00	9.00	7.00	1.00	**0.499**
TOTAL						**1.000**

Table 4. Result of AHP Calculation on Priority Towards Cost (CR% = 6%)

	SE	SC_1	SC_2	PA_1	PA_2	Priority
SE	1.00	3.00	3.00	3.00	0.50	0.257
SC_1	0.33	1.00	1.00	7.00	0.25	0.094
SC_2	0.33	1.00	1.00	1.00	0.20	0.089
PA_1	0.33	1.00	1.00	1.00	0.14	0.084
PA_2	2.00	4.00	5.00	7.00	1.00	**0.476**
TOTAL						**1.000**

Table 5. Result of AHP Calculation on Priority Towards Duration (CR% = 7.9%)

	SE	SC_1	SC_2	PA_1	PA_2	Priority
SE	1.00	7.00	7.00	5.00	5.00	**0.556**
SC_1	0.14	1.00	1.00	0.33	0.33	0.055
SC_2	0.14	7.00	1.00	0.20	0.20	0.047
PA_1	0.20	3.00	5.00	1.00	3.00	0.209
PA_2	0.20	3.00	5.00	0.33	1.00	0.133
TOTAL						**1.000**

Table 2 – 5 presented the results of pairwise comparison result of each alternative activity explained in Figure 1, following the other of the criteria ranking. Table 2, 3, and 4 indicate that the pairwise comparison ranked the Alternative of Place Attachment 2 as the first, while Table 5 shows that the Alternative of Socio-economical is the first rank. Afterwards, each value grade on priority is calculated to get the value of total priority, thus, resulted in the ranking of alternative solutions as presented in Table 6.

Table 6. Result of AHP Calculation on Total Priority

	Priority	Rank
Socio-economical	0.230	2
Socio-cultural 1	0.204	3
Socio-cultural 2	0.072	5
Place Attachment 1	0.109	4
Place Attachment 2	**0.386**	**1**

Based from the pairwise comparison of AHP, it is determined that the most prioritised alternative is **Place Attachment 2**: The government, private sector, and organization, establish cooperation in increasing public awareness of the quality of their living environment, through the reward/punishment method, e.g: giving incentive or public recognition when a settlement area is succeeded in managing its environment. The results of calculations with AHP show that this alternative gets the highest ranking in fulfilling 4 (four) criteria, namely: effectiveness, duration, cost, and sustainability. It means that this alternative method is likely to be preferred due to its economical, steadfast result, and sustainable nature. It is highly considered the possibility of the likeliness of the community to receive rewards and public recognition from the local and central government –especially in emerging cities like Jakarta– when each small event has the possibility of becoming viral on the Internet.

The result of the AHP ranking system that grade Place Attachment as the most effective solution to influence the pro-environmental behavior reinforced the results of previous studies that state how place attachment factors is one of the most crucial factors in the development of pro-environmental behavior [16],[17],[18] .

4. Conclusion

This study decided that, from 5 (five) alternatives in the socio-cultural, and settlement factors that were offered, Place Attachment Alternative 2 is the most suitable to be implemented to develop the pro-environmental behavior in Jakarta rural-urban area. Thus, this study recommends that the government, the private sector, and organization establish cooperation in increasing public awareness of the quality of their living environment through the reward/punishment method, since the results of calculations with AHP show that this alternative gets the highest ranking in fulfilling 4 (four) criteria, namely: effectiveness, duration, cost, and sustainability.

Acknowledgment

This work is supported by Hibah PITTA 2018 funded by DRPM Universitas Indonesia No. 2577/UN2.R3.1/HKP.05.00/2018

Reference

[1] R. Turaga, R. Howarth and M. Borsuk, "Pro Environmental Behavior Rational Choice Meets Moral Motivation", *Annals of the New York Academy Sciences*, vol. 1185, pp. 211-224, 2010.
[2] I. Chatterjee and S. Dey, "Some Correlates of Pro Environmental Behavior", Journal of The Indian Academy of Applied Psychology, vol. 41, no. 3, pp. 156-161, 2015.
[3] E. Fredericks, K. Stenner and E. Hobmen, "The Socio-Demographic and Psychological Predictors of Residential Energy Consumption: A Comprehensive Review", *Energies*, vol. 8, pp. 573-609, 2015.
[4] C. Manning, "The Psychology of Sustainable Behaviour: Tips for Empowering People to Take Environmetally Positive Action", *Minessota : Minnesota Pollution Control Agency.*, 2009.
[5] I. Putrawan, "Konsep Konsep Dasar Ekologi dalam Berbagai Aktivitas Lingkungan", *Bandung: Alfabeta*, 2014.
[6] "Indikator Perilaku Peduli Lingkungan Hidup 2013", *Bps.go.id*, 2013. [Online]. Available: https://www.bps.go.id/.
[7] M. Alibelli and C. Johnson, "Environmental Concern: A Cross National Analysis.", *Journal of Internation and Cross Culture Studies*, vol. 3, no. 1, pp. 1-10, 2009.
[8] R. Gifford and A. Nilsson, "Personal and social factors that influence pro-environmental concern and behaviour: A review", International Journal of Psychology, p. n/a-n/a, 2014. Available: 10.1002/ijop.12034
[9] G. Raj, S. Shilpa and R. Maheshwaran, "Revised Socio – Economic Status Scale for Urban and Rural India – Revision 2015. The Scientific Journal for Theory and Practices of Socio Economic Development", pp. 167-174, 2015.
[10] H. Welsch and J. Kuhling, "Pro Environmental Behavior and Rational Consumer Choice: Evidence from Surveys of Life Staisfaction. Journal WENKE, Vol.2 (8).", vol. 2, no. 8, 2015.
[11] S. Moser and S. Kleinhückelkotten, "Good Intents, but Low Impacts: Diverging Importance of Motivational and Socioeconomic Determinants Explaining Pro-Environmental Behavior, Energy Use, and Carbon Footprint", Environment and Behavior, vol. 50, no. 6, pp. 626-656, 2017. Available: 10.1177/0013916517710685
[12] E. Plombon, "Factors Affecting Pro Environmental Attitudes", *Journal of Undergraduate Research*, vol., 2011.
[13] S. Plavsic, "An Investigation on Gender Differences in Pro Environmental Attitudes and Behavior", *Honor Scholar Theses*, 2013. Available: http://digitalcommons.uconn.edu/srhonors_theses/404.
[14] L. Varela-Candamio, I. Novo-Corti and M. García-Álvarez, "The importance of environmental education in the determinants of green behavior: A meta-analysis approach", *Journal of Cleaner Production*, vol. 170, pp. 1565-1578, 2018. Available: 10.1016/j.jclepro.2017.09.214
[15] D. Makarenko and D. Štreimikienė, "Quality of life and environmentally responsible behavior in energy sector". *Journal of International Studies*, *7*(3), 179-192, 2014. https://doi.org/10.14254/2071-8330.2014/7-3/17
[16] Juneman and A. Rufaedah, "Influence of Five Types of Ecological Attachments on General Pro-environmental Behavior", *Procedia - Social and Behavioral Sciences*, vol. 85, pp. 535-543, 2013. Available: 10.1016/j.sbspro.2013.08.382.
[17] Halpenny, "Examining The Relationship of Place Attachment with Pro Environmental Intentions", *Proceedings of The 2006 Northeastern Recreation Research Symposium*, 2006.
[18] M. Şahin And H. Yurdugül, "A Content Analysis Study on the Use of Analytic Hierarchy Process in Educational Studies", *Eğitimde ve Psikolojide Ölçme ve Değerlendirme Dergisi*, pp. 376-392, 2018. Available: 10.21031/epod.373784
[19] M. Şahin And H. Yurdugül, "A Content Analysis Study on the Use of Analytic Hierarchy Process in Educational Studies", *Eğitimde ve Psikolojide Ölçme ve Değerlendirme Dergisi*, pp. 376-392, 2018. Available: 10.21031/epod.373784
[20] L. Scannell and R. Gifford, "The relations between natural and civic place attachment and pro-environmental behavior", *Journal of Environmental Psychology*, vol. 30, no. 3, pp. 289-297, 2010. Available: 10.1016/j.jenvp.2010.01.010.

Journal of Physics: Conference Series **1424** (2019) 012044 doi:10.1088/1742-6596/1424/1/012044

Application of Natural resource conflict management in resolving the Jakarta Bay reclamation

Elok Faiqotul Mutia, Herdis Herdiansyah[1,*]

[1]School of Environmental Science, Universitas Indonesia, Jakarta 10430, Indonesia

[*]herdis@ui.ac.id

Abstract. The reclamation of Jakarta Bay is a hot issue which lately still becomes a public debate. This debate is heated by environmental issues, marginalization of certain groups, authority regulation as well as corruption in this reclamation project. Some central and provincial government institutions take different stances on the issue. The objective of this research is to describe stakeholder perception in Jakarta Bay Reclamation. This research uses a mix-method approach which is quantitative and qualitative. Data is obtained from in-depth interviews with stakeholders who are 3 experts from central government and 3 experts from the provincial government. Furthermore, the data analysis is conducted with AHP to find out the perception of stakeholders about the causes of the complexity of the Jakarta Bay reclamation conflict. The results of the analysis show that environmental damage issues are not the first concern for the stakeholder. Central government considers a conflict of Jakarta Reclamation is mainly caused by the existing laws and regulations (27.6%). In contrast, the provincial government considers the main cause of the Jakarta Bay reclamation conflict rests on the management authority of the Jakarta Bay reclamation (28.7%).

1. Introduction

The reclamation of the Jakarta Bay is a hot issue which lately still becomes a public debate. The reclamation activity is based on the issuance of Presidential Decree no. 52 of 1995, on the Jakarta North Coast Reclamation. Although the presidential decree came out more than 20 years ago, the reclamation has not made any considerable progress. Rapid progress has only been seen since 2015 or the governance era of Basuki Tjahaja Purnama. The reclamation activities based on planning documents include the construction of 17 islands stretching from the eastern end of Jakarta bay in the Babelan region to the western end of Jakarta bay in the Kamal region. At the same time, the structuring of the coastal area of Jakarta as an integrated plan to create a flood-and-slum-areas-free Jakarta also takes place.

One of the issues in public debate about Jakarta Bay Reclamation is the impact it has. Research has been done, from academics and the government. Many studies that assess Jakarta Bay Reclamation will be bad. Jakarta Bay as a coastal area has a very diverse and interdependent ecosystem. One of the ecosystems that are threatened is the mangrove ecosystem. Mangrove forests in the Jakarta Bay play an important role in maintaining ecological balance[1]. Jakarta Bay reclamation adds pressure to water quality indicated by decreasing the water transparency, salinity and the value of diversity index of

phytoplankton and macrozoobenthic. It will eliminate the fishing area of a set net (fixed fishing gear), stationary lift net and green mussel trap nets, resulting in fish production loss[2]. When there is damage to the ecosystem, not only damage the environment but also resulted in decreased incomes of surrounding communities. A total value of estimated losses arising from the activities of Jakarta Bay reclamation perceived by fishermen and cultivators is Rp. 207.153.292.610, - / year, which is divided into the estimated value of loss for fishermen of Rp. 94.714.228.734, - / year, green shellfish cultivators Rp. 98.867.000.591, - / year and for pond farmers Rp. 13.572.063, - / year[3].
However, there are some studies assess that Jakarta Bay Reclamation does not significantly even not affect the current state. However, an increase in tidal current exists of about 300% to 600% in the future canals between those new islands. The tidals may result in erosive sea currents, which can be mitigated by lining the embankment of the future canals and present coastline. The construction of rubble mound revetment may avoid the occurrence of erosion. The reclamation and the existence of the gate do not pose a significant risk of bed profile change, and they cause sedimentation only of the order of cm in a year [4]. Besides, the coastal reclamation plan does not change the current pattern at Marina Ancol Beach, since primarily the current is generated by other factors (beyond all forms of artificial coastal processes such as coastal reclamation) such as wind, tidal, and density differences. The result of the model approach before coastal reclamation shows when the current tide moves from Northeast - West at a speed of 0.123 m / s and when the current tide moves from the West - Northeast at a rate of 0.005 m / s. As for the result of the model approach after coastal reclamation, when the current tide moves from the Northeast - West at a speed of 0.130 m / s and when the current tide moves from the West - Northeast with a speed of 0.004 m /s[5]. This different research becomes the basis for the argument for stakeholders in stating the attitude in the debate of Jakarta Bay Reclamation. In this study, it is esential to know the position of environmental damage issues in policy making of each stakeholder in the Jakarta Bay Reclamation

Conflict is a relationship between two or more parties (individuals or groups) who have or feel they have unaligned objectives [6]. Further, it is explained that conflicts arise because of an imbalance between relations between parties, including social inequalities, inequitable prosperity, differences in access to resources, discrimination and so on[7]. According to Mitchell et al., the roots of arising conflict include four aspects, namely: differences in knowledge and understanding, differences in values, differences in interests, as well as personal issues or historical background[8]. In this research, it focuses on resources conflict. Resource conflicts often occur due to differences in perception or interpretation of the resource is manifested in their attitude [9]. Conflict over natural resources can arise at various levels, from local, regional, to global. Conflicts can also occur between these levels. Conflicts occurring primarily in local contexts can extend the traditional, regional and worldwide l levels due to their specific legal relevance[10]. On the global level, resource conflict caused by gaps in demand and supply[11], which makes resources value increased and attract many actors to be involved. Not only gaps demand and supply, inequality access to resources can also lead to conflict[12]. In this research, resource conflicts occur due to high-value resources and different perceptions and interests of stakeholders at Jakarta Bay Reclamation.

The concept of natural resources is a comprehensive concept, so it has many categories and definitions from various disciplines. Most categorize natural resources into two, renewable resources and non-renewable resources[13]. Both of them requires a compatible management approach to avoid conflict[14]. In third countries, natural resource management is usually carried out by the Government, but this often leads to disputes in the absence of community access rights [15]. Furthermore, the government-centered administration has the risk of abuse of authority and corruption[16]. Several things need to be considered in the management of natural resources in the third country. First, analyses the actors involved, feel ownership and interest in the natural resources[17]. Second, regulation and authority to manage natural resources. Many countries have to implement decentralization policy for natural resources management[18]. Third, participation and involvement of the community in natural resources management, especially community that has relation interdependent on these resources[19].

In this Research, will analyses 6 stakeholders that involved in Jakarta Reclamation Bay. It will show the primary regulation of their statement and authority.

2. Method

The objective of this research is to describe stakeholder perception in Jakarta Bay Reclamation. This research uses a mix-method approach which is quantitative and qualitative. Data is obtained from in-depth interviews with stakeholders who are 3 experts from central government and 3 experts from the provincial government. Further, data analysis is done with the Analytic Hierarchy Process (AHP) to find out the stakeholder perceptions about the causes of the complexity of the Jakarta Bay reclamation conflict. AHP (Analytic Hierarchy Process) is a general theory of measurement used to find the ratio scale, either from discrete or continuous pairwise comparisons. AHP describes complex multi-factor or multi-criteria problems into a hierarchy[20]. With a hierarchy, a complex problem can be broken down into groups that are then organized into a hierarchical form so that the problem will seem more structured and systematic. Criteria are derived from an in-depth interview with stakeholder and literature study results such as media analysis. There are 8 criteria to be assessed by stakeholders (Figure 1). This assessment will show the comparison of causes that have the greatest to least impact on the current conflict. In addition, the comparison will also be drawn between the perceptions of central government and local government.

FIGURE 1. The hierarchy of Jakarta Bay reclamation conflict

3. Results and discussion

The analysis result of the AHP 3 participant from the central government, as shown in Table 1, indicates that the Jakarta Reclamation conflict is mainly due to the existing laws and regulations (27.6%). While the least influence is the power abuse (3.1%). This result is based on a consistency ratio of 5.3%. The central government's concern about the environmental damage and social problems as a source of conflict ranks 3 and 4.

TABLE 1. Central Government Result

No.	**Criteria**	**Weights**	**Ranking**
1	Environmental Damage	14.0%	3
2	Authority of Management	27.0%	2
3	Law and Regulation	27.6%	1
4	Power Lobbying	7.9%	5
5	Corruption	4.1%	7
6	Marginalization of Fisherman	8.9%	4
7	Economical Lobbying	7.5%	6

8	Power Abuse	3.1%	8

The analysis result of AHP 3 participant from the provincial government, as shown in Table 2, indicates that the Jakarta Bay Reclamation conflict rests heavily on the management authority of Jakarta Bay reclamation (28.7%). While the weakest reason is corruption. This result is based on a consistency ratio of 2.2%. The provincial government's concern about the environmental damage and social problems as a source of conflict ranks 5 and 6.

TABLE 2. Provincial Government Result

No.	Criteria	Weights	Ranking
1	Environmental Damage	7.5%	5
2	Authority of Management	28.7%	1
3	Law and Regulation	17.9%	2
4	Power Lobbying	15.9%	3
5	Corruption	4.9%	8
6	Marginalization of Fisherman	6.5%	6
7	Economical Lobbying	13.2%	4
8	Power Abuse	5.4%	7

Based on the above analysis, it is clear that the central and local governments have different perceptions about the causes of Jakarta Bay reclamation conflict. The central government considers the law and regulation (27.6%) to be the leading cause due to different legal basis used by the central and regional governments in viewing the Jakarta Bay reclamation project such as the Presidential Decree, the Ministerial Decree, the Regional Regulation and the overlapping Governor Regulation. Provincial governments are also aware that laws and regulations trigger the conflict, but they prioritize issues of management authority (28.7%) in the Jakarta Bay reclamation. For them, the legal and regulatory basis they use is so strong that this conflict is caused by the seizure of the Jakarta Bay reclamation authority. The central government places environmental (14%) and social issues (8.9%) in the third and the fourth cause respectively. This then reinforces the Ministry of Environment and Forestry's reason for conducting a moratorium on several Jakarta Bay reclamation islands and requests for full strategic environmental assessment. The Ministry of Marine Affairs and Fisheries firmly declared its rejection of the Jakarta Bay reclamation. In addition to their study of the adverse impacts that will occur on the coastal areas, the fate of fishermen in affected areas is also the primary concern of the Ministry of Marine Affairs and Fisheries. But for the provincial government, the environmental and social issues only rank 5 (7.5%) and 6 (6.5%). The results of in-depth interviews indicate that this is based on their beliefs that have accommodated the welfare of fishermen and conducted various studies to minimize environmental problems such as AMDAL and KLHS.

From the analysis results can show that the conflicts that occur in the Jakarta Bay Reclamation are a conflict of resource management. The Jakarta Bay Reclamation is a project involving many stakeholders and high economic value. This certainly makes these resources into many interests involved. The management authority of Jakarta Bay Reclamation became one of the most potent causes of conflict. The central and provincial governments use regulations that strengthen the authority over the Jakarta Bay Reclamation and ignore other weakening regulations. On this basis then each government has different attitudes and policies on issues that exist within the Jakarta Bay reclamation, such as environmental damage and fishermen's welfare. This conflict continues and focuses on the authority and regulation then ignore the environmental problems and the well-being of fishermen.

This conflict of "owning resources" also takes place in Manila, where the solar city reclamation project will be built in Manila Bay. This creates conflict because of the many stakeholders and interests involved. This project also became a public debate because of the threatened national heritage issue. Such conflicts are common in countries that still use the most important sources of economic resources such as third world countries. Projects that affect resources will involve multiple stakeholders and interests. However, not infrequently these interests focus on profit and ignore the sustainability of the supply itself. Conflicts involving many stakeholders and interest can be minimized by disclosure of information. This can reduce suspicion and equate perception among stakeholders. Information sharing can improve transparency, build trust, solve the fact problems, and distinguish them from eing issues. Previous studies conducted by the authors (Asteria and Herdiansyah) show that community engagement is essential in solving difficult environmental problems [21]. The process of conducting research and information sharing can also serve as a valuable tool for identifying and engaging multiple stakeholders. Information sharing and stakeholder identification represent the beginning of the conflict resolution process. Building trust and understanding together can begin with less controversial peripheral issues

4. Conclusion

This research concludes that the complexity of conflicts between central and provincial governments in the Jakarta Bay reclamation is the difference of interest seen in the agreement of natural resource management authority. This conflict continues and focuses on the authority and regulation then ignores the environmental problems and the welfare of fishermen. Disputes involving many stakeholders and interest like Jakarta Bay Reclamation can be minimized by disclosure of information.

Acknowledgment

This research is funded by the Grant of Indexed International Publication for Final Project of Students/*Publikasi Terindeks Internasional Untuk Tugas Akhir Mahasiswa* (PITTA) Universitas Indonesia 2018 with contract number 2579/UN.R3.1/HKP.05.00/2018.

References

1. Alikodra, H. S. Media Konservasi, 5(1) (1996).
2. Puspasari, R., Hartati, S. T., & Anggawangsa, R. F. Jurnal Kebijakan Perikanan Indonesia, 9(2), 85-94. (2018)
3. Ramadhan, A., Firdaus, M., Wijaya, R. A., & Muliawan, I. Jurnal Sosial Ekonomi Kelautan dan Perikanan, 11(1), 1-11. (2016).
4. Ajiwibowo, H., & Pratama, M. B. International Journal of Engineering and Technology, 9, 3763-3774. (2017).
5. Yuliasari, D., Zainuri, M., & Sugianto, D. Buletin Oseanografi Marina, 1(5), 1-9. (2012)
6. Fisher, S.. Zed books. (2000).
7. Stewart, F. (Ed.). Springer. (2016).
8. Mitchell, B., Rahmi, D. H., & Setiawan, B. Gadjah Mada University Press. (2003)
9. Cheng, A. S., Kruger, L. E., & Daniels, S. E. Society &Natural Resources, 16(2), 87-104. . (2003).
10. Talaue-McManus, L., Yambao, A. C., Salmo III, S. G., & Alino, P. M.. International Development Research Centre, 151-161. (1999).
11. Giordano, M. F., Giordano, M. A., & Wolf, A. T. Journal of Peace Research, 42(1), 47-65. (2005)
12. Homer-Dixon, T. F. 19(1), 5-40. (1994).
13. Grimble, R., & Wellard, K... Agricultural systems, 55(2), 173-193. (1997)

14. Reed, M. S., Graves, A., Dandy, N., Posthumus, H., Hubacek, K., Morris, J.& Stringer, L. C.. Journal of environmental management, 90(5), 1933-1949. (2009)
15. Heltberg, R.. Journal of Economic Surveys, 16(2), 189-214.(2002)
16. Le Billon, P.. Political geography, 20(5), 561-584. (2001).
17. Marfo, E. Powerful relations: the role of actor-empowerment in the management of natural resource conflicts: a case of forest conflicts in Ghana. (2006).
18. Ribot, J. C. Public administration and development, 23(1), 53-65. (2003)
19. Shackleton, S., Campbell, B., Wollenberg, E., & Edmunds, D. Natural resource perspectives, 76(1), 1-6. (2002).
20. Mu, E., & Pereyra-Rojas, M.. Springer, Cham. (2018)
21. Donna Asteria et al IOP Conf. Ser.: Earth Environ. Sci. **126,** 012146 (2018)

A Path Analysis: Study of Correlation Between Noise Intensity and Quality of Life of Workers In Textile Industry

Sumardiyono[1], Ari Probandari[1], Hartono[2], Reni Wijayanti[3]

[1]Department of Public Health, Faculty of Medicine, Universitas Sebelas Maret, Surakarta, Indonesia
[2]Department of Physiology, Faculty of Medicine, Universitas Sebelas Maret, Surakarta, Indonesia
[3]Occupational Health and Safety Program, Faculty of Medicine, Universitas Sebelas Maret, Surakarta, Indonesia

*sumardiyono@staff.uns.ac.id

Abstract. Noise intensity is a dangerous stressor in the working environment of the textile industry that can have an impact on the decrease in hearing loss of workers. The decrease in hearing loss will disrupt the workers' activities in their daily work. Textile industry workers exposed to noise while working every day will potentially be affected by a decrease in hearing loss. Activities that are constantly exposed to noise intensity can also result in a decrease in the quality of workers' life. This research was conducted to prove the correlation between noise intensity and quality of life through hearing loss using path analysis. Path analysis can explain the correlation between variables indirectly. The research method was used namely analytic survey with the cross-sectional design. The population of the study was workers in 5 textile industries in Central Java Province, Indonesia. The number of subjects was determined based on purposive sampling, each industry was 40 people thus, and the total number was 200 people. The correlation was included independent variables namely noise intensity, work period and age; the mediator variable was a loss of hearing loss, and the dependent variable was quality of life. Path analysis was used as the statistical analysis with Stata 13 software at a significance level of 5%. The research results showed that there was a direct correlation between hearing loss and decreased the quality of life of workers ($p = 0.026$). The decrease in the quality of workers' life was also influenced indirectly by noise intensity ($p = 0.000$) and age (0.012). The working period did not affect the decrease in quality of life ($p = 0.210$). Increased noise intensity can reduce hearing loss that decreases the quality of life.

1. Introduction

Noise intensity is a dangerous stressor that can cause health problems, including hearing loss(1). Long-term noise intensity can also cause stress (2), hypertension (3), diabetes mellitus (4), and decreased in quality of life(5). Noise intensity in the textile industry is interesting to study, especially in developing countries including Indonesia, because there are still many industries that use old models of engines with high noise levels.

Declining quality of life due to exposure to noise intensity will affect prosperity. The World Health Organization (WHO) defines quality of life as "individual perceptions of their position in life-based on the cultural context and the value system in which they live and it's related to their goals, hopes, standards, and concerns" (6). The quality of life concept is related to social, environmental, psychological and physical values (7).

Kim et al.'s research found a correlation between the daily lives of chronic stroke patients with quality of life, especially dimensions of social mobility and cognition (8). Serra et al. found a correlation between 6-minutes walking test and quality of life, but there was no correlation between the dimensions of emotional role, mental health, and vitality (9). Hong et al. found that quality of life was significantly reduced in the group of spinal osteoarthritis symptoms (10). In Indonesia, Rahmani and Andean found a correlation between 6WMT activity and quality of life in heart failure outpatients in General Hospital of Dr. Hasan Sadikin Bandung (11).

Previous researches described the quality of life more widely studied in patients and the elderly. We see that research on the quality of workers life that are exposed to hazardous factors from the work environment such as noise intensity has not been conducted. Nada et al. found that in the textile industry produce high noise levels that cause hearing loss for workers, which can disturb the quality of workers life (1).

This research was conducted by observing noise intensity as a hazardous risk factor for the labor force that is different from previous researches, and it analyzed using path analysis to find the correlation between noise intensity, hearing loss and quality of life of textile industry workers.

This research is useful to be a reference for noise intensity control in the working environment of the textile industry to improve the health status of workers.

2. Material and Methods

Research Design and Sampling

Design of the study was a cross-sectional study and sampling technique was used namely proportional purposive sampling. The population was workers in 5 textile industries in Central Java, Indonesia. Inclusion criteria were female gender, minimum work period was 1 year, and the minimum age was 20 years. The number of subjects was determined using the correlative analysis research formula (12). In this study determined $Z\alpha$ (5%) = 1.96, $Z\beta$ (10%) = 1.28. The value of r was determined based on the preliminary survey (r = 0.227). Calculations using online software from the Clinical & Translational Science Institute (13), obtained n = 200 people, taken from each company as much as 40 people.

Three instruments were used in this study, the first was the demographic characteristics of workers questionnaire, the second was the Short-Form Health Survey (SF-36) to assess the quality of workers' life that had been validated in the general population and various subpopulations(14)(15), and third was pure tone audiometry to measure hearing loss.

Measurement of noise intensity

Noise measurement used a sound level meter, GM1356 Sanfix type, which results were categorized into categories> 85 dBA and ≤ 85 dBA.

Measurement of age and work period

Age and work period are calculated in years using demographic data sheets. The age group is based on an average was 42 years, thus the age group was categorized as age > 42 years and ≤ 42 years. The

Journal of Physics: Conference Series **1424** (2019) 012045 doi:10.1088/1742-6596/1424/1/012045

average working period was 17 years, thus the working period is grouped into >17 years and ≤ 17 years.

Measurement of hearing loss
The level of hearing loss was measured using pure tone Audiometry towards both ears at frequencies of 500 Hz, 1000 Hz, 2000 Hz, and 4000 Hz, then the average was calculated (16)·(17). Measurements are made before the subject works (06.00-08.00). The average hearing loss of both ears = (monaural in the better ear x 5 + monaural in the worse ear x 1) / 6. The American Speech-Language-Hearing Association (ASHA) provides categories of hearing loss to Normal (-10 to 15 dB), slight (16 to 25 dB), mild (26 to 40 dB), moderate (41 to 55 dB), moderately severe (56 to 70), severe (56 to 90), and profound (91 dB or more). In this study, hearing loss was categorized as having hearing loss (≥ 26 dBA) and normal (<26 dBA) (18).

Measurement of Quality of Life
The category of quality of life according to Ware et al, a score of 100 is perfect, a score of 84 to less than 100 is very good, a score of 61 to less than 84 is good, a score of 25 to less than 61 is less good, and a score of 0 to less than 25 is not good. This study categorizes the quality of life as being Good with a score of ≥ 61 and less good with a score of <61 (15).

Statistical analysis
The descriptive analysis presents cross tables. Bivariate correlation analysis used Chi-Square test. Path Analysis using the Stata 13 Program analyzed the correlation between noise intensity, hearing loss and quality of life.

Ethical Clearance
Before the subjects participated, they have filled out informed consent. Subject information is kept confidential. The Health Research Ethics Commission, Faculty of Medicine, Sebelas Maret University issued an ethical clearance, by letter No. 63 / UN27.6 / KEPK / 2018, protocol number: 01/18/04/059 dated April 24, 2018.

3. Results
Information of Social-demographic Respondents
All subjects were women, Indonesians, and aged between 25 to 50 years with an average age of 42 years. Body mass index between 18.7 kg / m2 to 29.9 kg / m2 with a mean of 24.1 kg / m2. Subjects did not have a smoking habit (100%) and did not have a drinking alcohol habit (100%).

Expose noise intensity
The highest noise intensity was 104 dBA in the weaving section and the lowest was 64 dBA in the warehouse section. The threshold value standard is 85 dBA exposure in 8 working hours of a workday. The hearing protection device used by the subject was cotton. According to the Occupational Safety and Health Administration (OSHA), United States Department of Labor, cotton is not a good hearing protection device.

Bivariate statistics description of correlation between variable directly
The correlation between the variables of noise intensity, age, and work period and hearing loss through bivariate was presented in table 1.

Table 1. Correlation between noise intensity, age, and work period and hearing loss (n = 200).

Variable	Hearing loss						X^2	p	OR	CI95%
	Deaf		Normal		Total					
	n	%	n	%	N	%				
Noise intensity										
> 85 dBA	94	71,2	38	28,8	132	100	11,232	0,001	2,78	1,52 – 5,12
≤ 85 dBA	32	47,1	36	52,9	68	100				
Age (years)										
> 42 years	57	73,1	21	26,9	78	100	5,570	0,018	2,08	1,13 – 3,86
≤ 42 years	69	56,6	53	43,4	122	100				
Work period (years)										
> 17 years	78	71,6	31	28,4	109	100	7,530	0,006	2,25	1,25 – 4,05
≤ 17 years	48	52,7	43	47,3	91	100				

The correlation between the variables of hearing loss and quality of life through bivariate was presented in table 2.

Table 2. Correlation between the variables of hearing loss and quality of life (n = 200).

Variable	Quality of Life						X^2	p	OR	CI95%
	Less Good		Good		Total					
	n	%	n	%	N	%				
Hearing loss										
Deaf	70	55,6	56	44,4	126	100	4,996	0,025	1,94	1,08 – 3,48
Normal	29	39,2	45	60,8	74	100				

Path analysis

The correlation between variables and test results is presented in Figure 1.

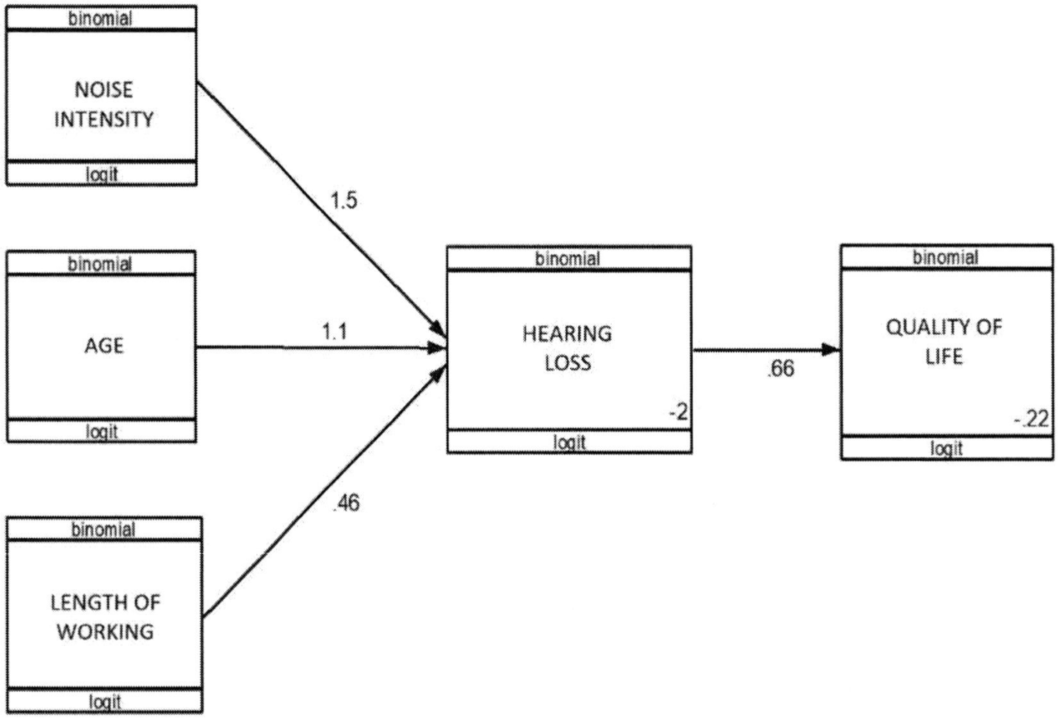

Figure 1. Correlation diagram between variables in the Path Analysis

The results of the correlation test between variables using Software Stata 13 is presented in table 3.

Table 3. Path Analysis Results of the correlation between Noise intensity, Hearing Loss and Quality of Life (n = 200)

Dependent Variable	Independent Variable	Cons	b	95%CI	p
Direct Effect		-0,22			
Life quality	← Hearing loss		0,66	0,08 – 1,25	0,026
Indirect Effect		-1,99			
Hearing loss	← Noise intensity		1,52	0,78 – 2,25	0,000
Hearing loss	← Age		1,09	0,25 – 1,94	0,012
Hearing loss	← Work period		0,46	-0,25 – 1,17	0,210
N observation = 200					
df= 6					
AIC= 520.52					
BIC= 540.31					

Discussion

This study revealed that there was a negative correlation between hearing loss and the quality of life of textile industry workers directly. Noise intensity and age indirectly reduced the quality of life of workers, while the work period did not correlate with quality of life. Thus, the stressors causing the decline in quality of life in textile industry workers were the intensity of noise intensity and age. The impact of noise intensity exposure on health has been previously studied by Afiah et al. who found that the subjects had dissatisfaction with knowledge, attitudes, and practices against the occurrence of hearing loss due to noise intensity, then appropriate action is needed to identify workers at risk of hearing loss due to noise intensity (19).

Hearing loss has been also studied in the mining industry, which was found that in the processing department (94 dBA), underground mining (102 dBA), and underground workshops (103 dBA) can cause hearing loss was 36.7% of workers; and will increase as workers age(20). In the tile and ceramics industry, the prevalence of hearing loss was 44% (21). In India textile industry, hearing loss occurred in 39% of workers who were exposed to noise intensity exceeding 87.3 dBA for 8-12 hours a day(22). Research in Malaysia found that hearing impairment costs were a very high financing burden, entrepreneurs need extra effort to carry out hearing conservation program in the workplace and the Malaysian government must enforce regulations effectively to reduce noise intensity exposure in the industry (23).

Hearing loss and quality of life have been studied by Carlsson et al., which found severe and very severe patient of hearing loss had higher anxiety and depression than general patients. Patients who had hearing loss were negatively related to the quality of life of patients (24). Hearing loss due to noise intensity ultimately causes a decrease in the quality of life of agricultural workers (25). The results of different studies stated that there was no correlation between hearing loss and quality of life with HIV patients (26).

Our research found that there was a significant negative correlation between age and quality of life of textile industry workers. This study is in line with Campos et al's study who found that adults have a good quality of life if they maintain personal health and do not have depression, whereas in women with poor physical activity tend to have a poor quality of life ($p = 0.022$) (27). Thus, increasing age, quality of life can also be maintained well by maintaining personal health, do not get depression and doing good physical activity. This is in line with the statement of Netuveli & Blane, who found that the main factor that influences the good quality of life in old age is depression (28).

This study was not found any relationship between the workers' working period and quality of life. There have not been many comparisons with our study, but von Bonsdorff et al's study found that men with long working hours and lack of sleep, his quality of life were not good at terms of physical function, vitality, and general health. Meanwhile, men with long working hours but regular sleep time had a poor quality of life or physical function in old age (29).

4. Conclusion

Noise intensity and age increase hearing loss, and hearing loss reduces the quality of life of textile industry workers, so the entrepreneur needs to carry out the hearing conservation program.

Conflict of Interest

We declare that there was no conflict of interest in this study.

Acknowledgments

The author would like to thank the Directorate of Research and Community Service, Directorate General of Strengthening Research and Development, Ministry of Research, Technology and Higher Education of the Republic of Indonesia and Head of Research Institute and Community Service of Sebelas Maret University who has given this research project.

References

1. Nada E, Ebraheem W, Sheta S. Noise-induced hearing loss among workers in a textile factory. Egypt J Otolaryngol [Internet]. 2014;30(3):243. Available from: http://www.ejo.eg.net/text.asp?2014/30/3/243/138484
2. Burns KN, Sun K, Fobil JN, Neitzel RL. Heart rate, stress, and occupational noise exposure among electronic waste recycling workers. Int J Environ Res Public Health. 2016;13(1).

3. Zijlema W, Cai Y, Doiron D, Mbatchou S, Fortier I, Gulliver J, et al. Road traffic noise, blood pressure and heart rate: Pooled analyses of harmonized data from 88,336 participants. Environ Res [Internet]. 2016;151(March):804–13. Available from: http://dx.doi.org/10.1016/j.envres.2016.09.014
4. Zare Sakhvidi MJ, Zare Sakhvidi F, Mehrparvar AH, Forrester M, Dadvand P. Association between noise exposure and diabetes: A systematic review and meta-analysis. Environ Res [Internet]. 2018;166(January):647–57.
5. Seidman MD, Standring RT. Noise and quality of life. Int J Environ Res Public Health. 2010;7(10):3730–8.
6. World Health Forum. What quality of life? The WHOQOL Group. World Health Organization Quality of Life Assessment. Vol. 17. 1996. p. 354–6.
7. Theofilou P. Quality of life: Definition and measurement. Eur J Psychol. 2013;9(1):150–62.
8. Kim K, Kim YM, Kim EK. Correlation between the Activities of Daily Living of Stroke Patients in a Community Setting and Their Quality of Life. J Phys Ther Sci [Internet]. 2014;26(3):417–9. Available from: http://jlc.jst.go.jp/DN/JST.JSTAGE/jpts/26.417?lang=en&from=CrossRef&type=abstract
9. Serra AJ, De Carvalho PDTC, Lanza F, De Amorim Flanders C, Silva SC, Suzuki FS, et al. Correlation of six-minute walking performance with quality of life is domain- And gender-specific in healthy older adults. PLoS One. 2015;10(2):1–8.
10. Hong JY, Han K, Shin DH, Chun EM. Quality of life analysis and smoking correlation in symptomatic spine osteoarthritis: A nationwide health survey analysis of an elderly population with EQ-5D. PLoS One. 2016;11(3):1–10.
11. Rahmani S, Erwinanto E AR. Correlation between Six-Minutes ' Walk Test and Quality of Life in Heart Failure Patients. Althea Med J. 2015;2(4):469–73.
12. Hulley, S.B., Cummings, S.R., Browner, W.S., Grady, D., Newman TB. Designing clinical research: an epidemiologic approach. 4th ed. Philadelphia: Lippincott Williams & Wilkins; 2013. 79 p.
13. Kohn M, Senyak J JM. Correlation sample size [Internet]. 2014. Available from: www.sample-size.net/correlation-sample-size/
14. Ware Jr. JE, Sherbourne CD. The MOS 36-Item Short-Form Health Survey (SF-36). I. Conceptual Framework and Item Selection [Internet]. Vol. 30, Medical Care. 1992. p. 473–83. Available from: http://www.ncbi.nlm.nih.gov/entrez/query.fcgi?cmd=Retrieve&db=PubMed&dopt=Citation&list_uids=1593914
15. Ware Jr, J.E., Snow, K.K., Kosinski, M., Gandek B. F36 Health Survey: Manual and Interpretation Guide. New England Medical Center, Boston, Massachusetts: The Health Institute; 1993. 6-21 p.
16. Beck RM, Ramos BF, Grasel SS, Ramos HF, Moraes MF, Almeida ER, et al. Comparative study between pure tone audiometry and auditory steady-state responses in normal hearing subjects. Braz J Otorhinolaryngol. 2014;80:35-40. Braz J Otorhinolaryngol [Internet]. 2014;80(1):35–40. Available from: http://linkinghub.elsevier.com/retrieve/pii/S1808869414500092
17. Al-Abri R, Al-Balushi M, Kolethekkat A, Bhargava D, Al-Alwi A, Al-Bahlani H, et al. The accuracy of IOS device-based uHear as a screening tool for hearing loss: A preliminary study from the Middle East. Oman Med J. 2016;31(2):142–5.
18. American Speech-Language-Hearing Association (ASHA). The degree of Hearing Loss [Internet]. Available from: http://www.asha.org/public/hearing/Degree-of-Hearing-Loss/
19. Afiah N, Zulkefli M, Farhan Bin Rusli M, Rahman AA, Zulkefli AM. Predictors of Knowledge, Attitude and Practice of Noise-Induced Hearing Loss among Workers in an Automotive Industry in Malaysia. Mal J Med Heal Sci Malaysian J Med Heal Sci [Internet]. 2017;13(1):61–8. Available from: http://www.medic.upm.edu.my/upload/dokumen/2017050314451408_MJMHS_Jan_2017_-_0011_Predictors_of_Knowledge.pdf

20. Chadambuka A, Mususa F, Muteti S. Prevalence of noise-induced hearing loss among employees at a mining industry in Zimbabwe. Afr Health Sci. 2013;13(4):899–906.
21. Mehrparvar AH, Heidari F, Sharifabadi MS, Javad M. Prevalence and Pattern of Noise-induced Hearing Loss in Tile and Ceramic Industry. 2017;60–5.
22. Ranga R, Ranga S, Yadav S, Yadav A, Yadav N. Prevalence of occupational noise-induced hearing loss in industrial workers. Indian J Otol [Internet]. 2014;20(3):115. Available from: http://www.indianjotol.org/text.asp?2014/20/3/115/136848
23. Narita T., Aljunid SM. Direct and Indirect Cost of Noise-Induced Hearing Loss among Manufacturing Industrial Workers in Malaysia. Malaysian J Public Heal Med. 2016;16:108–9.
24. Carlsson P-I, Hjaldahl J, Magnuson A, Ternevall E, Edén M, Skagerstrand Å, et al. Severe to profound hearing impairment: quality of life, psychosocial consequences, and audiological rehabilitation. Disabil Rehabil [Internet]. 2015;37(20):1849–56. Available from: http://www.tandfonline.com/doi/full/10.3109/09638288.2014.982833
25. Sena TRR de, Vargas MM, Oliveira CC da C. Saúde auditiva e qualidade de Vida em trabalhadores exposes an agrotóxicos. Cien Saude Colet [Internet]. 2013;18(6):1753–61. Available from: http://www.scielo.br/scielo.php?script=sci_arttext&pid=S1413-81232013000600026&lng=pt&tlng=pt
26. Duong N, Torre P, Springer G, Cox C, Plankey MW. Hearing Loss and Quality of Life (QOL) among Human Immunodeficiency Virus (HIV)-Infected and Uninfected Adults. J AIDS Clin Res [Internet]. 2016;7(12). Available from: https://www.omicsonline.org/open-access/hearing-loss-and-quality-of-life-qol-among-human-immunodeficiencyvirus-hivinfected-and-uninfected-adults-2155-6113-1000645.php?aid=83660
27. Campos ACV, Ferreira EF e., Vargas AMD, Albala C. Aging, Gender and Quality of Life (AGEQOL) study: Factors associated with good quality of life in older Brazilian community-dwelling adults. Health Qual Life Outcomes. 2014;12(1):1–11.
28. Netuveli G, Blane D. Quality of life in older ages. Br Med Bull. 2008;85(1):113–26.
29. Von Bonsdorff MB, Strandberg A, Von Bonsdorff M, Törmäkangas T, Pitkälä KH, Strandberg TE. Working hours and sleep duration in midlife as determinants of health-related quality of life among older businessmen. Age Ageing. 2017;46(1):108–12.

Journal of Physics: Conference Series **1424** (2019) 012046 doi:10.1088/1742-6596/1424/1/012046

Competitive Force Model for Indopipe Industry with Analysis of Customer Requirements

Rita Ambarwati[1*]; A G Fathurochman[2]; As'ad Rizal[1]

[1]Department of Magister Management, Universitas Muhammadiyah Sidoarjo, Majapahit 666B, Sidoarjo 61215, Indonesia
[2]Department of Management Technology, Institut Teknologi 10 Nopember, Cokroaminoto 12A, Surabaya 60264, Indonesia

*ritaambarwati@umsida.ac.id

Abstract. The pipeline industry is one of the manufacturing industries that support national development infrastructure. Increased efficiency and expanded brand market share with optimizing the opportunities available on the market is one of the efforts in increasing competitive advantage. The design of this study used interviews and data collection through questionnaires. The analytical unit is the engineering expertise of the Indopipe company at all manager levels. The result of this analysis is the determination of competitive strength strategies with Porter Five Forces in the pipe industry based on criteria of quality, cost, delivery, safety, and morale. The results of the analysis of the study show that strategies to address the threat of new entrants are a priority for managers in this industry. In addition, by improving the quality of pipe products and making this product brand as a market leader, it is the main strategy in increasing competitiveness.

1. Introduction

The company's strategic decision making that has a good impact on the company is greatly supported by good strategic management implications for the company [1]. Analysis of strategic management consists of an internal environment including the strengths and weaknesses of the company and the external en-vironment including opportunities and threats for the company [2]. In order to help plan strategies for developing an early warning system to avoid threats or develop strategies that can turn threats into benefits for the company [3]. Therefore, to be able to survive in the market in developing its business, a company or industry depends on five porter powers. A company will have a competitive advantage if the company is in a favorable condition of the five competitive forces, namely the entry of new entrants, the threat of substitute products, the bargaining power of buyers, the bargaining power of suppliers, and the threat of rival companies existing competitors [4]. Ideally a company can have all the competitive strength factors, but the company must work as much as possible to be able to maximize the five Porter's strengths [5].

Competition between similar companies is very sharp, new competitors can enter the industry relatively easily, and suppliers and customers can increase their bargaining power [6]. In order to win business competition, every company must have a competitive strategy [7]. The ultimate goal of a competitive strategy is to increase environmental strength in the interests of the company [2]. Competitive strategies are often also referred to as business strategies, focus on increasing the competitive position of company products and services in certain industries or market segments [8]. The competitive strategy is an effort to find a competitive position in an industry, the fundamental arena in which competition takes place [9]. The competitive strategy aims to foster a position that is profitable and strong in fighting the forces that determine competition in the industry [2]. Therefore, the competitive strategy is not only a response to the environ-ment but also an effort to shape the environment in accordance with the wishes of the company [10]. The purpose of this study is to analyze what strategies are being implemented by the Indopipe Company, the competitive climate of the pipeline industry in Indonesia, and alternative strategies that can be developed by the current Indopipe company using the Five Forces Porter model.

2. Method

This type of research is included in descriptive qualitative research. This study uses descriptive qualitative research because it is in accordance with the objectives of the study to find out and get the depth of information related to the topic. The subject in this study is the Indopipe Company in Surabaya and the object of the research was an indicator of the 5 strengths of Porter's competition. Data collection techniques were carried out by semi-structured observation and interviews. The validity test of the data is done using the triangulation test. The source triangulation carried out in this study was interviews with different sources. In this research stage, checking the intended information is directly obtained by collecting information in the field, and will be compared with information directly from the company itself. Furthermore, based on the analysis of Porter's five forces model, a test is conducted to determine the priority strategies in the analysis with Analitycal Hierarchy Process (AHP) . The survey of Porter's priority five forces model is conducted on 8 decision-making managers at Indopipe. Furthermore, data analysis techniques are carried out by: 1) Examining all data from various sources; 2) Conduct an assessment for each of the five porter powers to find out which external factors are most influential; 3) Conducting content analysis of the results of the interviews that will be used as the basis of the analysis of five porter powers; 4) Compile the company's marketing strategy; 5) Prepare conclusions and evaluations of the research that has been done.

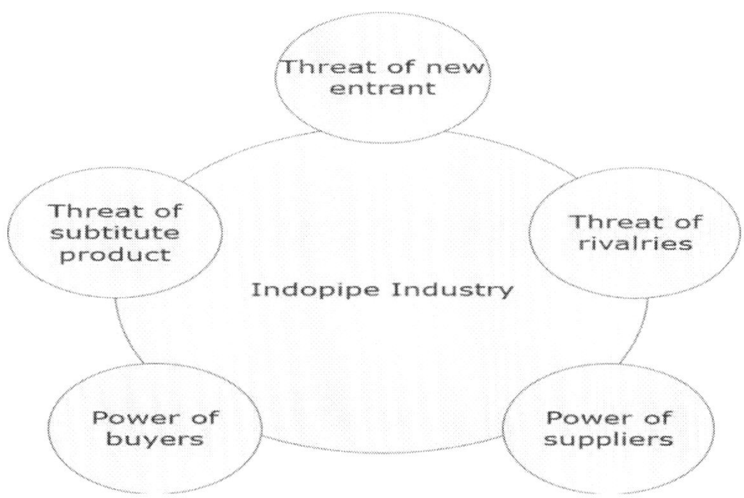

Figure 1. Competitive Five Force Model

3. Results and Discussion

Five Force Model Porter Analysis is used to analyse the company's external environment based on the threat of rivalries, the threat of new entrants, threat of substitute product, bargaining power of buyers, and bargaining power of suppliers .

Table 1. Analysis of Five Forces Model Porter

Journal of Physics: Conference Series **1424** (2019) 012046 doi:10.1088/1742-6596/1424/1/012046

No.	Indicator	Analysis	Conclusion
I.	**Analysis of the threat of Rivalries**		
1.	number of competitors	the number of competitors in the area of Surabaya and Sidoarjo is around 3-4 companies	low competition
2.	increasing number of competitors	annual increase of competitors around 1 company	Low competition
3.	annual industrial growth each year	industrial growth increased in line with increasing market demand, where market demand increased 15% and industrial growth also increased by 20%	High competition
4.	the impact of industrial growth on inflation	Industrial growth had an impact on inflation because inflation had an impact on purchasing power, so that the original sales of 15 tons to 11 tons	moderate competition
5.	product differentiation with competitors	the pipe industry in general has product differentiation reaching 3 products	High competition
6.	differentiation of raw materials	the majority of companies use the different raw materials	High competition
7.	production cost	non-fixed production costs with increases ranging from 3-5%	moderate competition
II.	**Analysis of The threat of new entrants**		
1.	Capital requirements	capital requirements to open a plumbing business are very expensive. the average capital for machines is at least 100 million with land capital, where the overall capital reaches an average of 500 million	Low threat
2.	How to meet capital requirements	sufficient capital needs through outside shareholders and sponsors. 0-10% of paid up capital, 10-15% medium, and 15-20% high	High threat
3.	Customer loyalty level	The level of loyalty depends on the needs and services provided to customers. Indopipe companies have customers above 60%.	Middle threat
4.	The formation of customer loyalty	Customer loyalty is formed from the service and good performance of the company. So far the company has been committed to always meeting and satisfying customer needs	Middle threat
5.	Access to distributiin channels	access to distribution channels using shipping fleets owned by the company itself and more than competitors	Middle threat
6.	Cost requirements to distribution channels	the costs used are relative, depending on tonnage and distribution distance	Middle threat
7.	Government policy	industrial growth is inseparable from government policy, because the government has facilitated the licensing and conditioning of the industrial sector to grow	High threat
8.	The impact of government policy	government policy has a considerable impact	High threat
III.	**Analysis of The threat of substitute products**		
1.	the level of need for a substitute product	the level of need for substitute products is not too high, where people only need up to 10%	Low threat
2.	ease of getting a substitute product	ease of obtaining a subtitute product of around 30%	Middle threat
IV.	**Analysis of The Bargaining power of buyers**		
1.	level of buyer dominance	buyers are dominated by construction projects.	bargaining power of buyers is high
2.	the dominance of buyers every year	the dominance of the buyer is not constant depending on the number of requests.	bargaining power of buyers is low

3.	the impact of product switching costs	the magnitude of the impact of product transition costs depends on the value that exists in the transition. transition costs <5 million are low, 5-10 million are moderate, and 10-20 million are high.	bargaining power of buyers is high
4.	reserve fund for product switching costs	reserve fund <5 million low, 5-10 million medium, and 10-20 million high. The company has a large reserve fund for product switching costs.	bargaining power of buyers is high
5.	level of clarity of product information	there is a sales person who informs the product to the customer.	bargaining power of buyers is high
6.	official permission from product information	product information does not have official permissions.	bargaining power of buyers is low
7.	growth in buyer market share each period	the market share growth is around 5-10%.	bargaining power of buyers is high
8.	buyers' market share every year	the buyer's market share is not constant depending on purchasing power.	bargaining power of buyers is high
V.	**Analysis of The bargaining Power of Supplier**		
1.	level of supplier dominance	suppliers are not dominated by only one supplier.	bargaining power of suppliers is low
2.	alternative supplier	the company has 1 high supplier, 2-4 medium suppliers, 4 high suppliers. Indopipe has 7-10 suppliers.	bargaining power of suppliers is low
3.	product supplier quality level	supplier product quality is good because the company has standards that must be met by suppliers.	bargaining power of suppliers is low
4.	Supplier product quality is always constant	Supplier product quality is not always constant because quality follows the purchase price.	bargaining power of suppliers is low
5.	supplier market competition level	competing suppliers prices by raising prices according to market conditions.	bargaining power of suppliers is middle
6.	supplier market competition is always neutral	supplier market competition is relative.	bargaining power of suppliers is middle
7.	the amount of the cost of switching to another supplier	switching costs depend on which supplier is used.	bargaining power of suppliers is high
8.	supplier turnover costs every year	supplier switching costs are not always constant each year	bargaining power of suppliers is high

The results of the Porter Five Force analysis show that the competitive climate of the pipe industry in Indonesia is quite low. It is known from the number of competitors in the Sidoarjo and Surabaya regions that currently only about 3-4 companies are engaged in the same field, namely the pipe industry. Furthermore, the increase in the number of the pipeline industry is also evidenced by the increase in competitors which produce around 1 to 2 companies each year. The increase in the number of the pipe industry has resulted in quite competitive competition with a marked market demand that has also increased by 10-15%. Higher competition can also be seen from the increasing number of companies that have product differentiation from other competitors. In this case, the Company has product differentiation on mixed products that can reduce production costs but still pay attention to the quality that exists, so that the company can provide benefits to consumers by providing cheaper and yet quality products [11]. For the main raw materials themselves, most of the companies do not have differentiation, so that competition in existing raw materials tends to be low. Indopipe's strategy in overcoming increasingly fierce competition among competitors is to expand demand or strengthen the level of product differentiation by improving and maintaining product quality.

In addition, the Company must also do good marketing, and receive advice from customers without changing existing quality standards. Customer loyalty is one factor that can affect the low threat of new entrants. In this case plastic companies have customers that are above 60% of total customers, so the threat of new entrants is not too high, because with customer loyalty it is expected that customers will continue to make purchases in the same company so that they will not switch to new competitors. However, to build customer loyalty is not easy because companies must have performance and provide maximum service to customers. This can be done by providing quality products, affordable prices and various forms of convenience provided to customers such as ease of ordering, ease of delivery, and ease of payment to make customers trust the company [12]. In addition, government policies will also pose a threat to the entry of new entrants, because government policies have contributed to the growth of industries such as the pipe industry. The government has a policy to facilitate licensing for industrial establishments and the government also provides support for industrial development in Indonesia, one of which is the pipe industry. This ease in industrial licensing will facilitate the entry of new entrants who can be a threat to the company.

In the threat of substitute products, Indopipe is not too worried because the level of customer needs for substitute products is not too high, which is around 10%, but in obtaining a replacement product it is still easy to find so it can be a threat. which needs to be considered by the company. Buyers in this company are dominated by construction projects, where the dominance of buyers every year is not constant depending on the number of buyer requests. In increasing bargaining power, company buyers can also provide clear information about the products they have. Pipeline companies have a sales team to offer and provide information about the company's products to customers, customers can ask directly about information that is clearly needed. The growth of buyers' market share can also increase the bargaining power of buyers, where market share growth has so far reached 10% per period, only the buyer's market share is not constant and depends on people's purchasing power. Adequacy of raw material requirements in the company is not dominated by one supplier. Indopipe has 7-10 suppliers, because if one supplier is in trouble, the company can use other suppliers and the production process can continue smoothly. The quality of products from suppliers is not always constant, this is because following the purchase price, the lower of selling price, the lower of quality [11]. However, the Company has suppliers that have quality raw materials, because the company already has standards that must be met by suppliers to be able to maintain the quality of the pipe products.

Table 2. Priority of Five Forces Model Porter

Porter 5 Forces	T. of New Entrants	B. Power of Suppliers	B. Power of Buyers	T. of Substitute Products	Threat of Rivalries	Priority Vector
Threat of New Entrants	0,40	0,44	0,43	0,36	0,36	0,40
Bargaining Power of Suppliers	0,15	0,16	0,23	0,21	0,13	0,18
Bargaining Power of Buyers	0,13	0,10	0,14	0,17	0,20	0,15
Threat of Substitute Products	0,15	0,11	0,11	0,14	0,16	0,13
Threat of Rivalries	0,17	0,18	0,10	0,13	0,15	0,14
SUM	1,00	1,00	1,00	1,00	1,00	1,00

In table 2, based on the results of data analysis, prioritizing five forces porter models shows that the threat of new entrants has the highest weight, which means that this threat for managers of Indopipe is considered as one of the most important problems that requires the right strategy to overcome the threat [13]. Furthermore, bargaining power of buyers occupies the second priority, it can be explained that this is an important factor in determining the strategy in the pipeline industry competition [14]. The threat of product subtitute, the threat of rivalries and the bargaining power of suppliers sequentially occupy a priority level based on Indopipe managers' perceptions.

4. Conclusion

The current strategy implemented by Indopipe based on a five force model analysis is to increase product differentiation; cheaper prices than competitors; better delivery system; easier payment system. The climate of industrial competition tends to increase due to the number of companies in the industry similar types in Sidoarjo and Surabaya as many as 3-4 companies, where there is an increase in competitors per year between 1-2 industries both small and large scale accompanied by market demand which also continues to increase by 10-15%. An alternative strategy that can be developed by Indopipe is the addition of capital to develop the business; add machine production tools; training program for employees to develop abilities and skills. The priority of the five forces porter model based on Indopipe's manager's perception is threat of new entrants. There needs to be an evaluation of the strategy that has been determined by the company so that it can excel in business competition

5. Acknowledgements

We hereby thank you to RISTEKDIKTI and Universitas Muhammadiyah Sidoarjo for supporting the publication of this research

References

[1] D. P. Pamungkas, "Analisis Competitive force dan Competitive Sistem Informasi Kuliner di Indonesia (Studi Kasus: Kulina.id)," *J. Electron. Informatics, Vocat. Educ.*, vol. 1, no. 2, pp. 118–127, 2016.

[2] P. H. K. Ho, "Analysis of Competitive Environments, Business Strategies, and Performance in Hong Kong's Construction Industry," *J. Manag. Eng.*, 2015.

[3] M. Jeremiah, B. Kabeyi, C. Moses, J. Barasa Kabeyi, and M. Kabeyi, *Michael porter's five competitive forces and generic strategies, market segmentation strategy and case study of competition in global smartphone manufacturing industry*. 2018.

[4] O. Nurlansa, "Analysis Porter's Five forces Model on Airbnb," *J. Electron. Informatics, Vocat. Educ.*, vol. 1, no. 2, pp. 83–94, 2016.

[5] G. E. Indiatsy, C.M.; Mwangi, M.S.; Mandere, E.N.; Bichanga, J.M.; George, "The Application of Porter's Five Forces Model on Organization Performance: A Case of Cooperative Bank of Kenya Ltd," *Eur. J. Bus. Manag.*, vol. 6, no. 16, pp. 75–85, 2014.

[6] A. Pratap, "Apple Inc. Five Forces Analysis (Porter's Model)," *Panmore Institute*, 2017. .

[7] J. Dedrick and K. Kraemer, "Intangible assets and value capture in value chains: the smartphone industry," *World Intellect. Prop. Organ.*, 2016.

[8] M. L. L. S. F. Johnson, "The five competitive forces framework in a technology mediated environment : do these forces still hold in the industry of the 21st century?," *3rd IBA Bachelor Thesis Conf.*, 2014.

[9] T. P. Pervan, Maja; Curak, Marijana; Kramaric, "The Influence of Industry Characteristics and Dynamic Capabilities on Firms' Profitability," *Int. J. Financ. Stud.*, 2017.

[10] H. J. Sirajuddin Omsa, Ibrahim H. Abdullah, "Five Competitive Forces Model and the Implementation of Porter's Generic Strategies to Gain Firm Performances," *Sci. J. Bus. Manag.*, 2017.

[11] J. G. Khan, R. S. Dalu, and S. S. Gadekar, "DEFECTS IN EXTRUSION PROCESS AND THEIR IMPACT ON PRODUCT QUALITY," *Int. J. Mech. Eng. Rob. Res*, 2014.

[12] E. P. Stringham, J. K. Miller, and J. R. Clark, "Overcoming Barriers to Entry in an Established Industry: Tesla Motors," *Calif. Manage. Rev.*, 2015.

[13] A. Mantovani, O. Tarola, and C. Vergari, "End-of-pipe or cleaner production? How to go green in presence of income inequality and pro-environmental behavior," *J. Clean. Prod.*, 2017.

[14] D. S. Shim, K. P. Kim, and K. Y. Lee, "Double-stage forming using critical pre-bending radius in roll bending of pipe with rectangular cross-section," *J. Mater. Process. Technol.*, 2016.

Workload Analysis by Using Nordic Body Map, Borg RPE and NIOSH Manual Lifting Equation Analyses: a Case Study in Sheet Metal Industry

Jacky Chin[1], Herlina[2], Hardianto Iridiastadi[3], Lin Shu-Chiang[4], Satria Fadil Persada[5]

[1]Department of Industrial Engineering, Mercu Buana University, Jakarta, Indonesia
jacky.chin@mercubuana.ac.id
[2]Department of Industrial Engineering, Universal University, Batam, Indonesia
[3]Department of Industrial Engineering, Institut Teknologi Bandung (ITB), Bandung, Indonesia.
[4]Department of Business Management, Texas Health and Science University, Texas, USA.
[5]Department of Business Management, Institut Teknologi Sepuluh Nopember (ITS), Surabaya

*jacky.chin@mercubuana.ac.id

Abstract. The purposes of this research were to identify the lifting complaints experienced by the painting operators in the sheet metal industry at west java, Indonesia. The lifting issues such as operators' lifting effort and risk were evaluated by combining the Nordic Body Map, the Borg Rating of Perceived Exertion and the NIOSH manual lifting equation tool. Specifically, two questionnaires that consist of Nordic Body Map and Borg Rating of Perceived Exertion and field observation were used as the research methodologies to evaluate respondents' lifting process when loading the panel box, where its body part, door part and base plate were examined. All of the 44 full-time operators in the painting department of the sheet metal company participated in this study. Their age ranged from 20 to 40 years old. Five complaints were identified by Nordic Body Map questionnaire. These complaints were pain, discomfort, and tingling at operators' neck, shoulders, and hips. Finding from Borg Rating of Perceived Exertion questionnaire revealed that the effort needed to lift the panel body parts were substantially higher than door parts and base plate. A new manual conveyor was then installed and the study's recommendations were proven to be successful by reducing the 1991 NIOSH lifting index value from 1.65 to 0.91 when lifting panel body part, from 1.3 to 0.72 when lifting panel door part, and from 0.93 to 0.51 when lifting panel base plate. The value of the Borg Rating was also decreased from 13.79 to 11.96 for panel body part, from 9.36 to 8.82 for panel door part, and from 8.82 to 8.25 for panel base plate, and the lifting complaints of the painting operators were reduced.

Keywords: Lifting, Nordic Body Map, BORG Rating of Perceived Exertion, NIOSH manual lifting equation.

1. Introduction

Indonesia, as one of the developing countries, requires a sufficient supply of electricity to be distributed to the manufacturing factories. In line with this, of course, it takes an enormous amount of electrical panels containing electrical components. The electrical panels are made of sheet metal material. There are several steps to be able to make the electrical panel. First, sheet metal is cut by means of punching and bending in order to get the box form. Once bent, then every corner of the parts connected by means of welding. Painting is the next step, where the system of painting is done using powder coating. After all parts painted finish, the final step is to assemble the panel. In the painting process, there are several steps that must be done, ie the process of pre-treatment, drying 1, powder coating process, ends with the drying 2 process.

The placement and position of the panel's parts are very essential, especially when the operators need to move them from one station to the other station. Operators need to maintain their body posture in order to lift the panel's parts. NIOSH Recommended Weight Limit (RWL) equation and Lifting Index (LI) are based on the concept that the risk of lifting related low back pain increases as the demands of the lifting task increase [1]. The equation is widely accepted and used throughout industry in setting acceptable lift limits for workers.

Manual handling and lifting are a major cause of work-related Low Back Pain (LBP) and impairment [2]. LBP also can occur by direct trauma, a single exertion or potentially as the result of multiple exertions ('repetitive trauma') [3]. Painting operators must lift parts from the first drying oven (static oven) to the powder coating process. Based on these observations, this study aimed to analyze the complaints experienced by the operators in the department of painting and the risk of their body. Hopefully this research can provide inputs for companies engaged in sheet metal fabrication to improve the efficiency of the operator.

More than thirty years ago, the National Institute for Occupational Safety and Health (NIOSH) recognized the growing problem of work related back injuries and published the Work Practices Guide (WPG) for Manual Lifting [4]. The NIOSH WPG (1981) contained a summary of the lifting-related literature before 1981; analytical procedures and a lifting equation for calculating a recommended weight for specified two-handed, symmetrical lifting tasks; and an approach for controlling the hazards of low back injury from manual handling [1].

On 1991, the revised lifting equation reflects new findings and delivers methods for evaluating asymmetrical lifting tasks, and lifts of objects with less than optimal couplings between the object and the workers' hands. The revised lifting equation also provides guidelines for a more diverse range of lifting tasks than the other equation [2].

About 1993, Waters et al suggest a new revised lifting equation [2]. The recommended weight limits derived from the revised equation are consistent with, or lower than the generally reported in the letter. Moreover, the proper application of the revised equation is more likely to protect healthy workers for a wider variety of lifting tasks than methods that relies only a single task factor or single criterion [13].

The concept of perceived exertion was introduced half a century ago and an operational definition presented with methods to measure different aspects of perceived effort, strain and fatigue [4]. One very common method is the RPE-Scale for "Ratings of Perceived Exertion" ("the Borg Scale") officially now: the "Borg RPE Scale".

2. Methodology

The aims of this study are to identify the complaints experienced by the operator using a Nordic Body Map questionnaire, to analyze how big the effort needed by using the BORG Rating Perceived Exertion, to analyze the risk incurred by using the NIOSH lifting equation, to propose an improvement plan for the department of painting by design tools [5, 6].

The study was conducted in the painting department from Contract Manufacturing, Sheet Metal Business Unit – Indonesia. The data needed for this research are the layout of the manufacturing plant, work process detail, effort of the operators, vertical distance, horizontal distance, subjective perception of the operator and time study for the lifting process [7].

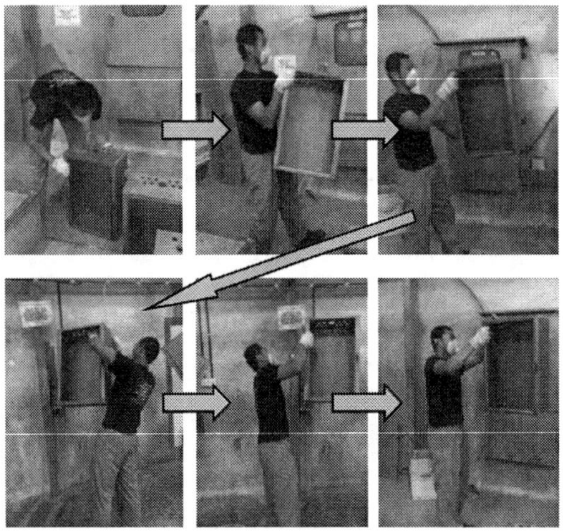

Figure 1. Operator Body Posture

3. Analysis and Result

There are 51 workers in the painting department, consist of 1 supervisor, 2 leaders, 8 coordinators, 32 powder coating operators, 4 treatment operators, and 4 HSE & Inventory operators. The Panel box divided into body, door, and base plate, with the size 600*400*180+20 mm. Figure 1 shows the body posture of the operator when lifting the body of the panel.

Table 1. Summary For Recommended Weight Limit

Multiplier	Formula	Parameter	Value
Load Constant (LC)	23 kg	23 kg	23 kg
Horizontal (HM)	25/H	H = 35 cm	0.714
Vertical (VM)	1- (0.003 |V-75|)	V = 160 cm	0.745
Distance (DM)	0.82 + (4.5/D)	D = 160 cm	0.850
Asymmetry (AM)	1- (0.0032 x A)	A = 30°	0.904
Frequency (FM)	Table Frequency multiplier	3 lifts/ min	0.550
		6 lifts/ min	0.270
Coupling (CM)	Good, Fair, Bad	Fair	0.9

From Table 1, the lifting frequency divide into two, the first one is to lift the body of panel which is 3 lifts/min, and the second is to lift the door and the base plate which is 6 lifts/min. From the calculation above, the Lifting Index for the three parts are more than 1.0. As we know, LI > 1.0 can increase the risk for low back pain or some fraction [8].

Table 2. Descriptive statistics BORG RPE Scale rated by 44 operators.

Parts	BORG RPE SCALE			
	Trial 1		Trial 2	
	Mean	Std. Dev	Mean	Std. Dev
Body	13.71	0.91	13.86	1.03
Door	9.43	0.76	9.29	0.83
Base plate	8.86	0.66	8.79	0.70

Table 3. Descriptive statistics lifting time

Parts	Lifting Time			
	Trial 1		Trial 2	
	Mean	Std. Dev	Mean	Std. Dev
Body	20.05	1.29	20.03	1.08
Door	10.97	1.13	11.01	0.96
Base plate	12.04	1.03	12.02	1.31

To examine the relationship between the Borg RPE scale and lifting time, then the correlation test using SPSS 18.0 was done. The result was there is a correlation between RPE Scale and lifting time for the first trial (0.992) and the second trial (0.996), as seen in table 2 and table 3. To obtain the subjective perception of complaints on certain body parts, the Nordic body map questionnaires were distributed to 44 operators. Questionnaires conducted over 26 working days. Figure 2 describes the percentage of complaints for 44 operators for almost 1 month. The operator was given an understanding of the Borg scale, the scale at which the operator provides a great perception of energy spent on lifting, ranging from no exertion at all to the biggest power is needed. This study was done twice and performed the time measurement when the operators lifted the panel, so the research is closer to the actual results [9-12].

From the Figure below, Upper neck, Waist and Back are the three biggest complaints among the other body parts. From the calculation of lifting index, the value of LI for the three parts is bigger than 1.0, which can increase the low back pain risk. From the Nordic Body Map Questionnaire, the biggest complaints rated by operators are upper neck, waist, and back. the conveyor table was made by using anthropometry data from West Java – Indonesia's workers as an improvement tool. The anthropometry data used are standing waist height and hands reaching. After the painting pre treatment process, panel parts (Body, door and base plate) will glide through the pipes of the conveyor. Based on this improvement tool, recalculate the RWL & LI must be done. Here is the summary of the calculation: The RWL is 8.37 kg for 3 lifts/min and 4.1 kg for 6 lifts/min. When the operators use the improvement tool, a lowering value of Lifting Index.

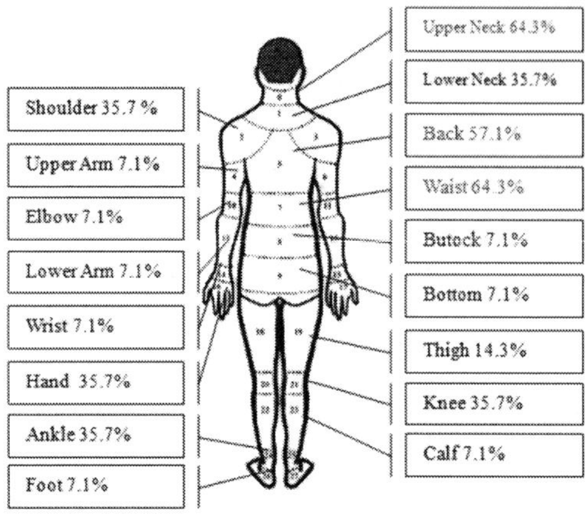

Figure 2.BORG RPE Scale rated by the operators.

Table 4. Percentage of Lifting Index After Improvement

Part	LI (Before Improvement)	LI (After Improvement)	%
Body	1.65	0.91	44.84
Door	1.30	0.72	44.61
Base plate	0.93	0.51	45.16

Table 5. Percentage BORG RPE Scale After Improvement

Part	BORG RPE Scale		
	Before	After	%
Body	13.79	11.96	15.27
Door	9.36	8.32	12.47
Base plate	8.82	8.25	6.93

4. Conclusion

Based on the results of Nordic Body Map Questionnaire, for the one month, the operators have complaints such as pain, discomfort, and tingling in the neck, back and waist. The Borg RPE scale is used to assess how big the effort to lift the parts. From Borg Rating of Perceived Exertion questionnaire revealed that the effort needed to lift the panel body parts were substantially higher than door parts and base plate. From NIOSH manual lifting equation measurement found that the biggest risk of injury was the operators' lower back Complaints submitted by the operator that pain in the waist, back and neck was indeed a great risk in line with the work. Using anthropometric dimensions of workers in West Java, the authors proposed an improvement tools such as conveyor table. Basic selection conveyor table this is a fairly low cost factor, and a fairly short time of manufacture. After testing by using a conveyor table, it is evident that these tools help to reduce the workload of the operator. as shows in table 4, calculation results lifting index value were reduced by 44-45%. While the Borg scale was also reduced to 6-15%, as seen in table 5. Proved by the existence of these tools, operator workload can be reduced. Design and manufacture of improvement tool can be proceeded with the manufacture of conveyor table that can adjust height, and the addition of wheels on the legs so that the conveyor table can serve as a lorry.

References

1. Waters, T.R., V. Putz-Anderson, and A. Garg, *Applications manual for the revised NIOSH lifting equation*. 1994.
2. Waters, T.R., et al., *Revised NIOSH equation for the design and evaluation of manual lifting tasks*. Ergonomics, 1993. **36**(7): p. 749-776.
3. Singh, S., et al., *Anthropometric measurements and body composition parameters of farm women in north Gujarat*. J Ergonomics, 2013. **3**(114): p. 2.
4. Borg, G., *Borg's perceived exertion and pain scales*. 1998: Human kinetics.
5. Osha, E., *The Study of Work US Department of Labor Occupational Safety and Health Administration*. 2000, OSHA.
6. Ray, G., S. Ghosh, and V. Atreya, *An anthropometric survey of Indian schoolchildren aged 3–5 years*. Applied ergonomics, 1995. **26**(1): p. 67-72.
7. Syazwan, A., et al., *Poor sitting posture and a heavy schoolbag as contributors to musculoskeletal pain in children: an ergonomic school education intervention program*. Journal of pain research, 2011. **4**: p. 287.
8. Okimoto, M.L.L.R. and E.R. Teixeira, *Proposed procedures for measuring the lifting task variables required by the Revised NIOSH Lifting Equation – A case study*. International Journal of Industrial Ergonomics, 2009. **39**(1): p. 15-22.
9. Chin, J., et al., *The investigation of consumers' behavior intention in using green skincare products: a pro-environmental behavior model approach*. Sustainability, 2018. **10**(11): p. 3922.
10. Chin, J. and S.-C. Lin, *A behavioral model of managerial perspectives regarding technology acceptance in building energy management systems*. Sustainability, 2016. **8**(7): p. 641.
11. Chin, J. and S.-C. Lin, *Investigating users' perspectives in building energy management system with an extension of technology acceptance model: a case study in indonesian manufacturing companies*. Procedia Computer Science, 2015. **72**: p. 31-39.
12. Mufidah, I., et al., *Understanding the consumers' behavior intention in using green ecolabel product through pro-environmental planned behavior model in developing and developed regions: lessons learned from Taiwan and Indonesia*. Sustainability, 2018. **10**(5): p. 1423.

Journal of Physics: Conference Series **1424** (2019) 012048 doi:10.1088/1742-6596/1424/1/012048

The Work Posture Assessment Towards Musculoskeletal Disorders in Coloring Activities of Indonesian Hand Drawn Batik

Siswiyanti[1], Rusnoto[2], Saufik Luthfianto[3], Nurjanah[4] and Arif Zainudin[5]

[1]Engineering Department, Universitas Pancasakti Tegal, Indonesia
[2]Engineering Department, Universitas Pancasakti Tegal, Indonesia
[3]Engineering Department, Universitas Pancasakti Tegal, Indonesia
[4]Fisheries Department, Universitas Pancasakti Tegal, Indonesia
[5]Goverment science Department, Universitas Pancasakti Tegal, Indonesia

*arifzainudin@upstegal.ac.id

Abstract. This study aims to determine how big the level of musculoskeletal complaints is, the fatigue and increase productivity of batik maker after body posture changes in coloring (dyeing) hand drawn batik cloth uses a machine in standing body posture. This study used Nordic Body Map musculoskeletal measurements and subjective fatigue, the research method used the experimental design of the same subject, namely the design that is variable observation carried out several times, the control subject also applied as the experimental subject. The results of processing data on 20 batik makers were concluded that the percentage of pain in the coloring activity (manual dyeing) by squatting and stooping body posture with a mean was about 65.18 ± 6.45%. The percentage of pain in the machine dyeing activity for standing posture was 50.73 ± 71.73%. The average difference between the level of musculoskeletal complaints on batik staining with squatting/stooping body posture compared to the staining of the standing body posture was 16.80 or a decrease of 19.34%. The results of subjective fatigue questionnaire obtained a decrease in fatigue of manual dyeing activity with dipping machines, namely an activity attenuation decrease from 90% to 62%, attenuation of motivation from 88% to 58%, physical attenuation from 78% to 64%. The average difference between the degree of fatigue in batik staining and stooping body posture compared to the staining of the body posture is 25,25 or there is a decrease in fatigue of 26.69%. The work posture improvement of standing position on batik dyeing machine that is designed using 3 roll (cylinders) which has a driving power of 0.190 HP, with the working concept of a sheet of fabric wrapped around a cylinder with a 72.5 rpm rotation, frame material made of elbow iron painted and cylinders made of data a the PVC pipe to avoid corrosion could actually increase work productivity by 0, 0190 or 29.19% .

1. Introduction

Batik is a painting on cloth which is a hereditary inheritance from Southeast Asia, especially in the region of Java Indonesia. The cloth is painted with a heated candle, then undergoes a coloring process or dyeing and washing the wax [21]. Batik technique is a way of holding a candle to be scratched in the form of a point or line and can be a motif of artistic value and beautiful [11].

Batik process starts from the process of making patterns, painting, coloring (removing wax by heating), washing and drying. Among the most difficult batik processes are coloring or dyeing process, which is to arrange the color composition to produce the desired blend of motifs. Coloring of fabrics that have been painted with hot wax must be evenly distributed so that the wax attached to the fabric does not break. Uneven coloring makes the painting become pale and faded [18].

The process of coloring batik in Kalinyamat Wetan Kota Tegal Indonesia is generally done by dipping (manual) with stooping and squatting work postures. The results of the initial study of 10 batik makers after doing batik coloring manually caused musculoskeletal complaints in back, lower neck, right shoulder, waist, left elbow, right elbow, right thigh, right ankle pain. Getting pain on the back of Occupational low back disorders (LBD) (lower back pain, lower back muscular tissue, lumbar disc disorder) is a socio-economic burden in many Industrial countries and often associated with physical factor which is related to works, such as directing energy, highly repetitive movements and continuous static posture, such as squatting and standing posture. LBD is caused by non ergonomic work postures such as stooping and standing [15]. This disorder occurs because of fatigue which leads to changes in the trajectory of the spinal movements which can increase the risk of injury.

Abnormality of Musculoskeletal often happens and has a major impact in health and quality as well as workers' health. Safe lifting techniques become a concern due to high risk of injury and low back pain (LBP) because of frequent lifting activity in the industry. The recommended compression strength limits for material handling often occur such as lifting, squatting and stooping. [1]. Analysis of squat and stoop dynamic liftings: muscle forcesand internal spinal loads. Biomechanical literature does not allow for squatting techniques as a means of preventing lower back pain [5]. Musculoskeletal complaints of batik coloring activities as shown in Figure 1 influence the level of batik makers and reduction of productivity. One of the efforts to reduce complaints is by creating a coloring machine for dyeing batik cloth.

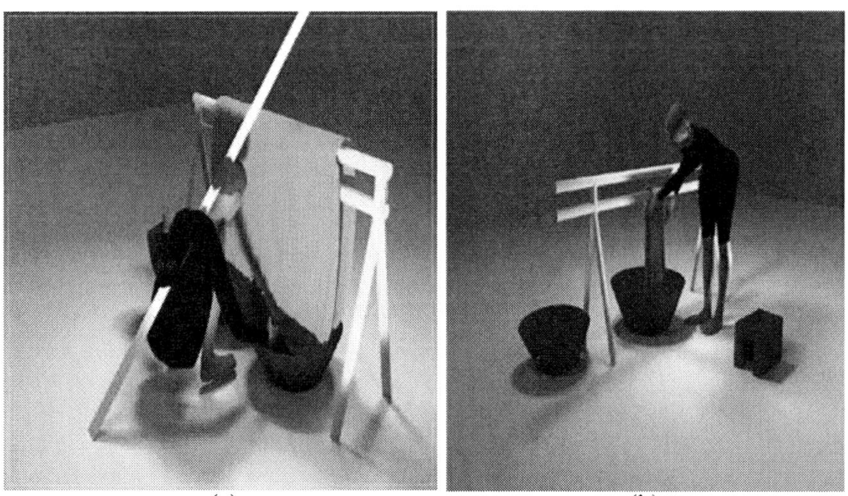

(a) (b)

Figure 1. The Animation of Hand Drawn Batik Staining Process with Manual Dyeing of (a) Squatting and (b) Stooping Body Posture

Musculoskeletal disorders and the risk of injury can be reduced by using an ergonomic work system design [6].Workers' tools help to reduce physical loading on the spine by reducing or eliminating some load during stooping or squatting posture. There are many types of mechanical work and worker assistance which have a potential to control the stooping posture as in agricultural production work so that It is needed the design / specification feature which can determine better height / size / installation location [14]. In designing the dye machine, height of the machine is adjusted to the height of the batik maker posture so that the machine can be adjusted up and down to adjust the length of the fabric.

The main factor of batik makers who get musculoskeletal is from work postures that stooping and squatting by using buckets and wooden frame during the coloring process. Squatting and stooping posture causes discomfort. The working conditions force workers to always be in an unnatural posture and work position. An unnatural, long-lasting / static attitude can cause injury [20]. The results of the analysis of hand drawn batik dye machine are designed using 3 roll (cylinders) that have a driving power of 0.190 HP, with the concept of sheet fabric work wrapped around the cylinder with a 72.5rpm rotation. The frame material is made of painted elbow iron and cylinders of PVC pipe to avoid corrosion. The assessment of the workers' posture in the coloring process (manual dyeing) is obtained by REBA scores of 5 to 8 at moderate to high musculoskeletal levels (need to be repaired immediately) . After applying the design results of coloring machines in the batik coloring process by workers, REBA scores obtained namely 2-4 at safe musculoskeletal levels to low (long-term repairs needed). It shows that there is a change in body posture after applying a staining machine, which is from a squatting and stooping body posture that is not natural to become a more natural standing body posture. The development of batik coloring tools/machines is advised by conducting socialization to workers to apply the dyeing method using dye machines so as to increase the productivity of batik artisans and reduce the risk of injury [12]. Paying attention to the body position in batik staining by squatting and stooping, the complaints of the musculoskeletal system will often be found by batik makers. Therefore this research was carried out to uncover musculoskeletal complaints and fatigue systems faced by batik makers after designing batik dye machines, wherein the machine operator worked with standing posture.

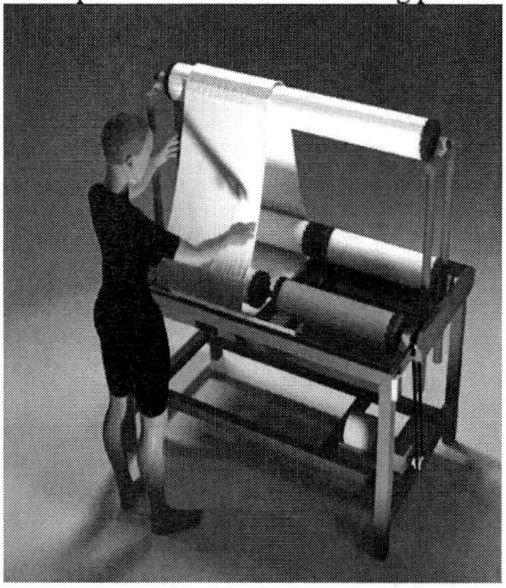

Figure 2. The animation of the Hand drawn Batik Coloring Process with the dyeing of standing body posture [12].

Problem Formulation: How big is the level of difference in musculoskeletal complaints, fatigue and increase in the productivity of batik maker after a change in standing posture on coloring (dyeing) of batik cloth using a machine?

2. Methodology

The subjects of this study were 20 female batik makers. The quantitative analysis of the subjects was done by calculating the mean and standard intersections for each criteria, namely age, height, weight, and work experience. The object of the research is work attitude or body posture for batik coloring activities with manual dipping positions, namely body posture squatting or stooping, and the dipping position of the machine with standing posture. The Research was located in Fitri Ayu batik group in Kalinyamat Wetan village, South Tegal sub-district, Tegal city, Indonesia.

Methods and data analysis : The measurement of Musculoskeletal complaints is done with NIOSH Nordic Body Map Subjective Filling, the measurement of fatigue with 30 Item Self Rating Questionnaire Industrial Fatigue Research Committee is from Japan Association Of Industrial Health. The data analysis is divided into three parts, namely descriptive analysis, normality test, and different test.

a. Quantitative analysis calculates the mean and standard intersections for each criterion, namely age, height and weight.
b. Test for normality: using the Kolmogorov-Smirnov Test (with a significance level of $\alpha = 0.05$)
c. Different test: using different test two groups in pairs with a significance level ($\alpha = 0.05$). If the data is normally distributed, then paired t test is used. If the data is not normally distributed, the Wilcoxon test is then used.

Nordic Body Map Qusisioner is a map of the body that analyzes the body parts of the muscles or joints that undergo complaints. Parts of the body from numbers 0 to 27 or from neck to feet.[19]. A biomechanical tool for assessing MSS subjectively is the Nordic body map questionnaire, developed by the Institute of Occupational Health. This tool has been widely used throughout the world. The Nordic body maps are equipped with body maps with a number of each part of the body. The person assessed must fill out the questionnaire by giving a sign of the level of discomfort in the intended body part.

Musculoskeletal disorders (MSD) that was experienced by workers caused by uncomfortable working conditions while practicing the operation of machinery and equipment. The findings are supported by several ergonomically designed workstation studies that can reduce MSD and increase productivity. Musculoskeletal disorders are obtained from repetitive work including injury or disruption of muscles, bones, tendons, tendon sheaths, ligaments, bursa, blood vessels, joints, intervertebral discs, etc. Other terms which are used to describe MSD including Repetitive Strain Injury, Musculoskeletal Injury, Cumulative Trauma Disorder, Occupational Overuse Syndrome, or Strain or Sprains [4]. to reduce the risk of working related to musculoskeletal disorders (WMSD) Finally, the prevalence of pain symptoms in various parts of the body has been found relating to age, work experience, work shift, and body mass index [6].

According to [10] fatigue is caused by a non-ergonomic work environment, batik makers who complain of fatigue while doing work activities. Housewives, have jobs other than as batik craftsmen who are able to maintain a family economy of more than 50% of their daily economic needs. To work more efficiently, Batik craftsmen are needed to ergonomically design batik workspaces.

In the study [8] showed that the health of spinning operators was greatly affected due to the body posture and imperfect workload. These postures require them to work in a non-neutral position which increases overall discomfort and pain in the lower back, neck and shoulders, non-ergonomic work postures and modification. New breakthroughs for health promotion in the workplace are still needed to increase K3 which are implemented in UKM Batik and innovative approaches are needed to get the attention of UKM workers [22] According to [13] in batik process is carried out manually for approximately 7 hours every day (08.00 - 15.00 WIB) so that it has the potential to cause various kinds of complaints experienced by workers, including fatigue and risk of injury. The study was conducted to evaluate the risk of muscle injury using the Posture Activity Tools Handling (PATH) method and the Ovako Working Posture Analysis System (OWAS).

In developing the industry, especially batik industry, which is a hereditary inheritance, the home industry must pay attention to healthy environmental conditions, such as the tools and machines used by the place to work. When batik makers feel pain in certain parts of the body, they cannot complete the work optimally so that they cannot fulfill the order as targeted. According to [17] that certain body aches cause work discomfort resulting in impaired health factors and lower productivity. The important aspect of work productivity and quality measurement is time. In human interaction, the cycle timer requires the work time of the worker and the operating time of the machine. The working time of the worker is influenced by the interaction of the mass work posture and the workplace dimension. This interaction produces comfortable work so that it will increase work productivity and quality of workers. [3] according to (Hutagalung) that work productivity is measured from the ratio of output where the amount of time produced divided by input, namely the average workload (the average work pulse) during working hours.

3. Result and Discussion

a. Subject Characteristics

The subjects of the study were batik makers who were coloring fabric with a total of 20 women. The subject description is shown in Table 1. It states that the age of the subjects was obtained with a mean of 46.65 years ± 9.82 years with a range of 34-67 years. Subject height obtained with a mean 1.52m ± 0.06m with a range of 1.40-1.60 meters. Subject weight gained mean of 51.65 kg ± 4.42 kg with a range of 45-58 kilograms. Subject work experience obtained with a mean 22.58 th ± 1.87th with a range of 18-24 years.

Table 1. Subject Description

Aspects	Women		
	Average	Standar Deviation	Range
Age (years)	46,65	9,82	34,00-67,00
Height (m)	1,52	0,06	140,00-160,00
Weight (kg)	51,65	4,42	45,00-58,00
The duration of work (years)	22,38	1,87	18,00-24,00

b. Body Posture of Batik Coloration Activity

1) The manual dyeing body posture is the body posture on batik coloration activities which uses a bucket (tub) with a body posture is squatting or stooping, both hands are entering the bucket (tub) by dipping a cloth that will be given dye so that the body position adjusts the material and tools on top of wooden frame then the body posture of the batik makers stoop to adjust to the material / tool being done. The position of the knees (feet) of the batik maker is bent. The body position of manual dyeing and dyeing of the machine can be seen in Figures 3 and 4.

Figure 3. The Body Posture of Batik Maker on Manual Dyeing (Squatting)

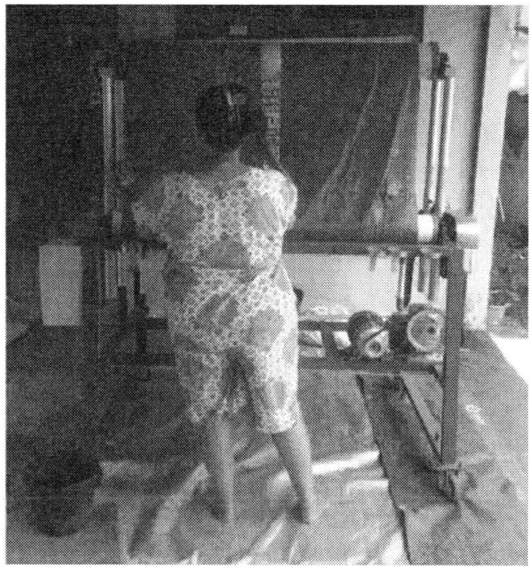

Figure 4. The body posture of Batik maker on machine dyeing (standing)

2) The body posture of machine dyeing is the attitude or standing body posture with both hands holding the cloth forward and upward with hand angle of 45 degree, the material that is done is placed in front of the body, or placed on the machine then the body posture is adjusted to the material / tool what is done. The knee (foot) position of the batik maker is straight.

c. Musculoskeletal systems of batik coloration in manual dyeing and machine dyeing

The questionnaire results of the level of musculoskeletal complaints decrease can be seen in Figure 5 below. The reduction in subjective complaints is feeling pain in upper neck from 55% to 50%, lower neck from 60% to 45%, left shoulder from 85% to 50%, right shoulder from 60% to 50%, left upper arm from 70 % to 45%, back from 65% to 45%, right upper arm from 65% to 60%, waist from 65% to 50%, buttocks from 70% to 60%, buttocks from 65% to 50%, left elbow from 55% to 40%, right elbow from 70% to 45%, left forearm from 65% to 50%, right forearm from 65% to 60%, left wrist from 55% to 50%, right wrist from 70 % to 65%, left hand from 65% to 45%, right hand from 70% to 45%, left thigh from 60% to 45%, left thigh from 60% to 55%, right thigh from 70% to 55%, left knee from 65% to 30%, right knee from 70% to 55%, left calf from 60% to 50%, right calf from 75% to 45%, left ankle from 65% to 20%, ankle from 60% to 45%, left foot from 65% to 60%, right foot from 65% to 60%.

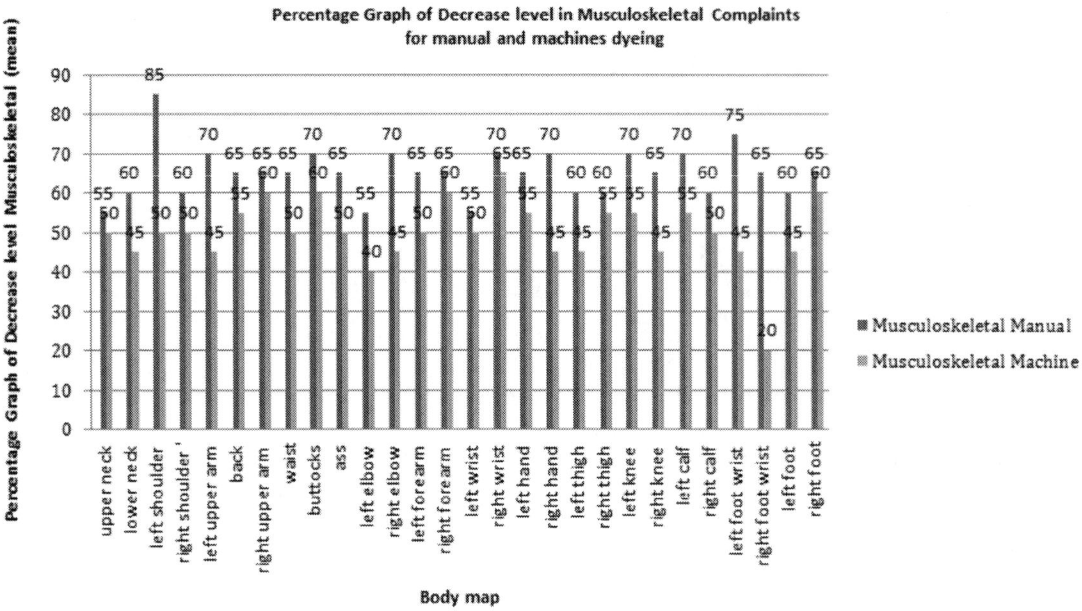

Figure 5. Percentage Graph of Decrease level in Musculoskeletal Complaints for manual and machines dyeing

d. Fatigue in manual and machine dyeing batik coloration

The fatigue measurement results using 30 questions questionnaires of activities attenuation, motivation, and physical can be seen in Figure 4 below. Most of the samples experienced a decrease in fatigue, namely a decrease in activity attenuation from 90% to 62%, weakening motivation from 88% to 58%, physical attenuation from 78% to 64%.

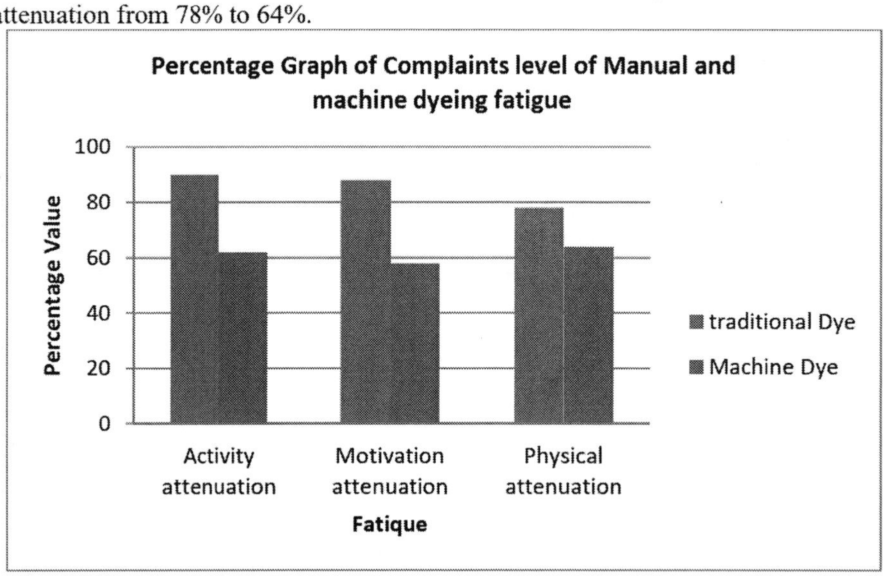

Figure 6. Percentage Graph of Complaints level of Manual and machine dyeing fatigue

e. Productivity of manual dyeing and machine dyeing batik coloration

Batik makers' productivity is the ratio between the number of products divided by the number of pulses. The productivity which is produced by 20 batik makers in dyeing machine has a mean of 0.07 ± 0.02. The average yield of pulse rate is 77 per minute, thus the batik coloring activity with squatting and stooping position has a light workload. The productivity which is produced from 20 batik makers in stooping and squatting postures on fabric staining using manual dye has a mean of 0.08 ± 0.01. The average yield of pulse rate was 75.03 per minute, thus batik activity with manual dyeing position has a light workload.

Table 2. The average difference of Control Groups and Experimental Groups

Aspect	Control group	Experimental group	Difference Remarks	%	Remark
muskuloskelatal	86,8500	69,6000	-17,25	-19,86	Decrease
Exhaustion	94,6000	69,3500	-19,00	-21,63	Decrease
Productivity	0,0650	0,0840	0,0190	29,19	Increase

f. Data Analysis

1) **Normality Tests on Musculoskeletal Complaints, Fatigue, and Productivity**

The Normality test is to test questionnaire data on the level of musculoskeletal complaints in the control group before activity, the level of musculoskeletal complaints in the experimental group before activity, the level of musculoskeletal complaints in the control group after activity, the level of musculoskeletal complaints of the experimental group after activity, fatigue of the control group before activity, exhaustion of the experimental group before activities, fatigue control group after activity, exhaustion of the experimental group after activity, productivity of the control group before activity, productivity of the experimental group before activity, productivity of the control group after activity, productivity of the experimental group after activities. This normality test aims to find out whether the sample comes from a population with a normal distribution of distribution. The normality test using Kolmogorov-Smirnov is shown in Table 3.

Table 3. The average Standard Deviation and Normality Test. Source. Author.

Aspect	Average	Standard Deviation	*p*
Musculoskeletal aspects of manual dip	86,8500	10,2406	0, 056
Fatigue aspect of manual dip	96,6000	3, 1522	0, 112
Productivity Aspect manual dip	0, 0650	0, 0202	0,128
Musculoskeletal aspects of machine dip	69,6000	5, 4134	0, 071
Fatigue aspect of manual dip	69,3500	11, 0895	0, 117
Productivity Aspect manual dip	0, 0840	0, 0127	0,127

p = probability value

Based on the calculations, it is obtained that the p value for all aspects is greater than 0.05 ($p > 0.05$) thus all data are normally distributed.

1) **T Test of Musculoskeletal Complaints, Fatigue, and Productivity**

The overall normality test results of the data are normally distributed, so the analysis used is the compare mean test by using paired sample T-Test. The results of the t test for the subject are shown in Table 4.

Table 4. Mean, Mean Difference, and T Test Between Control Groups and Experimental Groups

Variable	Group	Mean	Standard Deviation	Mean Difference	t count	P
Musculoskelet al	Control	86,8500	10,2406	-17, 25	7,48	0,000
	Experiment	69,6000	5, 4134			
Exhaustion	Control	94,6000	3, 1523	-19, 00	-9,363	0,000
	Experiment	69,3500	11, 0895			
Productivity	Control	0, 0650	0, 0202	0, 0190	-3,300	0, 004
	experiment	0, 0840	0, 0127			

Table 4 It states that the level of musculoskeletal complaints, fatigue and productivity in the sample obtained the probability value of 0,000; 0,000; and 0,004 ($p < 0.05$). Thus it can be concluded that there is a significant reduction between all variables in the control group and the experimental group. The average difference in the level of musculoskeletal complaints between the control group and the experimental group was 17.25 or a decrease in musculoskeletal complaints by 19.86%. The difference in average fatigue between the control group and the experimental group was 19.00 or there was a decrease in fatigue by 21.63%. The average difference in the level of productivity between the control group and the experimental group is 0.0190 or an increase in productivity is 29.19%.

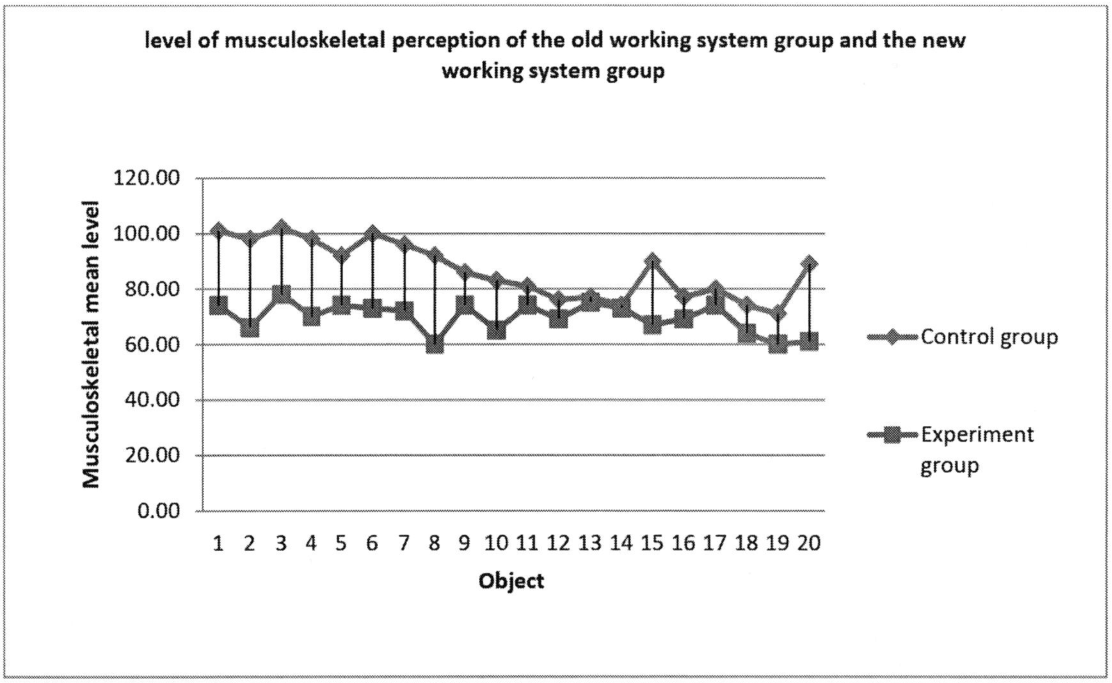

Figure 7. Graph of Musculoskeletal Complaints between Control Groups and Experimental Groups

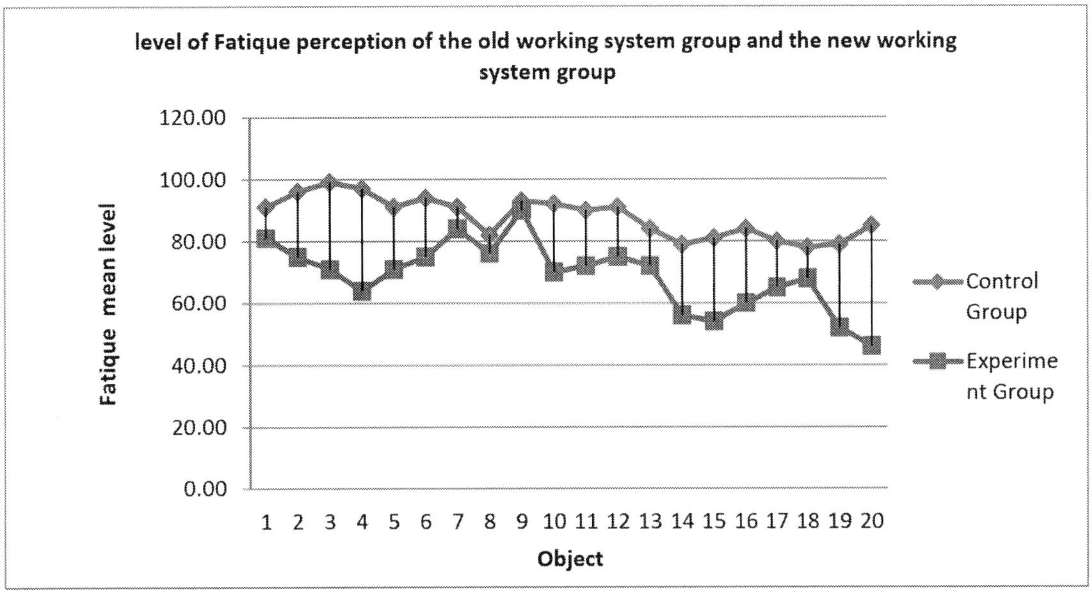

Figure 8. Graph of Fatigue Level between Control Group and Experimental Group.

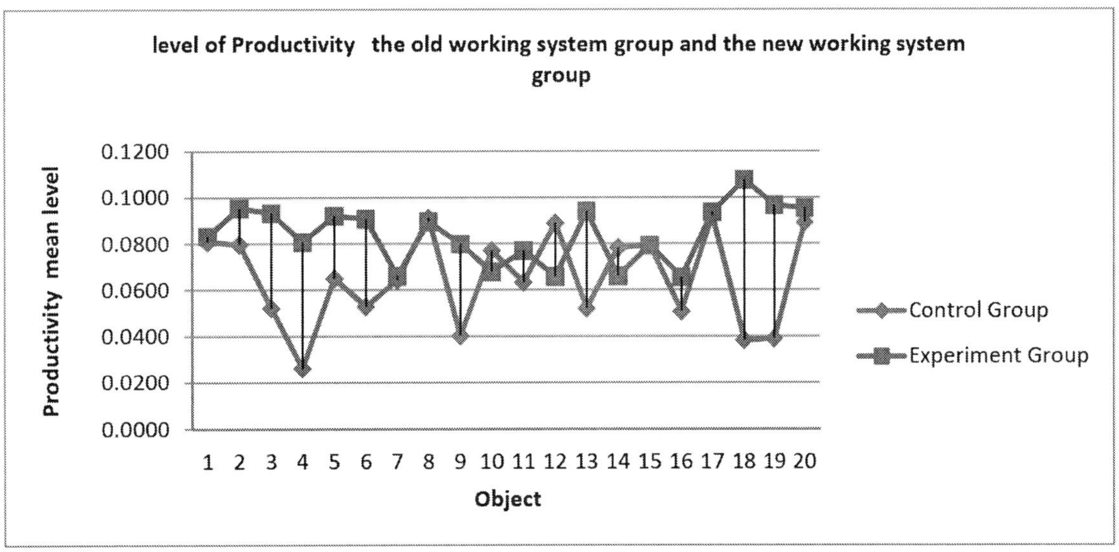

Figure 9. Productivity Level Chart between Control Group and Experiment Group.

The improvement of work attitude or staining of the body posture from the squatting position (stooping) into a standing position as a whole can reduce musculoskeletal complaints, fatigue and increase productivity of the batik maker. This matter is caused by the squatting work position which is depending on the ankle support which must support weight, [2]. Working posture by squatting has a higher blood pressure burden than standing posture, namely decreased blood pressure after standing from squatting [9].

According to [16] works with squatting and stooping techniques using maximum acceptable weight (MAW), perceptions of exertion (RPE), discomfort and preference. Mean (SD) MAW for squatting is lower than stooping (7.0 (2.2) kg vs. 8.5 (2.4) kg) and RPE for squatting is greater than for stooping (15.2 (1.5) kg vs 13.3 (1.5) kg). More subjects reported discomfort after squatting and preference for stooping. The result shown the limited support for the use of stooping techniques rather than squatting.

The application of ergonomics into the work system has been proven to be able to increase productivity, health, safety and comfort of work by the use of methods such as tools or practical and inexpensive tools for a long time which can achieve the desired results in reducing musculoskeletal disorders on a small scale and poor industries in developing countries [7].

4. Conclusion

The results of data processing and analysis can be concluded that the changes in squatting/stooping body posture to standing on coloring / dyeing of hand drawn batik cloth using machines makes the level of musculoskeletal complaints, fatigue decrease and productivity increases significantly. The Different mean of musculoskeletal complaints before and after the experiment is equal to 25.50 or there is a decreased level of musculoskeletal complaints by 26.26%. While the mean difference in fatigue complaints before and after the experiment was 25.25 or a decrease in complaints of 26.69%. The average difference in productivity achievement before and after the experiment was 0, 0 190 or there is an increase in productivity was 29.19%. It is necessary to be investigated further about the factors that cause differences in musculoskeletal system complaints, work fatigue and increase of the productivity of batik makers especially for batik coloring.

References

[1] Æ. A. S. (2007). Analysis of squat and stoop dynamic liftings : muscle forces and internal spinal loads, 687–699. https://doi.org/10.1007/s00586-006-0240-7
[2] Asuyama, T. A. K., & Akamoto, M. A. S. (2009). Ankle Joint Dorsiflexion Measurement Using, 195–199.
[3] Hartomo, H., Hassan, A., & Haron, C. H. C. (2005). Validation of Ergonomics Model for Estimating Work Productivity and Quality in Manual Handling Activities. *Jurnal Teknoin*, *10*(2). https://doi.org/10.20885/.V10I2.686
[4] Hazards, M. D. (2008). Musculoskeletal Disorders.
[5] Hoozemans, M. J. M., & Toussaint, H. M. (1999). Stoop or squat : a review of biomechanical studies on lifting technique, *14*.
[6] Islam, U., Islam, U., & Surakarta, B. (2017). THE USE OF MACRO-ERGONOMIC WORK SYSTEM DESIGNS TO REDUCE MUSCULOSKELETAL DISORDERS AND INJURY RISK IN TRAINING H. Purnomo 1 *, E. Giyono 1 & A.E. Apsari 2 ARTICLE INFO, *28*(May), 47–56.
[7] Jahangiri, M., Ali, S., Najarkola, M., Gholami, T., Jahangiri, A., Hesam, G., & Jalali, M. (2015). Ergonomics Intervention to Reduce Work-Related Musculoskeletal Disorders in a Lead Mine, *4*(4), 5–7. https://doi.org/10.17795/jhealthscope-29507
[8] Kolgiri, S. (2018). Work Related Musculoskeletal Disorders among Power-Loom Industry Women Workers from Solapur City , Maharashtra , India, (March).
[9] Krediet, C. T. P., Go-schön, I. K., Kim, Y., Linzer, M., Lieshout, J. J. Van, Wieling, W., … Lieshout, J. J. Van. (2010). Management of initial orthostatic hypotension : lower body muscle tensing attenuates the transient arterial blood pressure decrease upon standing from squatting To cite this version : HAL Id : hal-00479369.
[10] Maarif, U., & Latif, H. (2017). Redesign of Work Environment with Ergonomics Intervention to Reduce Fatigue, *12*(7), 1237–1243.
[11] Maulik, S. R., Bhowmik, L., & Agarwal, K. (2014). Batik on handloom cotton fabric with natural dye, *13*(October), 788–794.
[12] Siswiyanti, Studi, P., Industri, T., Pancasakti, U., Teknik, S., Universitas, M., & Tegal, P. (2018). Jurnal Optimasi Sistem Industri Penerapan Ergonomi pada Perancangan Mesin Pewarna Batik untuk Memperbaiki Postur Kerja, *1*, 75–85. https://doi.org/10.25077/josi.v17.n1.p75-85.2018
[13] Pratiwi, I., & Kartikasari, I. (2018). Evaluation of Work Posture for Non Repetitive Job in

Kampoeng Batik Laweyan Using PATH and OWAS Method, *20051*. https://doi.org/10.1063/1.5042907

[14] Proceedings, C. (2004). *Squatting Postures*.

[15] Spyropoulos, E., Kyvelidou, A., Stergiou, N., & Athanassiou, G. (2016). Quantifying Muscle Fatigue of the Low Back during Repetitive Load Lifting Using Lyapunov Analysis, *6*(6). https://doi.org/10.4172/2165-7556.1000180

[16] Straker, L., & Duncan, P. (2000). Psychophysical and psychological comparison of squat and stoop lifting by young females, 27–32.

[17] Susihono, W., & Istianah, I. (2018). Assessment of Musculoskeletal Disorders and Fatigue of Workers in Creative Industry of Carving Wood in PK Novi Cilegon , Banten, *2018*, 361–371. https://doi.org/10.18502/kls.v4i5.2567

[18] Tembalang, C. (2015). Measuring Efficiency of Using Resource in the Production Process of Making Stamped-Batik : A DEA Approach Aries Susanty Sri Hartini Diana Puspitasari, *6*(5), 318–327. https://doi.org/10.5901/mjss.2015.v6n5s2p318

[19] Wahyudi, M. A., Dania, W. A. P., & Silalahi, R. L. R. (2015). Work Posture Analysis of Manual Material Handling Using OWAS Method, *3*, 195–199. https://doi.org/10.1016/j.aaspro.2015.01.038

[20] Widyanti, A. (2018). Ergonomic Checkpoint in Agriculture , Postural Analysis , and Prevalence of Work Musculoskeletal Symptoms among Indonesian Farmers : Road to Safety and Health in Agriculture, *20*(1). https://doi.org/10.9744/jti.20.1.1-10.

[21] Wyvill, B. (1999). Rendering Cracks in Batik.

[22] Yunita, F. H., Kurniawidjaja, L. M., & Susilowati, I. H. (2018). Workplace Health Promotion Related to Occupational Safety Climate : A Case Study by the Government on Batik SMEs in Pekalongan Regency , Central Java , Indonesia, *2018*, 260–267. https://doi.org/10.18502/kls.v4i5.2558.

Learning of Medicinal Herbs Using QR Codes

Nur Yasmin Khairani Zakaria[1], Hafiz Zaini[1], Saedah Siraj[2,3], Melor Md Yunus[1], and Harwati Hashim[1]*

[1]Faculty of Education, Universiti Kebangsaan Malaysia
[2]Universiti Pendidikan Sultan Idris
[3]Universiti Malaya

*harwati@ukm.edu.my

Abstract. Rapid development of information technology today has widened the use of computers as well as mobile phones. With more sophisticated devices is equipped with high resolution color, wireless access and high-speed processing. The current development has evolved with the creation of the portal since the mid-90's and given that the potential of the portal, it has been used by various firms regardless of body governmental or statutory bodies. Various advantages that can be utilized by the use of portals as it is has a dynamic source of information from a variety of sources and various formats. The current study aims to discover effects of QR codes implementation on medicinal herbs learning. The findings of students' responses revealed that the QR Code facilitate the learning of medicinal herbs, whereby the mean from this analysis revealed that average majority of the students find the learning approach is easy to be employed.

1. Introduction

A mobile tag is a way of displaying familiar visual elements through smartphones and devices that has similar functions [1]. According to Janssen, this two-dimensional bar code can be found anywhere in printed media or virtual media whereby the users are able to scan the code using the camera on the smartphone and the code will continue to display to the user about the inside information online scanned. This two-dimensional code bar is better known as the QR Code. This development of this technology has presented a two-dimensional code that is also known as the QR Code was created and developed rapidly [2]. QR Code is used as a guide to identifying sign, help access information of an object and also work for intercept in achieving a portal or information and is used in various fields including logistics, management and so on.Although there are many advantages of the QR Code but its use in the field education is still preliminary [3]. According to Syed Ardi & Zadatun (2008), the use of technology in education is inevitable
due to the sophistication of technology [4].

Along with the rapid development of technology today, various ways or medium that can be used to provide information about knowledge to something [5]. Various mediums have been used for making it easier for you to access information is by using the web portal and the latest by using a two-dimensional code type QR Code as an intermediate medium to retrieve information. The development of technology has brought a new revolution in a whole world sophisticated, especially in education in Malaysia. These technologies have affected people in every aspect of life. The use of computer has also become very important tool and a big influence on daily basis. This is further enhanced by the existence of a portal for education [5]

A portal is a website or service that offers a variety of resources and services such as Electronic mail (E-mail), forums, search engines and purchases byonline or Electronic commerce (E-commerce) service. There are also news portal, research and statistics, discussions, newsletters, online tools, and many other services related to a particular industry [6]. According to Maloney (2004), websites or portals can be
defined as a web-based tool that provides an interface where it can be customized to display the collected information from various sources [7]. According to him, the portal was implemented as an interface based on software help that provides access integrated into the source of relevant information and services. Implementations of these portals are vary in terms of the application framework, tools and programming languages. The use of the portal has begun since the mid-90's and started to grow rapidly [8]. However, lately, with the presence of the use of smartphones, tablet computers and the advancement of technological advances communication, there is growing interest in the use of easy tags instead of using a kind of two-dimensional code named QR Code in various fields including in education.

However the implementation of QR code in higher education is low, and it is also less exposed in the context of teaching and learning [3]. The use of mobile tags can further promote activity and improve motivation to students towards a field [9]. Previous studies indicate that using mobile tags can have a positive impact to the student's interest in doing something but it depends on the benefit and satisfaction of the tool itself. However educators have realized that there is growing interest in use of mobile tags for educational purposes [4].

2. Methodology

This study is employed a survey involving 70 respondents from a public higher institution in Malaysia. Convenience sampling method was conducted for data collection. Respondents in this study were required to use camera and QR software in their smartphones or tablets to scan the QR Code found on the printed source determined. The mobile device will interpret the QR Code to retrieve the website that contain information about herbs required and information will be obtained in the form of multimedia information (QR Code has been produced will be placed on the printed information provided in accordance with the circumstances). On each face that has been accessed by the user, there is complete information about the herbs that they intend to know and its importance in daily life

3. Result and Discussion

The objective of this study was to determine the ease of use of the QR code for medicinal herbs learning by the students in a public university in Malaysia. The overall frequency was calculated from the data collected in the questionnaires distributed. Statistical Package for the Social Science (SPSS) 22 was used to analyse the descriptive frequency of the data on students' agreement on the posited statement on the use of QR code for medicinal herbs learning.

The Use of the QR Code for Herbal Learning

		Sangat Setuju (Strongly Agree)	Setuju (Agree)	Neutral (Neutral)	Tidak Bersetuju (Disagree)	Sangat Tidak Bersetuju (Strongly Disagree)	Min (Mean)
1.	Saya mendapati sistem pembelajaran Kod QR herba kesihatan ini tidak sukar untuk digunakan. *I find that this QR Code for herbal learning system is not difficult to use.*	N=65 (92.9%)	N=5 (7.1%)	N=5 (7.1%)	N=5 (7.1%)	N=5 (7.1%)	4.93
2.	Pembelajaran menggunakan sistem pembelajaran Kod QR herba kesihatan ini mudah untuk saya. *Learning using QR Code for herbal learning system is easy for me.*	N=61 (87.1%)	N=9 (12.9%)	N=0 (0%)	N=0 (0%)	N=0 (0%)	4.87

3.	Berinteraksi dengan sistem pembelajaran Kod QR herba kesihatan sering tidak mengecewakan. *Interacting using QR Code for herbal learning system often does not disappoint me.*	N=47 (67.1%)	N=23 (32.9%)	N=0 (0%)	N=0 (0%)	N=0 (0%)	4.67
4.	Saya merasa mudah menggunakan sistem pembelajaran Kod QR herba kesihatan dalam mendapatkan maklumat tentang herba kesihatan. *I find it easy to use the Herbal QR Code learning system to get information on health herbs.*	N=59 (84.3%)	N=10 (14.3%)	N=1 (1.4%)	N=0 (0%)	N=0 (0%)	4.83
5.	Sistem pembelajaran Kod QR herba kesihatan ini tidak kaku dan fleksibel untuk digunakan. *Herbal QR Code learning system is not rigid and flexible to use.*	N=58 (82.9%)	N=12 (17.1%)	N=0 (0%)	N=0 (0%)	N=0 (0%)	4.83
6.	Mudah bagi saya untuk mengingat bagaimana untuk menggunakan sistem pembelajaran Kod QR herba kesihatan. *It is easy for me to remember how to use the code learning system QR herbs.*	N=60 (85.7%)	N=10 (14.3%)	N=0 (0%)	N=0 (0%)	N=0 (0%)	4.86
7.	Berinteraksi dengan sistem pembelajaran Kod QR herba kesihatan adalah mudah. *Interacting with QR code learning system is easy.*	N=57 (81.4%)	N=13 (18.6%)	N=0 (0%)	N=0 (0%)	N=0 (0%)	4.81
8.	Interaksi saya dengan sistem pembelajaran Kod QR herba kesihatan ini adalah jelas dan mudah difahami. *My interaction with the code learning system QR herbs are clear and easily understandable.*	N=66 (94.3%)	N=4 (5.7%)	N=0 (0%)	N=0 (0%)	N=0 (0%)	4.94
9.	Saya mendapati tidak banyak usaha diperlukan untuk menjadi mahir dengan menggunakan sistem *I find that there is not much effort required to become proficient by using system.*	N=60 (85.7%)	N=8 (11.4%)	N=2 (2.9%)	N=0 (0%)	N=0 (0%)	4.83

Analysis from the data revealed that the students perceived medicinal herbs learning through QR codes is highly facilitating, interactive and beneficial for their learning process. Generally, the students feel that the QR codes are easy to use and suitable for the learning of medicinal herbs. From the data obtained, the highest mean of students' agreement is in the item "*My interaction with the code learning system QR herbs are clear and easily understandable" (M=4.94)* and *"I find that this QR Code for herbal learning system is not difficult to use." (M=4.93)*. Analysis from the data showed that the students believed that the use of QR codes for medicinal herbs learning is highly manageable and easy to use. Instructions presented on how to use QR codes for the learning is found to be clear and understandable for the students. Students responses in the aforementioned statements on ease of use of QR codes' is also correlated with the item "*Learning using QR Code for herbal learning system is easy for me"* whereby majority of the respondent expressed their high agreement (87.1%).

In terms of practicality and feasibility of QR codes of medicinal herbs learning, majority of the respondents agreed that the application is very useful and has the ability to facilitate their learning process. In the item *"Interacting using QR Code for herbal learning system often does not disappoint me."* and *"Herbal QR Code learning system is not rigid and flexible to use."*, analyses from the data obtained revealed a higher mean (M=4.83) for the items. The results reflected students' responses on the ability of QR codes in improving their learning in the most useful and easy way. In addition, the statements also revealed that the use of QR codes for medicinal herbs learning is flexible and not disappointing.

4. Conclusion

The findings of students' responses revealed that the QR Code facilitate the learning of medicinal herbs, whereby the mean from this analysis revealed that average majority of the students find the learning approach is easy to be employed. From this finding, researchers have found that every herb has a total of over 80% acceptance by the student. This is generally reflected by the students' agreement on the items in the questionnaires. The findings of this study have also shown that the use of the QR Code can affect the interests of respondents in learning. This is because most respondents stated that they are interested in this learning system based on number percentage and mean of the interest items given in the items in questionnaire. Therefore it is concluded here that learning using the QR Code system against herbs can have a positive impacts on the respondents' knowledge on medicinal herbs.

Acknowledgement

This research is supported by Universiti Kebangsaan Malaysia under research scheme no. GG-2018-001.

References

[1] Janssen C, Langager C, Murphy C. Technical Analysis: Indicators And Oscillators. IAC,[Online]. Available: http://www. investopedia. com/university/technical/techanalysis10. asp.[Accessed 1 April 2016]. 2016.

[2] Ramsden A. The use of QR codes in Education: A getting started guide for academics. Accesible desde< http://zabdulkhaaliqcapstone. yolasite. com/resources/getting_started_with_QR_Codes. pdf. 2008 Nov;28(10):2014.

[3] Latif LA, Fadzil M, Azzman TA, Ng MS. Can The Use Of QR Codes Enhance M-Learning In A Blended Learning Environment?. Journal Lifelong Learning Society. 2011;8(2):1-20.

[4] Kamal SA, Tasir Z. Pembelajaran masa depan–mobile learning (m-learning) di malaysia. InSeminar Penyelidikan Pendidikan Pasca Ijazah 2008 (pp. 25-27).

[5] Nordin, A., Husain, W., & Hazwane, W. N, "Pembangunan Portal Web Berasaskan Moodle Bagi Tajuk Jadual Berkala Unsur. Pembangunan Portal Web Berasaskan Moodle Bagi Tajuk Jadual Berkala Unsur," Fakulti Pendidikan, Universiti Teknologi Malaysia, Skudai, 2010

[6] Rao S, Malhan IV. Portal approach to library websites: libraries need to discover new integrated platforms.

[7] Maloney K. Portals in libraries: library technology and planning for change. en ligne]. Bulletin of American society for information science and technology. 2004 Oct;31(1).

[8] Gant JP, Gant DB. Web portal functionality and state government e-service. InProceedings of the 35th Annual Hawaii International Conference on System Sciences 2002 Jan 10 (pp. 1627-1636). IEEE.

[9] Kopf S, Scheele N, Winschel L, Effelsberg W. Improving activity and motivation of students with innovative teaching and learning technologies.

Technology Development of Salt Products Using Geomembrane Thread Filter Technology in Kajhu Village Baitussalam, Aceh Besar

Wahyu Lestari[1*], Ratu Fazlia Inda Rahmayani[2], Adhisyahfitri Evalina[3], Mulia Saputra[4], and Ade Irma Suryani[5]

[1]Faculty of Medicine Syiah Kuala University, Banda Aceh
[2]Faculty of Teacher Training and Education Syiah Kuala University, Banda Aceh
[3,4,5]Faculty of Economics and Business Syiah Kuala University, Banda Aceh

*wahyu_lestari2000@yahoo.com

Abstract. The purpose of this community service is to increase sales turnover and expand the thread filter technology technique. The method of implementing the service activities in this program aims to improve the welfare of the people who work as salt farmers who previously processed salt using traditional techniques. Therefore, this service activity desires to develop and apply the latest technology, namely Thread Filter and/or Geomembrane (plastic), to provide solutions to salt farmers in developing their business in an easier and more modern way which increases the quantity and quality of the salt produced. Besides, through this service salt farmers are also trained on how to process salt into nanoparticles, as a good product for skin care. Thus, the salt production process will be easier and faster and the benefits obtained by salt farmers will be higher. In this service activity, it will focus on traditional salt farmers to be able to switch and apply the technology of thread filter (TUF) or Geomembrane (plastic) as a process of making salt easily and competently for the community. The results showed that using Thread Filter Technology (TUF) was able to produce results of four to five tons after using geomembrane. Previously without using TUF farmers from the village were only able to produce three tons of salt. The prospect of salt farm business is profitable when applied the techniques of TUF or geomembrane. Finally, the salt farmers can earn income significantly.

Keywords: Sea sand, salt, iodized, thread filter, skin care

1. Introduction

1.1. Situation Analysis

The Salt Industry in Indonesia, especially in Aceh is part of the creative industry which is currently promoted by the government to create prosperity and increase employment opportunities for the Acehnese people. The demand of salt in the daily lives of people is high, it makes the salt industry not only act as an economic commodity but as a fulfillment of the consumptive of housewives as a complement to spices. Salt is one of the most needed ingredients for the Acehnese people. So, it is not surprising that many salt industries in Aceh are commodities that are profitable to the sustainability of human resources.

However, from these various potentials, it can be observed that the competitiveness of various salt products in Aceh is still not optimal, such as packaging design which is less attractive and seems cheap or less trendy, it is a result of minor innovation and creation and quality of products. Therefore, to maintain its sustainability, Aceh's food production, especially salt processing, must be touched by innovation, aesthetics, creativity and improving the quality and diversification of products. Furthermore, innovation, aesthetics, creativity, quality and product diversification can be realized from the variety of products, a combination of raw materials, high level of precision/product neatness, and the effectiveness of product accessories. Finally, these salt products will be more attractive and attractive and easily marketed, both as a local and national community.

1.2 Salt Production Process

The salt production process is conducted in a general way, namely by evaporation of sunlight, usually, the salt farmers make salt using the method of plots for evaporation. Salt farmers directly vaporize the flow of seawater in partitions to produce levels of Baume (it is the density of liquid/density/thickness) in salt. The results obtained at each salt processing are very dependent on the sun's weather. If the dry season, salt farmers can process and produce a lot of salt, and vice versa if the rainy season and there is no solar heat for evaporation, the farmers stop production until the weather improves again. In the final stages of the packaging process, salt is packaged in a simple manner with transparent white plastic and then distributed to small kiosks in Aceh Besar.

Hence, salt farmer produces 200 kg of salt in a day. For example, Mr. Muslem Yusuf managed the business for around 9 years. All production activities are held with his family. The salt processing that he is working on is still fairly traditional, starting from cooking using a traditional stove, and the process of crystallizing salt is very influential on the weather, even if the rainy season arrives, Mr. Muslem Yusuf is forced not to produce salt until the weather returns to light. During the dry season, salt production can reach up to 3 tons of salt per month, while in the rainy season it is only 1.5 tons. With this production capacity, Mr. Muslem Yusuf obtained an average income of Rp. 1.600.000 per month depending on the level of weather and the production process.

Mr. Muslem Yusuf salt production has not used proper packaging and does not competitively in the market because he still uses plain white plastic for its packaging, it does not reach the market extensively. The selling price per package is Rp8.000 per kg, depending on demand, if consumers buy salt that has not been packaged with large quantities it will cheaper. The development of products that are still low and lack of expertise in the aspect of marketing their products, namely they sell their products by depositing salt in every kiosk in the area of Aceh Besar because according to him this method is an effective way to market the product.

To improve the quality of salt production which is currently a market demand, salt farmers must be willing to add the available facilities. Because at present the production of salt is considered to not meet SNI requirements, namely the low NaCl value, the brownish color, and fragile color. Therefore to overcome the existing problems geomembrane technology is currently being developed. In the geomembrane technology, the entire crystallization table is coated with plastic sheeting to ensure the cleanliness of salt production.

Moreover, Geomembrane technology is suitable for cultivating salt during the salt production period, the farmers can harvest the salt continuously. The farmers do not need to worry about the quality of the salt produced because the salt crystals do not come into contact with the ground, it will result in white, clean and weighty salt crystals. In addition to the crystallization table coated with plastic sheeting also on the old water intake channel from the old water reservoir to the crystallization table, it needs to be coated with plastic tarpaulin, this is intended to prevent soil sludge in the intake channel from being brought into the crystallization table, divide the entry of old water into the crystallization table.

1.3 The Problem of Salt Industry

There are several priority issues to be resolved in the salt production business becomes more developed in the aspects of production, management aspects, and business marketing aspects. There are the following issues:

a. Production Aspect

- Branding the product of salt business to be easy for marketed widely and no longer use ordinary plastic.
- Training the ability of group members to combine other ingredients in producing salt, such as adding iodine.
- Improved production equipment that is more adequate (for example More sophisticated cauldron burners and other equipment).

b. Management Aspect

- Increase salt quality and quantity
- Management training in business improvement.
- Control the salt industry in order to develop.

c. Marketing Aspect

- There are many marketable salt products it is caused local salt products to be less attractive in the market.
- The marketing process is not conducted effectively.

2. Methodology

The community service activities held included product creativity and increased management capacity/partner business organizations. Therefore, the implementation of science and technology was carried out through a comprehensive and multidisciplinary approach. The program also involves the community of Sira Kajhu "Acehnese Salt" group as beneficiaries of the program.

The training method is designed using a participatory approach with reference to the andragogy learning process, through an approach:

1) Presentation of material equipped with PowerPoint Presentations and simulations
2) Discussion: limited, group
3) Practice, and
4) Feedback from business actors in the form of understanding the material provided.

Production capacity improvement is carried out through two pieces of training including:

1) Training in innovation and creativity of Aceh salt products,
2) Procurement of combustion engines and other processing equipment, both of which are accompanied by assistance.
3) Mentoring is done by involving experts in culinary and salt flavors Aceh.

To improve capabilities in marketing, partners will be given material related to marketing techniques such as determining target consumers, analyzing the market or consumer needs, creating unique products and having design creativity, product marketing through the internet (web), making brand and product logos, and market development techniques and market breakthroughs.

Development in the field of HR is carried out through training and development of worker expertise, increased collaboration among group members. Besides, there is also a development of strategies for creating a group environment situation to brainstorm or create ideas in a short time, and provide motivation to the Sira Kajhu group to generate enthusiasm for producing innovative products, creative and obtained a high market value.

Training and mentoring preparation of financial reports accompanied by a prepared format. The implementation of the financial recording cycle is provided in accordance with the needs of business actors and financial institutions that will help businesses obtain business capital loans. This activity starts with collecting transaction evidence. Training and mentoring the preparation of financial reports accompanied by a prepared format. The application of the financial recording cycle is provided according to needs.

Monitoring and evaluation were conducted during the activities using data that had been collected in this service activity, including diaries, FGDs (discussions with the Sira Kajhu group), and observations to produce conclusions and recommendations on the overall activities.

The community service program includes product creativity and capacity building for partner management/business organizations. Therefore, the implementation of science and technology is carried out through a comprehensive and multidisciplinary approach. Increasing production capacity is carried out through the stages of salt processing creativity training, by utilizing other parties who have expertise in related fields.

3. Community Service Result

The purpose of this community service is to increase the income of salt entrepreneurs by introducing the technique of Geomembrane Thread Filter (TUF) of salt production. In the process of making salt using Geomembrane, it requires modification of farmland by adding threads in the initial stages with the aim of accelerating the aging process of seawater.

In this technique, the thread filter is made in the form of a partition of a winding soil pool with an uneven base to create a natural flow of water so that the evaporation process is assisted by sunlight and wind. Then, the presence of a thread filter it is expected to accelerate the aging of seawater, the production process is shorter than 40 days of land preparation until production becomes 25 days.

The problems faced by salt entrepreneurs have been going on from the beginning, starting from the process of salt production, packaging, and marketing. Inadequate production equipment and appropriate science and procedures in the process of making salt in a modern way are still lack of understanding by salt entrepreneurs. Moreover, salt entrepreneurs still rely on the knowledge they learn on their own based on the experience they have gotten from previous farmers and surrounding communities who have the same business.

However, during the dry season, salt production can reach 3 tons every month, while the rainy season is only 1.5 tons. With this production capacity, Mr. Muslim Yusuf as a salt farmer obtained an average income of Rp1.600.000 per month depending on the level of weather and the production process. The selling price of salt per 1 kg is Rp8000 because the packaging of the salt product is attractive which influence the income is less profitable then the opportunity to expand the business cannot be achieved.

Regarding the current problems, in order to increase the production and quality of salt farmers, they need to be involved with technology for the salt farmers. Thus, for increasing production, it is necessary to arrange existing land, namely changing land from traditional to semi-intensive, because on traditional land it generally consists of light reservoirs, ponds, crystallization tables while old water reservoirs are only around the crystallization table with trenches. In semi-intensive land consists of light reservoirs, ponds, threaded pools, venerable water reservoirs, and crystallization tables. From these differences in semi-intensive land, old water will be quickly obtained, namely, by adding a screw pool, and to increase salt production, the crystallization table is expanded. It is not necessary to worry about a shortage of venerable water because venerable water stocks are available in venerable water storage ponds.

In this activity, it is more focused on the technique of developing the salt industry using Geomembrane Filter (TUF) Technology. The main principle of this technology is to accelerate the process of making venerable water (20° Be) by extending the flow of water and still maintaining clean water. The process of maintaining clean water is done by installing a filter on the water channel. Finally, the last step is to put the iodine into salt.
, there was no iodine in it.

However, after finishing the program, there are other activities held, it is making the salt into iodized salt. The purpose is that salt in the village can also be a healthy salt that is useful for people who consume it. Then, the new packaging design is provided. The design uses a new logo and it is in the form of thin plastic and thick plastic. The aim is to attract buyers to buy salt in the village.

Thus, the next is conducting counseling. Service extension activities were attended by salt business partners, namely partners in the salt production business group and salt marketing business group. Besides, this counseling activities also participated by village officials and communities who are interested in salt production.

In this event, there three invited speakers from Syiah Kuala University, Mr. Fahrizal., STP.,MA who discussed salt processing and production using Geomembrane and Thread Filter Technology (TUF) and the second speaker was from the Department of Chemistry of the Syiah University, she is Ratu Fazli Inda Rahmayani, S.Pd., M.Sc, which talks about processing salt through nanoparticles into cosmetic products. Then, the last speaker is Mr.Dr. Mulia Saputra., SE., M.Sc., Ak, who speak about how to make profits from salt production and how to streamline the efficient and profitable marketing process.

Furthermore, this activity will continue with nanoparticle filtering into cosmetics. These cosmetic examinations will be held at Balai POM until the licensing process in order to be legal in selling the salt for cosmetics.

Regarding these service activities, the results obtained were double the number of salt processing production, and selling quantity of sale has been increased. This is because of implementing an efficient marketing system. Then cosmetics produced from these salts can be bought and sold for skin health. This activity generated in the increase the marketing of salt makers increased production and raise profits for salt farmers.

4. Conclusion

This community service introduces the pattern of producing salt from traditional into the modern system by using thread filters technology (TUF) and it results that salt entrepreneurs find more and cleaner salt products than the previous production. This activity has presented by a speaker who talked about the salt production process and then filtered it using nanoparticles to produce cosmetics product for skin care. Then, the last it was the process of marketing the product to increase the benefit for the salt farmers group. Finally, the speaker discussed how to achieve maximum profits from the sale of salt and cosmetic products and also introducing efficient and effective marketing flow.

References

1. Clark, V.L.,2009, Food Ingredient, http://www.vlclark.com/foodaddive.html, accessed on 25 February 2011 Kurniawan, T., Azizi, A and Yanti B.V.I. 2011. Revenue Analysis of Salt Business in Pati District. Proceedings of the National Seminar on Research and Virtue on Marine and Fisheries Social Economy.
2. Chairunnisa. 2011. The Effect of Using Iodized Salt on Short Nutrition Nutrition status in Amuntai Tengah District, Huli Sungai Utara Regency. Undergraduate Thesis: STIMIK Husada Borneo Banjar Baru.
3. Department of Health RI 2000.Guidelines for Monitoring Iodized Salt at the Community Level. Jakarta: Depkes
4. Huda Nailul. 2013. Analysis of the Local Salt Industry in Rembang District(Structure Conduct Performance Approach). Undergraduate Thesis: Faculty of Economics and Business, Diponegoro University, Semarang.
5. Lee, J.D., Concise Inorganic Chemistry, Mc Graw Hill Company. New York:2002.
6. Mittiatur, Edma. Pressure filtering technique.Paper: Department of Pharmacy, Faculty of Medicine and Health Sciences, Jenderal Soedirman University.
7. Purbani, D. The Process of Forming Salt Crystallization.Marine area research center and non-biological resources. Jakarta. 2001.

8. Salt Institute,.2011. http://www.saltinstitute.org/Production% industry/Production.html Accessed on 20 February 2011.
9. Sanusi. 2012. Salt Production Process with Threaded Filter Technology. Salt Farmers of Flowen Aras, Ambulu Losari, Cirebon.
10. Wiendy, 2010.http://wiendy89.blogspot.com/2010/01/fungsi%manajemen%strategis.html
11. Lestari, Wahyu, 2017. www.kba.one/news/mendulang-harapan-di-ladang-garam /index.htmlAccessed on 10 October 2017.

Application of Environmental Awareness and Environmental Management System towards Behavior at Palm Cooking Oil Company, Bekasi City

Muhammad Yusuf Amartha[1], **Udi S Hamzah**[1], **Herdis Herdiansyah**[1,*]

[1]School of Environmental Science, Universitas Indonesia, Jakarta 10430, Indonesia

[*]herdis@ui.ac.id

Abstract. Environmental pollution generally occurs because of the behavior of workers who are not environmentally friendly. The purpose of this research is to analyze the relationship of environmental awareness to the behavior of workers of palm cooking oil company; examine the relationship of implementation of ISO 14001 environmental management system to the behavior of workers; analyzing the relationship of implementation of ISO 14001 environmental management system and environmental awareness simultaneously to the act of workers. The results were processed by univariate, bivariate, and multivariate methods. The results of the analysis show a weak and unidirectional relationship between environmental awareness and employee behavior with correlation value (R=0.23). Different results on the ISO 14001 environmental management system with worker behavior. The results show no relationship between the ISO 14001 environmental management system and workers' behavior with correlation values (R=0.39). While the correlation between environmental awareness and ISO 14001 environmental management system to worker behavior showed a strong correlation with correlation value (R=0.56). The conclusion that can be drawn from the results of the analysis is the awareness of the environment plays an essential role in the behavior of workers in the eenvironment. Corporate programs such as the ISO 14001 environmental management system can not run effectively without the support of environmental awareness.

1. Introduction

Sustainable development is one of the alternatives undertaken to ensure environmental sustainability to meet the needs of present and future generations. An essential component of sustainable development consists of three main pillars, namely environmental pillars, economic pillars and social pillars. The implementation of EMS ISO 14001 provides a gradual, systematic, efficient and sustainable environmental improvement [1]. Positive benefits of the system approach are the anticipation of short-term and long-term impacts of environmental aspects of environmental conditions and business activities, as well as relationships with governments and consumers [2]. Lack of awareness and worker participation in the environment at work that elicits behavior does not care about environmental management [3]. On the description formulated research questions as follows:

Is there a relationship of environmental awareness to the behavior of workers?

Is there any relation of implemented ISO 14001 environmental management system to worker behavior?

Is there a relationship between implemented ISO 14001 environmental management system and environmental awareness simultaneously with workers behavior?

The purpose of this research is to analyze the relationship of environmental awareness to the behavior of workers of palm cooking oil company; investigate the relationship of implementation of ISO 14001 environmental management system to the behavior of workers; analyzing the relationship of application of ISO 14001 environmental management system and environmental awareness simultaneously to the behavior of workers.

Environmental psychology is a branch of psychology that discusses human behavior with the surrounding physical environment, the process of human perception and cognition is more important than the study of behavior is visible (overt behavior). Human behavior is more caused by processes of perception. Objects, perception, and settings are a unity in the perception process [4]. Environmental consciousness is defined as a state of conscious mind capable of controlling the mind and being able to choose such things as good and bad. Another definition of environmental awareness is the condition in which there is a feeling for the environment, and each shows it through action and behavior. Binding communication can be useful for raising awareness [5]. Environmental awareness is critical in realizing environmental sustainability [6]. Studies conducted by the author show that the development of awareness in Indonesia is now growing significantly in various fields, especially in the field of educational institutions [7], and especially in the industrial sector. Greater environmental awareness leads to greater involvement in land management programs [8]. The effort involves every citizen in growing and nurturing awareness to preserve the environment based on the values, that is the value of the environment itself with the philosophy of living in peace with the natural environment [9]. Definition of Environmental Awareness on this research is thought that underlies the behavior to preserve the environment in the company.

Environmental Management System (EMS) ISO 14001 is a document of environmental management system specification. This Standard contains elements to be met by companies seeking certification for the implementation of the ISO 14001 EMS standard. All elements specified in ISO 14001 must be implemented and documented to obtain certification from a third party certification body. The documentation of all elements that have been applied is used as the responsibility of the certification body that the organization or company has implemented and implemented the EMS ISO 14001 suitably [11]. EMS ISO 14001 on this palm cooking oil company used to comply with regulations and responsibilities to support environmental protection, prevent pollution, and gain economic benefits through improvements in overall environmental performance [10]. Definition of EnvironmentalManagement System on this research is a program designed to compensate for environmental problems that occur as a result of operations in the company. The differences in the psychological and behavioral variables related to respondents' awareness of EMS [11]. EMS factors such as regulatory agency bureaucracy have a firm rule for that support pro-environmental behavior [12].

The notion of behavior is a collection of one's life experiences that encompasses the interaction of the individual with his or her environment embodied in the form of the individual's knowledge, attitude, and actions [13]. Behavior is a response to the stimulus received by the individual from the outside [14]. This response is divided into two, namely:

a. Reflexive is the response generated by spesific stimuli. Commonly called eliciting stimulation which results in a relatively fixed response;
b. The operant response is the response that arises and develops then followed by a certain stimulus or stimulation. This stimulation is called reinforcing stimulation.

Behavior can also be said as an individual reaction to the stimulus received. Individual reactions to various stimuli are two things: active reactions and passive reactions. Active reactions occur when the individual receiving the stimulus reacts in the form of real action.
The behavior can be divided into two:

1. Internal factors, namely individual characteristics are innate, for example the level of intelligence, emotional level, and others.
2. External factors, ie environmental influences, whether physical or non-physical, socio-economic, and cultural. Environmental factors are generally the main factors affecting one's behavior.

While the passive reaction is when individuals give a reaction such as thinking, opinion, attitude. Behavioral forms are divided into three domains with terms of knowledge, perception, and action [15].

Knowledge is generally derived from the sense of sight and hearing. Educational factors significantly affect one's consciousness. This issue is due to the higher a person's education, the more expert knowledge he or she possesses [16].

The definition of perception is a person's interpretation of the stimulus received by the person through the sensing device [17]. Workers who do not yet have a good perception of environmental awareness. Judgment is formed from the observation of an object, which is subsequently concluded in various forms that depend on the attention of the individual who made the remarks.

The higher the attention of the individual in observing something, the better will be in response to the object [18]. Based on the development of action is a mechanical response to the stimulus received, whereas behavior is an active and creative mental process [19]. The behavior of individuals and groups is influenced by the social, cultural and personal systems of each. The pro-environmental atmosphere and a consistent sustainability policy might significantly foster pro-environmental behaviors [20]. In the process of behavioral change, there is a theory of learning Classic conditioning and Operant Conditioning [21]. Definition of worker behavior in this research is all activities of the workers during the operational activities in the company.

2. Method

This research uses a quantitative approach with the method used in this research is a mixed method between quantitative and qualitative methods. The research location is in the working area of the Palm cooking oil company located in the City of Bekasi, West Java Province, Indonesia. The sample takes by simple random sampling by using the Slovin formula:

$$n=\frac{N}{1+N\alpha^2} \quad (1)$$

Information:
n = Number of samples
N = Number of Population
α = Errors 5%

With a total population of 314 Workers, this calculation uses a percentage error of 5% in the calculation. The result of the calculation resulted in the minimum number of sample to be the respondent is 176 people, to anticipate the unusable result from the existing sample, the number of samples plus 10% from the minimum sample amount. The total sample size for this research is 194 people. This research was conducted to get correlation value between environmental awareness variable, environmental management system, and worker behavior. Before performing correlation analysis, data collection will be performed for the validation test and reliability test on the questionnaire to be used.

3. Results and discussion

Validation and Reliability

The research begins by performing validation and reliability tests on the instruments used. The validity test of environmental awareness instrument with 18 items of the question was obtained by 18 items of a valid question. the reliability test was for this environmental awareness instrument obtained alpha value of 0.934> 0.361. In the validity test of EMS ISO 14001 instrument obtained 8 valid question items from the number of 8 questions. Reliability test which for instrument EMS ISO 14001 obtained alpha value 0.896> 0.361. Invalidity test instrument obtained 8 items valid question from the number of 8 questions. Reliability test where for the behavior instrument obtained the alphavalue of 0.810> 0.361. The results of validity and reliability of the items on the questionnaire said to be reliable or reliable as a means of collecting data on research.

Environmental Awareness Relation to Worker Behavior

The result of correlation test of environment awareness variable to a behavioral variable to 194 sample that chosen by random has obtained the value of significance equal to 0.001 which means calculation value <0.05 so concluded that there is a relation between environmental awareness and behavior variable. Based on the value obtained from the calculation of the value of the Pearson correlation of 0.23. A value of 0.23 provides a clarification of the known relationship test if the environmental awareness and behavioral variables have weak relationships. As for the direction of the relationship is a positive relationship, which means if the value of environmental awareness variables increases, then the value of behavioral variables also increase.

Table 1. Environmental Awareness Correlation Test against Behavior

Korelasi		Environmental Awareness	Behavior
Environmental Awareness	*Pearson Correlation*	1	0.230**
	Sig. (2-tailed)		0.001
	N	194	194
Behavior	*Pearson Correlation*	0.230**	1
	Sig. (2-tailed)	0.001	
	N	194	194

Workers' knowledge of the environment is derived from the formal and non-formal education of such workers such as counseling, discussions, exercises, and others. The worker's knowledge comes from the observation of the worker using his or her senses, such as the eyes and ears. In line with the notion that knowledge is acquired by man through the means of his thoughts, especially the eyes and ears [22].

Behavior as desire, interest, will, knowledge, emotion, attitude, motivation, reaction, etc. and other factors such as experiences, beliefs, physical means, socio-cultural society and so on. Thus the low relation between environmental awareness and worker behavior can be caused by several factors [23]. Factors that affect environmental awareness consist of ignorance, poverty, humanity, and lifestyle. The lack of socialization and direction towards workers in the field leaves workers with no long-term thinking about the impact of their actions that pollute the environment [24].

Relation of Implementation of the environmental management system to Worker Behavior

The result of correlation test of the variable of EMS ISO 14001 to behavioral variable got the value of significance equal to 0.339 which mean calculation value> 0.05 so concluded that there is no relation between the variable of ISO 14001 and EMS behavior.

Table 2. EMS ISO 14001 Correlation Test against Behavior

Korelasi		EMS ISO 14001	Behavior
EMS ISO 14001	*Pearson Correlation*	1	0.069
	Sig. (2-tailed)		0.339
	N	194	194
Behavior	*Pearson Correlation*	0.069	1
	Sig. (2-tailed)	0.339	
	N	194	194

Based on the results of a correlation test between the variable EMS ISO 14001 with the behavior of workers. Value test results correlation variable EMS ISO 14001 to behavioral variables obtained the value of significance of 0.339 which means the calculation value>0.05. If the value of significance <0.05, then correlated. If the value of significance>0.05 then not correlated. Based on the test results obtained conclusion there is no significant relationship EMS ISO 14001 with employee behavior, so Ha rejected, and H0 accepted.

The problem of implementing EMS ISO 14001 in an organization or company so not getting the benefit as it should is the lack of commitment from the top that makes the employees are not eager to run the system. The lack of socialization of management on all elements in the company, so that knowledge workers are inadequate to run the system. Lack of awareness and worker participation in the environment at work that leads to uncaring behavior towards environmental management [25].

For This Company, the absence of a relationship between the ISO 14001 EMS variable and the worker's behavior may be the lack of training or socialization of each worker and the lack of attention, awareness, and the participation of workers towards its application. Therefore, the high workload with limited time targets makes workers tend to focus their attention to complete the work and assume EMS ISO 14001 is only a formality that must be met. So in full awareness, they understand about EMS ISO 14001, but not apply it in everyday life.

Relation of Environmental Awareness and Implementation of the environmental management system to Worker Behavior

Relationship of environmental awareness and EMS ISO 14001 simultaneously to workers behavior, used multiple regression testing. The level of trust used in this test is 95% to test the hypothesis "There is an influence of environmental awareness (X1) and EMS ISO 14001 (X2) together on Behavior (Y)" under the following conditions:

1. If sig value <0,05 then there is the influence of variable X Simultaneously to Y;
2. If the value of sig> 0.05 then there is no influence of variable X Against Y.

Table 3. Multiple Regression Test of Environmental Awareness and ISO 14001 EMS against Behavior

	Model	Sig.
1	*Regression*	0.004^b
	Residual	
	Total	

The value of significance for the effect of X1 and X2 on Y is 0.004> 0.05 so that it can be concluded that the hypothesis is accepted which means there is the influence of X1 and X2 simultaneously to Y.R square value of 0.56 this means that the effect of variables X1 and X2 simultaneously affect 56% of variable Y. This shows the better environmental awareness and understanding of workers to EMS ISO 14001, will make the behavior of workers better too. Environmental awareness variables form the basis of changes in workers' behavior to care for the environment, it is better to be collaborated with a good understanding of ISO 14001 EMS to support environmentally friendly behavior. The relationship between environmental awareness and implementation EMS ISO 14001 to the behavior of workers is the better the environmental awareness and understanding that workers have of EMS ISO 14001, the better the worker's behavior. A worker who has good environmental awareness and good understanding implementation EMS potentially have better behavior to the environment. If the worker has proper knowledge, good perception, and good action will support worker to have good behavior toward the environment. Good environmental awareness and good implementation ofEMS ISO 14001 will further support changes in employee behavior better in their work.

4. Conclusion

There is a weak relationship between environmental awareness of the behavior of workers. There is no relationship between SML ISO 14001 with the behavior of workers. There is a relationship between environmental awareness and SML ISO 14001 applied simultaneously to the behavior of workers. This

Journal of Physics: Conference Series **1424** (2019) 012051 doi:10.1088/1742-6596/1424/1/012051

study concludes that there is a weak relationship between environmental awareness of the behavior of workers if it stands alone.In contrast to the ISO 14001 EMS relationship with employee behavior when applied separately to give no significant correlation. Furthermore, environmental awareness and EMS ISO 14001 applied simultaneously will affect on the behavior of workers of a Palm cooking oil company on Bekasi City

References

[1] ISO 14001:2004, , British Standard Institution (2004).
[2] A. Martinis; S. Mazi; C. Minotou, , Global Business & Finance Review, 132-146, (2016..
[3] A. Fisher, P.A.Bell, & A. Baum, Holt, Rinehart,&Winston (1984).
[4] A. Fisher, P.A.Bell, & A. Baum, Holt, Rinehart,&Winston (1984).
[5] P. Aymeric, P. Alexandre, J. Milena, K. Myriam, F. Marie-Line, and G. Nicolas, Environment and Behavior 339 –353. (2017).
[6] H. Herdiansyah et al, IOP Conf. Ser.: Earth Environ, Sci, **30,** 012017 (2016)
[7] H. Wu &L. Mweemba, Springer Environ Dev Sustain, 727-744. (2010).
[8] A. Neolaka. . Rineka Cipta Jakarta, (2008).
[9] ISO 14001:2004. British Standard Institution. (2004).
[10] S. Keraf,: PT. KOMPAS Media Nusantara, (2002).
[11] M. Andersson, O.Eriksson, and C. Borgstede, Sustainability. 1292-1308. (2012).
[12] C. Kola-Lawal, M. Wood, B. Alo, and A. Clark, Journal of Sustainable Development of Energy, Water and Environment Systems 408-421, (2014.
[13] S. Sarwono,; Rineka Cipta, (2004).
[14] S, Notoatmodjo, PT. Rineka Cipta(2010).
[15] Z. Ali,,Jakarta: Widya Medika (2002).
[16] S. Notoatmodjo, PT. Rineka Cipta (2007).
[17] A. Sugiharto, Universitas Diponegoro Semarang: Disertasi (2007).
[18] J. Rakhmat, Rosdakarya, (2007).
[19] D. Jones, Delaprasta (2009).
[20] L. Fu, Y. Zhang, X. Xiong, Y. Bai, ,Science and Technology Education. 427 –445. (2018).
[21] Geller & Scoot, *The Psychology Of Safety Handbook,* USA: Lewis Publisher. Delaprasta,(2001).
[22] S. Notoatmodjo, PT Rineka Cipta,(2007).
[23] Z. Ali, Widya Medika, (2002).
[24] ISO 14001:2004, British Standard Institution, (2004).

Compressive Strength Analysis On Geopolymer Paving By Using Waste Substitution Of Carbide Waste And Fly Ash

Mochammad Qomaruddin[1*], Han Ay Lie[2], Arif Hidayat[2], Sudarno[3], Anik Kustirini[4],

[1]Department of Civil Engineering, University Islamic of Nahdlatul Ulama, Indonesia
[2]Department of Civil Engineering, University of Diponegoro, Semarang, Indonesia
[3]Department of Civil Engineering, University of Tidar, Magelang, Indonesia
[4]Department of Civil Engineering, University of Semarang, Indonesia

*qomar@unisnu.ac.id

Abstract. The method to make paving sample in this research was done by mixing fly ash, carbide waste, sand, and alkali activator in the form of sodium silicate (Na_2SiO_3), 8M and 12M sodium hydroxide (NaOH) with ratio 1 : 2 as the binder, then it was pressed by using hydraulic press machine. The paving was made in the size of 20 cm x 10 cm x 6 cm with mix composition of 70% fine aggregate, 30% binder and activator from paving weight. The percentage of binder and activator was 65% and 35%, respectively. The curing was done by using wet burlap sack for about 28 days. After that, the compressive strength test was done. From the result of the compressive strength test of geopolymer paving sample can be concluded that the mixed test of carbide waste and fly ash (10 %: 90%) produced 34,6 MPa, and 39,8 MPa for its compressive strength.

1. Introduction

This Paving is one of the building material product that is often used for road pavement construction. The use of paving supports green construction because rain water falling on the paving surface will enter the soil, so it can maintain the ground water balance. In general, conventional paving blocks made of fine aggregate, cement and water are then molded in a solid state [1]. Past research suggests that portland cement material is not environmentally friendly, because the portland cement production process generates carbon dioxide gas that causes global warming. This carbon dioxide gas (CO_2) is generated during the process of calcination, fuel combustion and electricity consumption during chemical reactions. In the calcination process produced 0.407 tons CO_2 / ton clinker. The amount of carbon dioxide emissions is proportional to the average yield of cement produced by 0.77 tons CO_2 per ton of cement [2]. Carbide wastes include toxic and hazardous materials, if they are large in the soil surface, then rain splashing will soak into the soil, causing groundwater pollution. To reduce waste of carbide and fly ash wastes, as well as maintain environmental balance, both materials are used as raw material for paving polymer. Paving polymer is made of fly ash as an alternative to cement and sodium silicate (Na_2SiO_3) and sodium hydroxide (NaOH) instead of water, serves as activator[3,4].

2. Materials and Methods

2.1. Fly Ash

Fly ash used in this research is from unit 3 and 4 of Electric Steam Power Plant (ESPP) Tanjung Jati B Jepara categorized as fly ash type F. Before being used, EDX analysis test is done to know the chemical characteristic of fly ash.

Table 1. The Chemical Composition Result of Fly Ash

No	Parameter	Unit	Test Result
1	SiO_2	% Wt	46.70
2	Al_2O_3	% Wt	25.01
3	$Fe_2\,O_3$	% Wt	9.43
4	TiO_2	% Wt	0.98
5	CaO	% Wt	6.26
6	MgO	% Wt	3.65
7	K_2O	% Wt	2.19
8	Na_2O	% Wt	2.96
9	SO_3	% Wt	0.81
10	MnO_2	% Wt	0.07
11	P_2O_5	% Wt	0.41
12	Moisturecontent	% Wt	Nd
13	L O I	% Wt	1.05
14	Oil Content	% Wt	0.05
15	Unburned Carbon	% Wt	0.65

Figure 1. Fly Ash

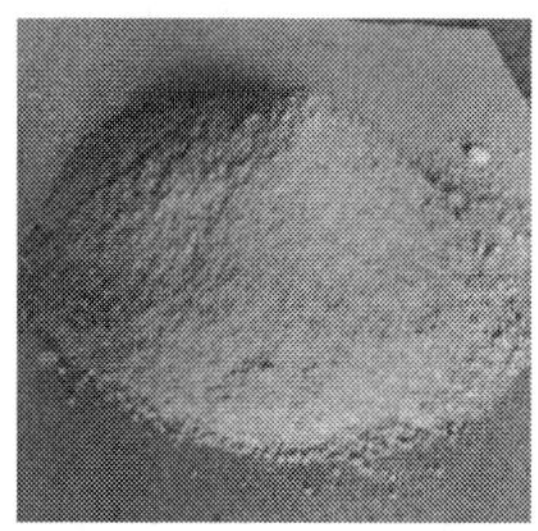

Figure 2. The Carbide Waste

2.2. The Carbide Waste

The carbide waste used in this research is originally from weld-shop in Jepara regency, Indonesia. Before being used, the carbide waste is dried first, because it still wet and the chemical composition of carbide waste by using Scanning Electron Microscopy-Energy Dispersive X-Ray (SEM-EDX) isanalyzed which is aim to know the content in it. The chemical composition of carbide waste can be seen in table 2.

Table 2. The Chemical Analysis Result of Carbide Waste

Component	Composition (% weight)
Carbon, (C)	23,06
Alumina, (Al_2O_3)	4,83
SilicaDioxide, (SiO_2)	1,97
Sulfite, (SO_3)	5,15
CalciumOxyde, (CaO)	63,95
Zinc (II) Oxyde, (CuO)	1,03

2.3. Fine Aggregate

The sand that is used in this research, for example, Muntilan sand. The aggregate passes through 4.76 mm sieve. Before being used, the inspection of material characteristics test is done which consists of: The specific weight of sand based on ASTM C128-01,from the experiment was obtained that the specific weight of sand was 2.43 gr/cm² so that the sand condition has fulfill the needed requirement[5,6,7,8]. The testing of mud content based on PBI 1971(Concrete Regulation Indonesia) N.1-2 analysis on the sand mud content, the result was good which has a maximum limit of 5% in the experimental sample. From the above conclusion, it is found that the sand meets the requirements due to the mud content of 5%.

2.4. The Alkali Activator

This study uses alkali activator in the form of sodium hydroxyde (8M and 12M) derived from white crystals dissolved with clean water and sodium silicate in the form of liquid and viscous. Sodium silicate functioned as a catalyst or mixer of sodium hydroxyde[9,10].

2.5. Mixed Design

Mixed design used in making 8 M and 12 M geopolymer sample will be explained in figure 3:

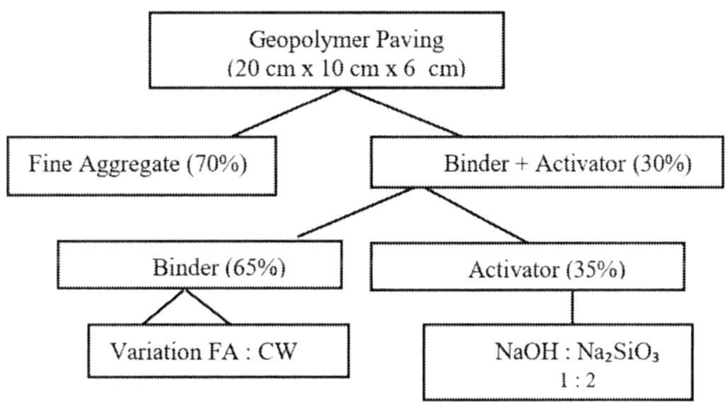

Figure 3. The Flow Diagram about Mixed Design of Geopolymer Paving

3. Result and Discussion

The Compressive Strength Test Result of Geopolymer Paving

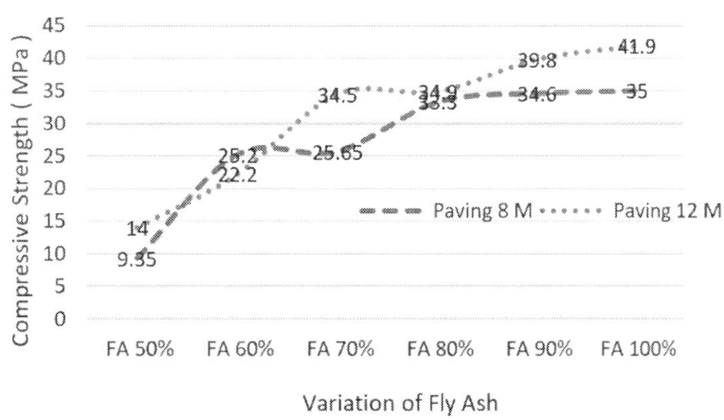

Figure 4. The Compressive Strength Graph of 8M and 12M Geopolymer Paving

Journal of Physics: Conference Series 1424 (2019) 012052 doi:10.1088/1742-6596/1424/1/012052

From the analysis of the figure 4, the compressive strength of 8M geopolymer paving in the age of 28 days with the variation in mixed design of 100% FA has the higher compressive strength equal to 35 MPa. It is different with the ratio variation of FA : CW (50% : 50%) that has the lower compressive strength result equal to 9.35 MPa. The most increasing tendency of FA variation was the increasing started from 60% to 100% variations of FA and significantly increased in variations of 50% to 60% FA that was equal to 9.35 MPa to 25.2 MPa with difference in the compressive strength of 15.85 MPa. Accompany with the increasing of carbide, it was influence the compressive strength of the geopolymer paving. According to[13] and SNI 03-0691-1996, based on the quality of the compressive strength above, most of them were categorized in type A and B with the compressive strength result equal to 35 MPa to 25.1 MPa which is used for roads and parking lots. Figure 4 shows that when on 12M geopolymer paving was more increasing of FA, it will make the compressive strength produced by the geopolymer paving higher[14]. It was seen from the mix ratio of FA 50% to 60% which increased significantly with the compressive strength equal to 14 MPa and 22.2 MPa with the difference of the compressive strength point equal to 82 MPa. It is different in the variation of FA 70% to 100% which the increasing with an average of compressive strength point was 34.5 MPa. The highest point was the mixture of FA 100% in the absence of carbide waste mixture with the compressive strength point equal to 41.9 MPa. The quality of paving blocks based on SNI 03-0691-1996were classified into the A quality category with an average of the compressive strength point was 40 MPa with a minimum of 33 MPa which is enabled for the road.

3.1. The Comparison of Geopolymer Paving

The classification of paving usage as the building material has its own quality especially on 8 M and 12 M geopolymer paving. The 12 M geopolymer paving has higher compressive strength point than 8 M geopolymer paving because of its molarity was bigger than 8 M geopolymer paving although both of them used the same fly ash mixture by using carbide waste. There is the classification table about the use of paving based on SNI-03-0691-1996 towards the compressive strength of 8 M and 12 M geopolymer paving.

Table 3. The Compressive Strength Result of 8M and 12M Geopolymer Paving

The Percentage of Carbide Waste in Each Mix	Compressive Strength (MPa)	
	8 M Paving	12 M Paving
FA50 %	9,35	14
FA60 %	25,2	22,2
FA70 %	25,65	34,5
FA80 %	33,3	34,9
FA90 %	34,6	39,8
FA100 %	35	41,9

- Paving block quality category D, used for gardens, with an average compressive strength of 10 MPa with a minimum of 8.5 MPa.
- Paving block quality category C, used for pedestrians, with an average compressive strength of 15 MPa with a minimum of 12.5 MPa.
- Paving block quality category B, used for parking lots, with an average compressive strength of 20 MPa with a minimum of 17 MPa.
- Paving block quality category A, used for roads, with an average compressive strength of 40 MPa with a minimum of 35 MPa.

From the each mixed design analysis on 8M and 12M geopolymer paving started from the percentage of fly ash (50%, 60%, 70%, 80%, 90%, 100%)with the percentage ratio of carbide waste (50%, 40%, 30%, 20%, 10%, 0%) seen from the graph above, the more increasing of fly ash will make the compressive strength point of geopolymer paving higher and the more increasing of carbide waste will make the compressive strength point decreased. It can be known from these variations that produce a high compressive strength in the 0% variable or do not use weld mixture at all. In the composition of (60% fly ash: 40 % carbide waste) and (50 % fly ash: 50% carbide waste) on 8M and 12M geopolymer paving produced 22 MPa compressive strength. It can be seen that from the graphic above, the higher percentage of the carbide waste mixture and the greater molarity of the geopolymer paving used was resulted in a very drastic reduction[15].

3.2. The Water Absorption on Paving

The water absorption test aims to be an indicator of water absorption into the soil and the saturation or maximum peak absorption rate in the material itself was the size of water absorption generated depends on the density and the number of cavities contained in the paving block. When water is saturated, it can cause humidity in the paving. In the calculation of water absorption with the reference[16] and SNI 03-0691-1996, the maximum in the A quality is 39%.

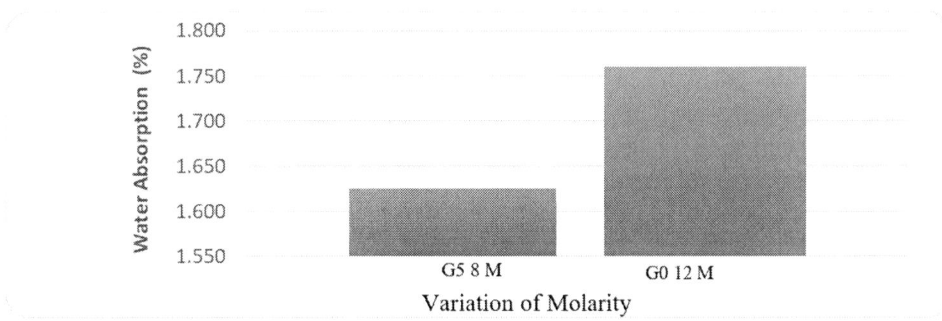

Figure 5. Water Absorption

From the calculation Figure 5 show, the result of the higher absorption was on 12M geopolymer paving that equal to 1.76% and the lower absorption of 8M geopolymer paving was 1.63%. It is known that the geopolymer paving that used a large molarity was more absorbed and the water absorption was less when the paving only used small molarity[14].The result of both types of paving fulfilled the standard. It is known that when the pressing process has well condensation, the pores on the paving will reduce so that the water absorption will be less.

4. Conclusion

Based on the research that has been conducted in the increasing of carbide waste ash towards the geopolymer paving, the conclusions can be taken as follows:

a. The more increasing ratio of fly ash in the geopolymer paving mixture produces higher compressive strength point.
b. The more increasing ratio of carbide waste variations with the percentage in the geopolymer paving mixture produces lower compressive strength point. So, it can be concluded that the use of carbide waste is only be used as filler.
c. The increasing of carbide waste cannot support the compressive strength towards the geopolymer paving.
d. The higher water absorption was in 12 M geopolymer paving which is equal to 1.76% and the lower water absorption of 8M geopolymer paving was 1.63%.

References

[1] Qomaruddin, M., Sudarno, "Influence of Bottom-Ash Mixed with Gypsum as Concrete Bricks for Wall Construction Material," Journal Of Applied Environmental And Biological Sciences; 8(4): 109-114, 2018.
[2] Sudarno S., Purwanto P., Pratikso P.," Life Cycle Assessment on Cement Treated Recycling Base (CTRB) Construction",Waste Technology; 2(2): 31-40, 2014.
[3] Sathonsaowaphaka, A., Chindaprasirt, P., Pimraksab, K,"Workability and Strength of Lignite Bottom Ash Geopolymer Mortar,"Journal Of Hazardous Materials; 68: 44-50, 2009.
[4] Chindaprasirt P., Jaturapitakkul C., Chalee W., RattanasakU.,"Comparative Study on The Characteristics of Fly Ash and Bottom Ash Geopolymers", Waste Management; 29: 539–543, 2009.
[5] ASTM C128-01, Standard Test Method for Density, Relative Density (Specific Gravity), and Absorption of Fine Aggregate, ASTM International, West Conshohocken, West Conshohocken, PA 19428-2959, United States,2001.
[6] American Society for Testing and Materials, ASTM C1437-01, " Standard Test Method for Flow of Hydraulic Cement Mortars", Annual Book of ASTM Standards, Philadelphia, 04.01., pp. 611–612, 2005.
[7] ASTM C618. (2010). Standar Specification for Coal Fly Ash and Raw or Calcined Natural Pozzolan for Use. US : ASTM Internasional.
[8] American Society for Testing and Materials, ASTM C109/C109M-02: Standard Test Method for Compressive Strength of Hydraulic Cement Mortars (using 2-in. or [50mm] cube specimens), Annual Book of ASTM standards, Philadelphia, 04.01., pp. 76–81, 2005.
[9] Khale, D., Chaudhary R., "Mechanism of Geopolymerization and Factors Influencingits Development": a review, Journal Material Science; 42: 729–746, 2007.
[10] Bernal SA, Rodríguez ED, Mejía de Gutiérrez R, Gordillo M, Provis JL, "Mechanical and Thermal Characterisationof Geopolymers Based on Silicate-Activated Metakaolin/Slag Blends",Journal Material Science; 42:5477–5486,2011.
[11] Sofi M, van Deventer JSJ, Mendis PA, Lukey GC,"Bond Performance of Reinforcing Bars in Inorganic Polymer Concrete (IPC)",Journal Material Science; 42: 3107-3116, 2007.
[12] Nath P, Sarker PK,"Geopolymer Concrete for Ambient Curing Condition", Proceedings ofAustralasian Structural Engineering Conference. Perth, Australia; 2012.
[13] Phoo-Ngernkham T, Chindaprasirt P, Sata V, Sinsiri T,"High calcium fly ash geopolymer containing diatomite as additive",Indian Journal Engineering Mateial Sciences; 20(4):310-318, 2013.
[14] Chindaprasirt, P.De Silva,P.Crentsil, SG.Hanjitsuwan,S., "Effect of SiO_2 and Al_2O_3 on The Setting and Hardening of High Calcium Fly Ash-Based Geopolymer Systems",Journal Material Science; 47:4876–4883, 2012.
[15] Pathak, A.,Kumar Jha, V.,*,"Synthesis of Geopolymer from Inorganic Construction Waste,"J. Nepal Chem. Soc;30: 45-51,2012.
[16] Lie, AH, Gan, SB.,Suryanto, B.,Priastiwi, AY.,"Influence of Inclusion Stiffness Modulus on Compressive Strength of Concrete" *The 3rd SCESCM Conference,*Bali-Indonesia, 2016.

Analyse Characteristics Physical Properties of Rock Manifestations Geothermal Semurup District of Kerinci Jambi With Hight Temperature

Nova Susanti[*]**, Larlen Nixolas, Didik Aryanto, Failosofah**

Department of Physics, Universitas Jambi, Jambi, Indonesia

Abstract. There were some changes of mineral components, internal structure and moisture content of the rocks constituent after the thermal annealing treatment was given at 400, 500, 600 ^{O} C. Physical of testing rocks like; density, porosity, specific gravity, Absorption have a relationship with temperature. The test results showed that the average density decreased 0.7 g/cm^3, porosity increased 26%, Density unchanged and Absortion increased 35%. The test of rock minerals components X-Ray Fluorescence (XRF).

1. Introduction

Geothermal energy is a renewable energy source that is environmentally friendly (clean and sustainable energy) and is one of the best solutions to overcome the limitation of energy. It makes geothermal energy is so favored by many countries that it has been used as a source of energy, particularly as a geothermal power plant (Rachel, 2015). One of the locations in Jambi that potentially has a geothermal source is hot water Semurup, subdistrict in Kerinci. Therefore it needs a surface manifestations to know the existing of geothermal energy.

Visible manifestations in the area of hot water Semurup Water District, Kerinci, Jambi are the hot springs, rocks and hot mud. Manifestation is an early indication of potential geothermal (Manfred, 2008). To determine the earliest manifestation of the physical characteristics of the subsurface needs first. It was done to at the rocks. The rocks in Semurup area show that the area around the hot spring is composed of andesite. Andesite rocks have special characteristic that is grayish while color and consists of very small grains (Khant, 2013). Thus, this andesite sample should be investigated to know more the physical parameters of andesite so to find the characteristics of the physical properties of rocks though at treatments different annealing temperatures 400, 500, and 600 ° C in a furnace (Weiqiang, 2016)

The study of the physical properties of rocks is very important to be done in a geothermal area because it can show the process of the geothermal system. Accordingly, researchers examined the physical properties of rocks that aims, to determine the impact of thermal effects, density, porosity, specific gravity and absorption (Weiqiang, 2016) using the Law of Archimedes as the basis of the research conducted and characterized the content using *X-Ray Fluorescence* (XRF) (Khant, 2013)

2. Experimental And Methods

The study started from January to March 2017. The testing of the rocks was conducted at the Research Center for a group of materials engineering Lembaga Ilmu Pengetahuan Indonesia (LIPI), it was a test of the physical properties of rocks at different annealing temperature at 400, 500, and 600 ° C in a furnace. The density, porosity, specific gravity and absortion rocks were calculated to know at relationship the effect of temperature given. Then, characterized crystal structure of rock *X-*

Ray Diffaction (XRD) in ~~Laboratarium Physics LIPI~~. The testing of content using *X-Ray Fluorescence* (XRF) was done at the Science Building, State University of Hasanuddin (UNHAS). The siteof the of the research object is the source of the hot springs in Semurup, district Kerinci, Jambi.

Temperature Test
To know the test effect of thermal annealicy, same steps were done as follows :

1. Rock samples in was cleaned using water, then dried with paper towels.
2. Rocks sample was in to cut rectangular pieces using burrs.
3. To get a clean sample parameters, the researcher used ultrasonic cleaner with a solvent mixture of water 80% + 20% solution of Acetone in the glass beaker.
4. A mixture of water and acetone was used 20 times in ultrasonic cleaning. While at the 5th, 10 th, 15 th, 20 th process, the mixture of water and acetone is replaced, Up to 4 times of cleaning, the water looked clearer.
5. Next is the drying process (using oven sample) at 60 ^{0} C for 24 hours
6. Rock sample mass was weighed with Digital Balance GF- 600 and put sample plastic bag and labeled.
7. The last step is postheating (annealing) with a temperature variation of 400 ° C, 500 ° C and 600 ° C respectively for 1 hour.

Physical properties of rocks Test
Tests conducted on the physical properties of rocks at the different annealing temperature 400, 500, and 600 ° C was done as, the following:

1. Rock sample masses was measured in beaker glass with distilled water at variation of temperature (400 ° C, 500 ° C and 600 ° C) three time.
2. To gain the weight + water at without temperature variations (400 ° C, 500 ° C and 600 ° C) the sample was processed in Ultrasonic Cleanerd for 12 minutes, then was dried using a tissue, The dry sample then was weighed again with Digital Balance GF-600.
3. The researcher than calculated the physical properties. They are density, porosity, absorption, spesific grafity of rocks.

3. Results And Discussion

This study was conducted by collecting the data through experiments that was carried out in same stages. Through the provision of thermal effects at 400, 500 and 600 ^{0} C in the rocks to measure the weight and water, we got the results of the physical properties of rocks in the form of density that can be seen below :

Table 1. Relationship between density and thermal effects

Temperature C ^{o}	V = M-S	Density B = D / V g / cm ^{3}
None	0978	0.774
400	0922	0.773
500	3089	0.748
600	1,509	0.740

Table 1 Relationship of in density without temperature of 400, 500 and 600 ° C can be explained based on Figure 1

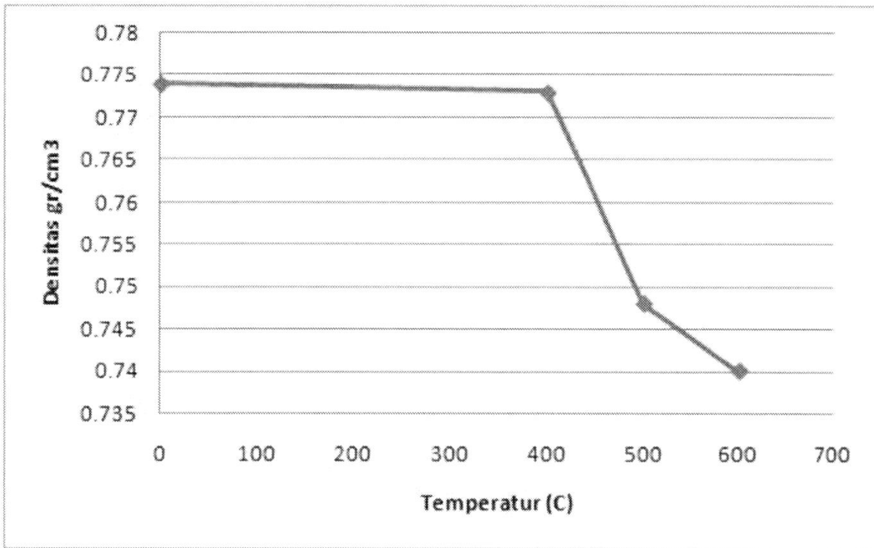

Figure 1. Relationships between density and thermal effects

Test standard that can be used in determining the density, porosity, spesific gravity, and absorption density is ASTM C373-88. In the calculation it should be noted that 1 cm^3 of water weighs 1 g. This applies to all three parts of 1000 g of water at room temperature. Then the equation (1) on the outside volume (V) in units of cm, as follows

$$V = M\text{-}S \tag{1}$$

Calculating the open pore volume (V_{op}) and the watertight part Vip in units cm, the equation (2), as follows

$$Vop = M - D \tag{2}$$

$$Vip = D - S \tag{3}$$

Calculating the density of rocks (B) in units of gr/cm^3 of the object test is the result of the air mass or dry divided by the outside volume, including the pores of rocks. Calculating the density of rocks (B) was the equation (4), as follows

$$B = \frac{D}{V} \tag{4}$$

From equation (4) then, the results showed that the density decreased at the 1st test without giving the temperature to 0.774 g/cm^3, The second test with the provision of thermal annealing at 400 °C, the density is 0.773 g/cm^3. The 3rd test with the provision of thermal annealing of 500 °C, the density is 0.748 g/cm^3. The 4th test with the provision of thermal annealing of 600 °C density of 0,740 g / cm $^{3.}$ Based on the value of density in table 1, the greater the temperature given the density will be smaller.

Table 2. Relationship porosity against thermal effects

Temperature (°C)	V = MS	Porosity (P = [MD) / V] x100%)
There is no	0978	23 619
400	0922	23 752
500	3089	25 509
600	1,509	26 640

Table 2 Relationship of porosity with no variation in temperature, 400°C, 500°C and 600°C can be explained based on Figure 2 below, as follows;

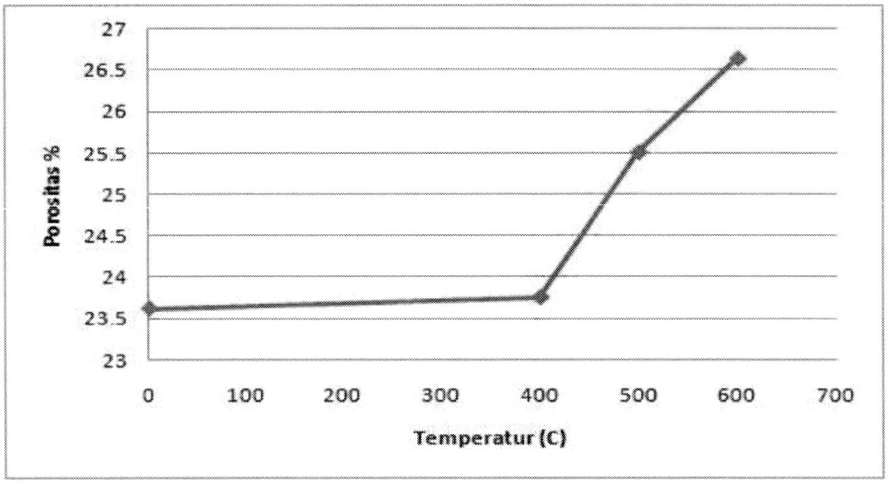

Figure 2. The relationship Porosity against thermal effects

Calculating porosity appears (P) expressed as a percent, so we get the volume of affairs outside. Calculating Porosity looked stated equation (5), as follows;

$$P = \frac{M - V}{D} x100 \tag{5}$$

From equation 5 then, the results showed that the porosity increases on testing to-1 without giving a temperature porosity of 23,619%, on a test-2 with the provision of thermal *annealing* of 400 ° C porosity of 23,752%, on a test 3 with the provision of thermal *annealing* 500 ° C porosity of 25,509%, in the 4th test by administering thermal *annealing* at 600 ° C porosity of 26,640%, Based porosity values in table 2, the greater the temperature given then the greater porosity.

Table 3. Relationship Absorption rocks against thermal effects

Temperature (°C)	Absorption (A = [(MD) / D] x100%)
There is no	30 515
400	30 715
500	34 097
600	35 989

Absorption of Table 3 Relationship with variations without temperature, 400 ° C, 500 ° C and 600 ° C can be explained based on Figure 3 below, as follows;

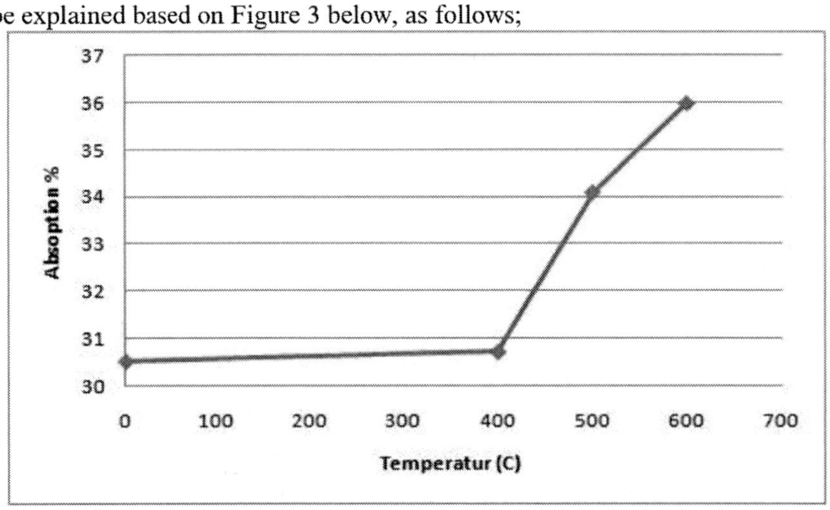

Figure 3. The relationship Absorption rocks against thermal effects

Calculate water absorption (Absorption) expressed as a percent, so it has a relationship once absorbed water mass and the mass of the air or dried. Calculating water absorption is expressed equation (6), as follows;

$$A = \frac{M - D}{D} x100 \qquad (6)$$

From equation 6, the results showed that the porosity increases on testing to-1 without giving a temperature of absorption 30,515% on the test-2 with the provision of thermal *annealing* of 400 ° C absorption amounted to 30,715% on the test 3 with the provision of thermal *annealing* 500 ° C absorption of 34,097%, in the 4th test by administering thermal *annealing* at 600 ° C 35,989% porosity, by absorption values in table 3.3, the greater the temperature given then absorption will be even greater.

Table 4. Relationships density of rocks against thermal effects

Temperature (°C)	Specific gravity T = D / (D-S)
There is no	1.013
400	1.014
500	1.004
600	1.009

Table 4 Relationship of density with no variation in temperature, 400 ° C, 500 ° C and 600 ° C can be explained based on Figure 4 below, as follows;

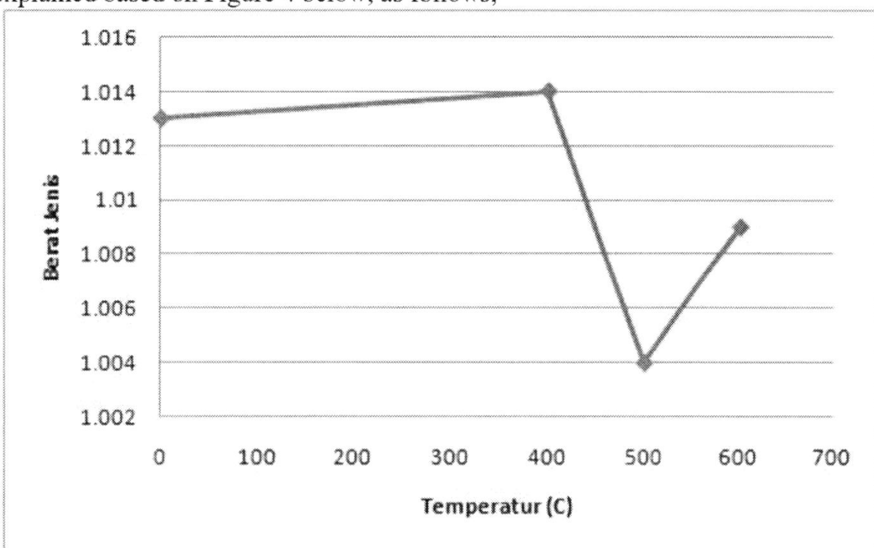

Figure 4. Relationships density of rocks against thermal effects

Calculating Density (T) is a test object is resistant to water, compute density (T) to be stated equation (7), as follows;

$$T = D(D - S) \qquad (7)$$

From equation 7 then, the results showed that the density does not change. In the 1st test without giving a specific gravity of 1.013 temperatures, the 2nd test by administering thermal *annealing* of 400 ° C gravity of 1.014, the testing of all three with the provision of thermal *annealing* of 500 ° C gravity of 1.004, in the 4th test with the provision of thermal *annealing* of 600 ° C for 1,009 porosity, density Based on the values in table 4, the temperature does not affect the density because it does not change the mineral components in the rocks.

Conclusion

By knowing the thermal effects on the physical properties of rocks (density, porosity, absorption and density) by providing thermal *annealing* of 400 ° C, 500 ° C and 600 ° C, then concluded:

1. The greater the temperature given the density will be smaller.

2. The greater the temperature given the porosity and absorption will be even greater.
3. Temperature does not affect the density because it does not change the mineral components in the rocks.

References

1. Zhang Weiqiang, et.al., Experimental Study on the Variation of Physical and Mechanical Properties of Rock After High Temperature Treatment, Applied Thermal Engineering, 98, 197-1304 (2016).
2. Jorand Rachel, et.al., Statistically Reliable Petrophysical Properties of Potential Reservoirrocks for Geothermal Energy use and their Relation Tolithostratigraphy and Rock Composition: The NE Rhenish Massif andthe Lower Rhine Embayment Germany, Geothermics, 53, 413-428 (2015)
3. Khanti Win, et.al., Geochemical Characteristics of Host Rocks of Polymetallic Epithermal Quartz Veins at Soripesa Prospect Area, Sumbawa Island, Indonesia. Procedia Earth and Planetary Science, 6, 30-37 (2013)
4. P.Hochstein Manfred, Sayogi Sudarman, History of Geothermal Exploration in Indonesia from 1970 to 2000, Geothermics, 37, 220-266, (2008)
5. B. Schmidt Roman, et.al., Reactivity of Geothermal Reservoir Rocks Under Temperature Conditions Found in the Upper Rhine Graben Germany, Procedia Earth and Planetary Science, 17, 881-884, (2017)
6. Okazaki. K, et.al., Permeability, porosity and pore geometry evolution during compaction of Neogene sedimentary rocks, Journal of Structural Geology, 62, 1-12, (2014)
7. Duchkov A.D, et.al., Thermal conductivity of the sedimentary-cover rocks of the West Siberian Plate in relation to their humidity and porosity, Russian Geology and Geophysics, 55, 784-792, (2014)

The Effect of Laboratory Based Learning in Developing Physics Teacher Candidates' Skills on Applying Measurement Instruments

Indrawati Wilujeng*[1], Faiz Hasyim[2], Indhah Permatasari[3]

[1,2,3]Physics Education Department, STKIP Al Hikmah Surabaya

*indrawati.physics@gmail.com

Abstract. Laboratory based learning is one kind of blended learning that combines the aspects of knowledge, skills, and attitudes. Simultaneously, students can apply the knowledge they have possessed, practice the skills, as well as growing their scientific attitude. Laboratory based learning method has been conducted for studying about measurement instruments, up to measurement data processing. This research aims to determine the effect of laboratory based learning in developing physics teacher candidates' skills on applying measurement instruments. The design of this research is qualitative in nature involving one class of undergraduate students of Physics Teacher Education Department at STKIP Al Hikmah, which is consist of six students, as the research subjects. Data collecting is obtained through interview result with physics teacher candidates and their laboratory report. Subjects showed higher improvement skills in using measurement instruments through laboratory based learning. Physics teacher candidates have a deeper understanding about measurement instruments that includes using measurement instruments, taking and analysis data, estimating measurement errors, and communicating the result both in written and oral. Laboratory based learning also improves concepts understanding (knowledge aspects) and students' scientific attitudes.

1. Introduction

One of the main skills of a physics teacher is to be able apply measurement instruments properly. Almost all courses of physics education in college will include measurement instruments skill as their subjects and must be taken by students. However, the lecture material is often limited to knowledge only. Students only know about measurement instruments from the books and explanations given by the lecturer. This kind of thing will only be memorized by them. Students should be practicing the knowledge they have received directly, so that they can interact directly with the measurement instruments. How can a physics teacher candidate have good skill on applying measurement instruments if he only gets a theory. Therefore, giving direct experience to physics teacher candidates to apply measurement instruments is important. Students' skills on applying measurement instruments will be develop if classroom learning can be integrated with laboratory activities. Students will interact with those measurement instruments so that their knowledge will be stronger because of direct experience.

Laboratory Based Learning

Laboratory based learning is often applied to science based learning [1]. It is because by applying laboratory based learning, students will have direct experience in finding knowledge. Almroth explains some advantages of laboratory based learning, such as trigger the curiosity of students and students can learn in small groups so that social interaction can be intertwined [2]. Besides, students who have less understanding can ask their friends in one group who have more understanding so that peer teaching will be happen. Laboratory based learning can also develop technical skills, for example applying measurement instruments. Furthermore, Almroth added, that by applying laboratory based learning, the learning will achieve several things at once, such as conceptual understanding, scientific thinking, general understanding of science, and laboratory skills [2].

Laboratory based learning proved can increase learning interest and success in science studying [3]. Laboratory based learning is appropriate to make students think, discuss, and solve the real problems. In addition, laboratory based learning is the best learning model in developing students' skills at once, including: academic skills such as critical thinking and problem solving, communication skills both written and oral, data handling, self management such as time management and planning, and interpersonal skills such as teamwork and leadership [4]. Therefore, laboratory based learning is one kind of blended learning that combines the aspects of knowledge, skills, and attitudes. Simultaneously, students can apply the knowledge they have possessed, practice the skills, as well as grow their scientific attitude.

Applying Measurement Instruments Skills

Applying measurement instruments skills is important to develop because it is a basic skill when conducting experiments in the laboratory. Especially for physics teacher candidates who will give an example for their students. The applying measurement instruments skills in this research include:

- **Calibration**
  Calibration is the first thing to do before taking measurement. Calibration aims to make sure whether the measurement instrument is still in standard conditions or not.
- **Smallest scale value determination**
  The smallest scale value of each measurement instrument depends on the number of lines on the main and nonius scale.
- **Measurement**
  Measurements using measurement instrument should be done carefully. Each measurement instrument have different sensitivities. For example, a thermometer should never be touched by a hand or a fluid container as it will affect the measurement result.
- **Measurement data reading**
  Measurement data reading should be done perpendicularly to avoid parallax errors. Furthermore, we must pay close attention to the measurements shown by the main and nonius scales carefully, along with the smallest scale values of the instrument (especially for vernier calliper and micrometer secrup).
- **Measurement data writing**
  Measurement data writing should always be accompanied by the measurement error. The measurement error is half of the smallest scale value of the measurement instrument.
- **Measurement data processing**
  Measurement data processing shall correspond to the type of measurement, whether single or repeated measurement, and whether direct measurement or not. On a single measurement, we simply write down the measurement result along with the measurement error. However, if repeated measurements must be searched for its mean and standard deviation first.

2. Methodology

This research is a qualitative research that aims to determine the effect of laboratory based learning in developing physics teacher candidates' skills on applying measurement instruments. The subject of this research is one class of undergraduate students of Physics Teacher Education Department at STKIP Al Hikmah, which is consist of six students. The research design consists of three stages. The first stage is the description stage. At this stage, the research subjects are observed during do measurement activities in the laboratory. All activity aspects are observed, that is cognitive, psychomotor, and affective. The second stage is the reduction stage. At this stage, the information that has been obtained in the first stage is reduced to determine the research focus. The selected research focus is the skill of applying measurement instruments. The last stage is the selection stage. At this stage, the research focus that has been selected in the previous stage is described deeper based on the interview results and laboratory reports. So, the description of applying measurement instruments are: how the subject calibrate the instruments, determine the smallest scale value, do measurements, read, write down, and analyze the measurement results. According to [5], the design of qualitative research generally can be described in the Figure 1 below.

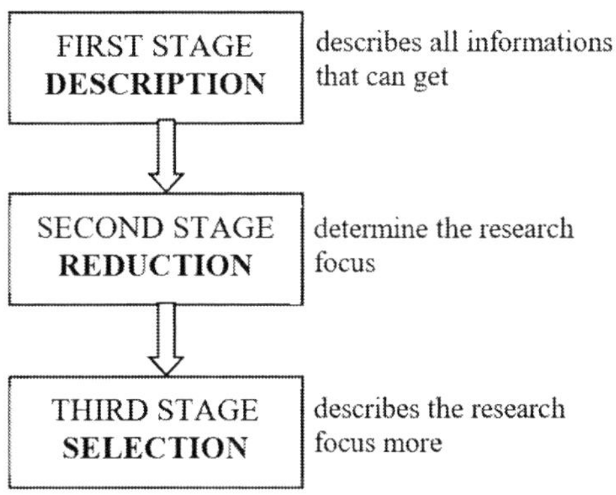

Figure 1. Research design

The research instruments used are observation sheet, questions list to be asked during interview, worksheet and analysis sheet as laboratory report. The data obtained in this research are the observation result, interview result, and worksheet and analysis sheet as laboratory report. Further research data will be analyzed inductively.

3. Result and Discussion

The data obtained are the observation result during the student do measurement activities in laboratory, the interview result, and the laboratory report consist of worksheet and analysis sheet.

Interview Result

Interview was conducted to each student after they work in the laboratory. Interview aims to confirm subjects' understanding of what they have done when do work in laboratory. Further, confirming whether their understanding is due to laboratory based learning or not. In the interview session, the information that explored and confirmed are calibration capability, determination of the smallest scale value, the way subjects do measurements, and the way subjects read the measurement data.

Calibration

Based on the observation, 67% of subjects were able to calibrate all measurement instruments correctly. They understand the calibration concept of each measurement instrument and can do it appropriately. Based on the interview, it is known that initially they only gain knowledge through reading references. Then, their understanding grew stronger as they practiced the knowledge directly.

While 33% of other subjects are only able to perform a calibration 4 of 5 measurement instruments correctly. This is because they do not perform the calibration of the fifth measurement instrument at the time in laboratory. The fifth measurement instrument is multimeter. Their understanding only knowledge aspect that they get from references but has no direct experience using multimeter in laboratory.

Smallest scale value determination

Based on the observation, 67% of subjects were able to determine the smallest scale value of all measurement instruments correctly. They understand the smallest scale value determination concept of a measurement instrument by watching the smallest scale of major scale (for the Ohaus's balance, thermometer, and multimeter) and the number of lines on a nonius scale (for vernier calliper and micrometer secrup). While 33% of other subjects were only able to determine the smallest scale value 4 of 5 measurement instruments correctly. The fifth measurement instrument is vernier calliper. It is because they only get the knowledge from references that generalizes the value of the smallest scale value of a vernier calliper is 0.01 mm without understanding the basic concept. Whereas the vernier calliper used in the laboratory is vernier calliper with the smallest scale value 0.05 mm, so when confirmed in the interview session they can not show the way of determining the smallest scale value correctly.

Measurement

Based on the observation result, all subjects were able to do measurement using all measurement instruments correctly. It is because they have understood the concept of using the measurement instruments. Based on the interview result, shows that their good ability to do measurement caused by they conducted their own measurements based on the knowledge they already had. There were some subjects who initially did not understand how to use a certain measuremet instrument, but by working together in groups they were able to help each other so that all subjects were able to do measurement using all of measurement instruments correctly. It shows that laboratory based learning not only develop the knowledge and psychomotor aspect, but also their affective aspects such as collaboration, honesty, and precision.

Measurement data reading

Measurement data reading can be known by observation and interview result. Based on the observation, all of subjects were able to read the measurement results for each measurement instrument appropriately. They also have observed several things in measurement results reading, such as seeing the measurements perpendicular to avoid parallax errors, not touching the thermometer to the body or the container so as not to affect the measured temperature, and pay attention to the smallest scale value in reading measurement result of vernier calliper and micrometer secrup. When confirmed through interviews, 17% of subjects not only rely on the knowledge they get through references (using formulas), but also use their logic in reading the measurement results. They calculate the nonius scale by comparing the smallest scale values to the lines that coincide on the nonius scale.

Laboratory Report

Each research subject writes down laboratory report that consist of worksheets and analysis sheet. From the worksheet can be known subjects understanding in writing the measurement results. While from the analysis sheet can be known subjects understanding in processing the measurement results.

Measurement data writing

Measurements data writing can be known from the worksheet of each student. Based on the worksheet, 50% of subjects were able to write the measurement results for each measurement instrument appropriately, while 33% of subjects were only able to write the measurement results 4 of 5 measurement instruments appropriately. They write down the measurements along with the measurement error that each measurement instruments has. Previously, they only had knowledge from references, but through laboratory based learning they were able to understand that the correct measurement results writing is write the measurement error also. In addition, they must also ensure that the number of significant figure behind comma on the measurement results and the measurement uncertainties are same. They must also ensure that the unit between the measurement results and the measurement uncertainties are same. Examples of measurement data writing by subject are shown in Figure 2.

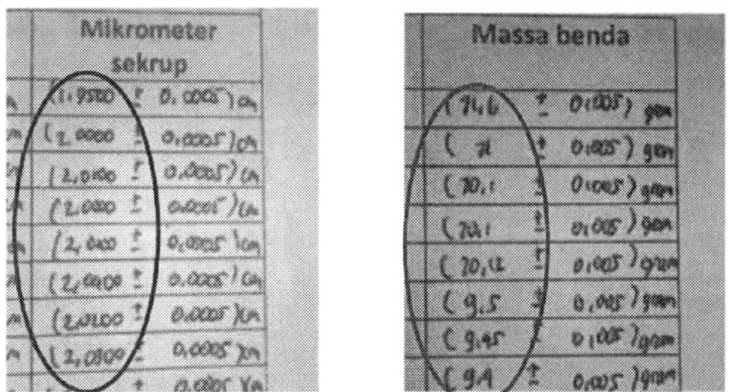

(a) (b)

Figure 2. (a) Example of correct measurement result writing, (b) Example of incorrect measurement result writing

Based on interview results, known that subjects write the measurement results incorrect because forget to equate the number of significant figure behind comma on the measurement results and the measurement uncertainties.

Measurement data processing

Measurement data processing can be known from analysis sheet each subject. The measurement that they did in the laboratory is a repetitive measurement. Furthermore, subjects are required to analyze the measurement data as direct and indirect measurements according to their knowledge from references. Based on the analysis sheets, it was found that 83% of subjects were able to analyze the direct measurements correctly, while only 33% of the subjects were able to analyze the indirect measurements correctly. Through further analysis by the researchers, it was found that their inaccuracy in analyzing is because they have not deeply understood yet the equations they used in analyzing, so it caused errors.

4. Conclusion

The effect of laboratorium based learning is physics teacher candidates have a deeper understanding of the measurement instruments that includes using measurement instruments, taking and analysis data, estimating measurement errors, and communicating the result both in written and oral. Laboratory based learning also improves concepts understanding (knowledge aspects) and students' scientific attitudes. The results of this qualitative research can be applied to other research as long as the condition of the research site is not much different from the conditions at this research site.

References

[1] I. Hughes and T. Overton, "Key aspects of learning and teaching in science and engineering," in *A handbook for teaching and learning in higher education*, 3rd ed., London: Kogan Page, 2008, pp. 424–448.

[2] B. C. Almroth, "The importance of laboratory exercises in biology teaching ; case study in an

ecotoxicology course," *Pedagog. Dev. Interact. Learn.*, no. september, pp. 1–11, 2015.

[3] B. Bayrak, U. Kanli, and Ş. K. İNGEÇ, "To Compare the Effects of Computer Based Learning and the Laboratory Based Learning on Students' Achievement Regarding Electric Circuits," *Turkish Online J. Educ. Technol.*, vol. 6, no. 1, pp. 30–37, 2005.

[4] I. Jamie, P. Karuso, and D. Scott, "Laboratory-Based Teaching and Learning : Developing Quality Assurance Processes," *UniServe Sci. Sch. Inq. Symp. Proc.*, pp. 74–75, 2001.

[5] P. Sugiyono, *Metode Penelitian Kuantitatif Kualitatif & R & D.* Bandung: Alfabeta, 2011.

Journal of Physics: Conference Series **1424** (2019) 012055 doi:10.1088/1742-6596/1424/1/012055

Kampung versus Climate Change: The Dynamics of Community Empowerment through the Climate Village Program (ProKlim)

Dodi Faedlulloh[1], **Retnayu Prasetyanti** [2], **and Bambang Irawan**[1]

[1]Universitas 17 Agustus 1945 Jakarta, Indonesia
[2]STIA LAN Jakarta, Indonesia

*dodifaedlulloh@gmail.com

Abstract. This paper aimed to contribute in explaining the implementation of community empowerment through the Climate Village Program (ProKlim) in Jati Village, East Jakarta and Kebon Kosong Village, Central Jakarta. Basically, both locations have demonstrated good practice of empowering and implementing ProKlim. But if faced with the problem of climate change which substantively has a global character –the village as the subject is only able to carry out limited adaptation and mitigation to climate change. Therefore, the structural narrative is crucial in overcoming the problem of climate change, namely by re-presenting the role and intervention of the state through macro policy.

1. Introduction

The problem of climate change has become a global phenomenon. When the world does not committed to take concrete steps to rehabilitate its activities towards nature, climate change will become a time bomb for the inhabitants of the earth. The Intergovernmental Panel on Climate Change (IPCC) report states that the rise in global surface temperature (global) ranges from 1.35oC and is expected to continue to increase between 1.5-2oC in the next 30 years. The increase of earth's temperature increases the threat to the risk of climate-related disasters such as floods, landslides, droughts, crop failures, biodiversity damage, sea level rise and degradation of the quality of human health. Indonesia particularly as an archipelagic country vulnerably responds to the effects of climate change which mostly occured in the form of hydrometereorological disasters (Dewi & Istiadi, 2016).

Humans and ecology live dependently, so that humans must consider the sustainability of nature. The Indonesian constitution after the amendment in 2002, incorporates movements towards becoming green. The spirit of the green constitution is stipulated in Article 28H clause (1) and Article 33 clause (4) of the Constitution. Article 28H Clause (1) states that *"Every citizen has the right to achieve social welfare in both physiologically and spiritually, have decent place to stay, have access to a good and healthy environment, and right to health care"* (Prasetyanti, Retnayu, 2017). Yet, contradictions often occured; humans commonly override environmental damage while persistently polluting nature. The latent problem of waste management in Indonesia considered not to be over. There are 64 out of 470 watershed areas (DAS) in Indonesia facing critical situations due to waste pollution (Luciana, 2017). In the forestry sector, 59 million hectares area affected by land conversion, fire, and illegal logging –with an average reduction of 1.9 million hectares per year (BPS, 2017). Factually, Forest damage caused major natural disasters, from landslides, floods, to air pollution.

In governance era, collaboration is mandatory, the responsibility of climate change does not only rely on governmental bodies. The urgency of community initiatives is recently trending, so that all actors including community must be actively involved. Responding to this problem, in 2012, Indonesia created the Climate Village Program (ProKlim) which focused on the basis of community empowerment in carrying out local actions to increase resilience to the effects of climate change and reduce greenhouse gas emissions. This program was developed by the Ministry of Environment and Forestry (KLHK). The Climate Village Program (ProKlim) is a national-scale program managed by the Ministry of Environment and Forestry in order to increase the involvement of the community and other stakeholders to strengthen adaptive capacity to the impacts of climate change and reduce GHG emissions and provide recognition of climate change adaptation and mitigation efforts that has been done that can improve welfare at the local level in accordance with regional conditions.

The actual implementation of ProKlim refers to the Minister of Environment and Forestry Regulation Number 84 of 2016 on The Climate Village Program, in which the main components; the terms of proposal; assessment; and program category are contained. In the Ministerial Agreement, it is also mentioned that ProKlim could be developed and implemented in the lowest administrative area at RW or Hamlet and Kelurahan or Village at the highest level. In 2013, there were 180 ProKlim in 69 districts/cities. In 2016, the number of ProKlim at the village level was 290 ProKlim. This number is predicted to increase to 2000 ProKlim which is targeted to be functioned optimally at the beginning of 2019 (Kompas, 2017).

Action for empowerment has been increasing within communities around issues of access to resources and entitlements, capacity building, the nurturing of leadership and local initiative and sustainable development. In ProKlim, empowerment considered as strategy to attain goals of local sustainable development. Through empowerment, community are expected to be able to enhance capacity to adapt and mitigate climate change. In practical realm, community empowerment, somehow, has led development programs into success, especially in rural area. ProKlim, therefore, is managed in attempt to achieve both community independence and sustainability.

By conducting research on cummunity based sustainable program namely ProKlim, the author aimed to analyze the dynamics of the ability of community in implementing the principles of sustainable development in local area. This paper concerns on how global problems such as climate change can (at least) be minimized by the collaborative efforts at the level of the village. This paper is intended to contribute to the theoretical conceptual explanation in the process of implementing ProKlim along with the inherent nature of ProKlim. The researchers also used empirical study data conducted in Jati Village –a Kampung in Rawamangun District, the City of East Jakarta and Kebun Kosong Village –a Kampung in Kemayoran, the City of Central Jakarta (Faedlulloh & Irawan, 2018).

2. Research Method

The fundamental purpose of this research is to analyze the implementation of ProKlim in the perspective of community empowerment and sustainability. In line with that, a qualitative research technique is chosen for this research to present comparative analysis between Jati

Village and Kebon Kosong Village. The use of qualitative method assessed to be appropriate because of the nature of research is the exploration of "what is going on" in specific situation regarding the issues of sustainability and empowerment in urban/rural development. Besides that, qualitative research was able to clarify "what went wrong" and define "what should government do" as response for development process (Alipour, Vaziri, & Ligay, 2011). This qualitative method is also precise to examine the phenomenology research dealing with people, organizations, and groups, such as ProKlim.

The author have conducted in-depth interview and observation in both research sites to collect information regarding activities of ProKlim, community involvement, constraining factors, also achievement. The primary data used to ensure facts and empirically strengthened the research argument (Moleong, 2004). To support the results of field research, researchers also used a library study approach with secondary data-based research. Literature study is a technique of collecting data by conducting study studies of books, literature, notes, reports that have to do with problems that are to be solved (Nazir, 2009).

3. Analysis

a. Community Empowerment, Sustainable Development and ProKlim

Empowerment has become the mechanism through which sustainable development goals can be achieved both locally and nationally (Pranarka & Vidhyandika, 1996).. Particularly in this paper, the concept of empowerment affirms the need to build the capacity of communities to respond to a fast-changing environment by inducing appropriate change-internally as well as externally through initiatives, innovation and commitment to a local sustainable development goals. Empowerment for sustainable development means giving to people and communities the true capacity to cope with the changing environment, for increased social awareness, higher levels of social and economic participation and the utilization of new insights on ecological processes of change and self renewal, as societies and communities strive to enter the transition towards sustainable patterns of development.

To achieve the requires, community based environment program –namely ProKlim is targetted to enhance the effectiveness of the action plan of the sustainable development priority programs. In this case, there are three means to implement community empowerment (Kartasasmita, 1996). *Firstly*, creating an atmosphere and climate that allows the potential of the community to well develop. This condition is based on the assumption that each individual and society has unique potential and self-encouragement to create change. In this context, we can see the process of the implementation of ProKlim in Kebon Kosong Village and Jati Village whose have different potential and social capital in managing ProKlim. In the case of Kebon Kosong Village, there was major role of the village apparatus as a creative minority who became the initiator to encourage community participation in the implementation of ProKlim.

The practice of community empowerment in Jati Village, meanwhile, supported by the financial assistance through CSR in 2015. The program was started by the pilot project *Bank Sampah* (Faedlulloh & Irawan, 2018). ProKlim in Jati Village has leveled up community awareness on nature. Through ProKlim, though in a small scale, there was impact on the change of microclimate, which indirectly decreased micro-temperature. Besides that ProKlim in Jati Village and Kebon Kosong Village considered effective in maintaining ground water, securing food stock, conserving biodiversity, providing green space for recreation/eco-tourism and even increasing social cohesion.

One of the exemplary activities of ProKlim is the reforestation process in Kebon Kosong. ProKlim did not stop at reforestation; the village apparatus pushed forward the collective achievement by organizing organic waste management. By conducting a self-help mechanism, local apparatus jointly worked with community in developing *Bank Sampah* at the end of 2013 to increase impacts on the mitigation activity. *Bank Sampah* turned out advantageous; this most well-known community based waste management program has given significant economic benefit, which used by community to make cultivation gardens with four variants –fish cultivation, fruit orchards, farm gardens, and ornamental plant gardens. This situation continues to be encouraged by the construction of environmentally friendly waste management locations in 2017. This is an affirmative action led by the Central Jakarta Environmental Agency as part of community based waste management program.

Secondly, strengthening the community potential by implementing concrete steps, accommodating various inputs, providing infrastructure and facilities –physically and socially, which can be utilized by community. In this case, the implementation of ProKlim needs to be implemented with the scheme of bottom-up, so that people can actively participate in determining the direction of the climate change adaptation and mitigation process in their neighborhood. The two sites of research successfully conclude that climate change relates to routine activities and community behavior. It is not abstract issue; climate change is more a matter of how awareness turns into small action.

Basically, the role of village apparatus in the two places has supported the program very well. The appointed authorized parties also provide education and socialization on ProKlim. Somehow, in the perspective of bottom-up approach, the actual implementation of socialization and education is not sufficient; it requires practical training and assistance from certified facilitator. The observation and interview results concluded that Kebon Kosong Village has consistency in mentoring activities. This attempt has made the Kebon Kosong Village received awards related to the environment and the success of ProKlim (2017), which one of the indicators is the process of mentoring activities. Apparently, this is what needs to be encouraged more by Jati Village; the enhancement of community potential through facilitation and mentoring activities.

Thirdly, empowering community in the sense of protecting the weak and defending the interests of the community. It means that the process of community empowerment bases on the implementation of social equity values which provide equality for all various parties. Certainly, the objective of ProKlim is basically providing protection for all citizens without the exception of the negative effects of climate change.

b. Presenting *Kampung* and Representing the Best Role of Government

Kampung is a small demographical and geographical entity. Demographically, *Kampung* describes a low-medium economic level with a (quite) strong social cohesion. Geographically, the term of *Kampung* defined as a group of local community with unique local wisdom located in rural, sub-urban, and urban area. *Kampung* has become the identity of Indonesia. The characteristic of *Kampung* is distinctive; depend on the development of modernization. Generally, *Kampung* is also located in urban area, such as Jakarta, but, the scope is surely small; only in several places where low-medium community lives and maintains local culture. Due to its social inclusiveness, community *Kampung* can practice and manage empowerment – a bottom up based program.

In the context of ProKlim implementation, community defined as the subject of development who owns small group in small scope –a *Kampung* scope. Therefore, ProKlim cannot be implemented in a top-down framework. Without participation, various steps taken by the government will be futile; unsustain. The expectation of a national climate protection program named ProKlim is to create awareness of environmental protection and management while also sharing equal benefits of green and healthy environment to communities. However, things need to be noted that the public will be responsible if the government also has the same commitment and willing to uphold the responsibility. Factually, the problem of climate change is a global problem that is also structural. Thus, the adaptation and mitigation steps need to be structurally dominant as well. This is not meant by denying non-structural roles in the effort to control the environment, but rather in the aim that several problems need to be presented in accordance with the right portion. Regarding the empirical practices that have occurred, for example what happened both in Kebon Kosong Village and in Jati Village, the community efforts of climate change adaptation activities were more non-structurally managed. Indeed, there were real impacts, however still limited.

Tabel 1. Activities of Climate Change Adaptation in Jati Village and Kebon Kosong Village

Location	**Activities**		
	Drought – Flood – Landslide Control	**Food Security Enhancement**	**Climate Change Disease Control**
Kebon Kosong Village	The development of city park as flood control	- The availability of small herbs/vegetables garden - Local food plants are used to support family needs - Research and development are managed to enhance food security	Habituation of clean and healthy living according to the program of KLHK
Jati Village	Flood control is not available	- The availability of small herbs/vegetables garden - Local food plants are used to support family needs	Habituation of clean and healthy living according to the program of KLHK

Source: Faedlulloh & Irawan (2018)

The fundamental difference from the structural approach with non-structural approach is the existence of state intervention through policy, to address public problems. For example, a mitigation and protection against flood certainly needs strong commitment of local authorities instead relying flood control facilities initiated by communities. Surely, government must uphold law enforcement on environmental protection to ensure not only the environmentally friendly behavior of community, but also the violation on land conversion/use by commercial parties. Representing government role in community empowerment means strengthening the policy processes and outputs in a sustainable urban development.

In mitigation process, the role of government is also crucial. Mitigation is a series of activities in attempt to reduce the level of greenhouse gas emissions as a form of climate change mitigation efforts. Mitigation activities can be simple to complex things. Examples are energy saving, green transportation, waste management, tree planting and control of land and forest fires involving the community (Ministry of Environment and Forestry, 2017). Practically in ProKlim, mitigation efforts are limited to waste management and reforestation. The availability of facilities, green knowledge, and technology is limited. In this case, mitigation efforts need to be managed by government through a macro level policy that also provides soft-approach to persistently educate community. The implementation of mitigation activities managed by community are explained in the table below.

Tabel 2. Climate Change Mitigation Activities in Jati Village and Kebon Kosong Village

Location	**Activities**			
	Solid Waste Management and Utilization	**Liquid Waste Management and Utilization**	**The Use of New/Renewable Energy and Conservation Energy**	**Vegetation Control**
Kebon	- The development of	Composting	Socialization	Reforestation

Kosong Village	*Bank Sampah Selaras Mandiri* - The development of an integrated solid waste management location			
Jati Village	The development of *Bank Sampah Asri Hijau*	-	Socialization	Reforestation

Source: Faedlulloh & Irawan (2018)

The damage that occured to the environment is the impact of global development that overrided human sustainability. The report entitled The Limits to Growth, written by Meadows to the Club of Rome (Project on the Predicament of Mankind) describes various problems that afflicted many countries and has become global attention, namely the progress of industrialization, population growth, food shortages, damage to the un-renewable natural resources, and environmental pollution (Supriadi, 2010). In substance, ProKlim is an innovative program which is urgent and prominent to sustain. However, the *Kampung*-community impossibly survive on its own capacity. The structural role is necessary to guarantee the balance of sustainable development impact. The state must settle the economic impact of development as the basis of community welfare. Therefore, the representing the role of government in particular climate change issue, must cover all aspects of development –economy, socio-culture, ecology, even technology through the development of macro policy.

4. Conclusion

In substance, the two locations have demonstrated good practice of community empowerment. Moreover, there are parties who get awards related to the success of ProKlim. But when faced with the problem of climate change which basically has a global character, the village or *Kampung*-community is only a subject who can carry out limited adaptation and mitigation to climate change. In fact, in a bigger perspective, the contributors to pollution that affect global climate change are actually companies, not villages. Therefore the structural narrative is needed again particularly in environmental protection program by representing the role and intervention of the state through macro policy. Thus, the existence of ProKlim is a supporting program, not a fundamental policy to address climate change.

References

[1] Alipour, H., Vaziri, R. K., & Ligay, E. (2011). Governance as Catalyst to Sustainable Tourism Development: Evidence from North Cyprus. *Journal of Sustainable Development*, *4*(5), 32–49. https://doi.org/10.5539/jsd.v4n5p32

[2] Dewi, I. K., & Istiadi, Y. (2016). Mitigasi Bencana pada Masyarakat Tradisional dalam Menghadapi Perubahan Iklim di Kampung Naga Kecamatan Salawu Kabupaten Tasikmalaya. *Manusia Dan Lingkungan*, *23*(1), 129–135.

[3] Faedlulloh, D., & Irawan, B. (2018). *Pengendalian dan Perlindungan Lingkungan Hidup Yang Berkelanjutan Berbasis Pemberdayaan Masyarakat: Studi Komparasi Pada Program Unggulan Kampung Iklim (ProKlim) di Kelurahan Jati Jakarta Timur dan Kelurahan Kebon Kosong Kemayoran Jakarta Pusat*. Universitas 17 Agustus 1945 Jakarta (Tidak Dipublikasikan).

[4] Kartasasmita, G. (1996). *Pembangunan untuk Rakyat: Memadukan Pertumbuhan dan Pemerataan*. Jakarta: PT. Pustaka Cidesindo

[5] Kementrian Lingkungan Hidup dan Kehutanan. (2017). *Road Map Program Kampung*

Iklim (ProKlim) Gerakan Nasional Pengendalian Perubahan Iklim Berbasis Masyarakat. Jakarta: Direktorat Adaptasi Perubahan Iklim Direktorat Jenderal Pengendalian Perubahan Iklim Kementerian Lingkungan Hidup dan Kehutanan.

[6] Kompas. (2017). Sebanyak 2000 Kampung Iklim Ditargetkan Berfungsi di 2019. Retrieved from http://nasional.kompas.com

[7] Luciana, S. (2017). Komunikasi Persuasif Duta Lingkungan dalam Meningkatkan Gaya Hidup Ramah Lingkungan pada Program Kampung Iklim Badan Lingkungan Hidup Kota Pekanbaru. *Jurnal Online Mahasiswa Fakultas Ilmu Sosial Dan Ilmu Politik Universitas Riau*, *4*(2), 1–18.

[8] Moleong, L. J. (2004). *Metode Penelitian Kualitatif.* Bandung: PT. Remaja Rosdakarya.

[9] Nazir, M. (2009). *Metode Penelitian.* Jakarta: Ghalia Indonesia.

[10] Pranarka, A. M. ., & Vidhyandika. (1996). *Pemberdayaan: Konsep, Kebijakan dan Implementasi*. Jakarta: Center of Strategic and International Studies.

[11] Prasetyanti, R. (2017). Ecocracy - Ecology Based Democracy: Pursuing Local Goals of Sustainable Development. *JKAP (Jurnal Kebijakan dan Administrasi Publik). 1*(1), 1-12.

[12] Supriadi. (2010). *Hukum Lingkungan di Indonesia Sebuah Pengantar.* Jakarta: Sinar Grafika.

AUTHOR INDEX

AUTHOR INDEX

Institute of Physics
Dirac House, Temple Back
Bristol BS1 6BE UK

ISSN: 1742-6588
ISBN 978-1-7138-1328-6

CURRAN ASSOCIATES INC.
proceedings.com